Peter D. Schumer
Number Theory

Also of interest

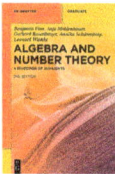

Algebra and Number Theory.
A Selection of Highlights
2nd Edition
Benjamin Fine, Anja Moldenhauer, Gerhard Rosenberger, Annika
Schürenberg and Leonard Wienke, 2023
ISBN 978-3-11-078998-0
e-ISBN (PDF) 978-3-11-079028-3

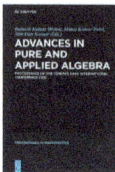

Advances in Pure and Applied Algebra.
Proceedings of the CONIAPS XXVII International Conference 2021
Ratnesh Kumar Mishra, Manoj Kumar Patel and Shiv Datt Kumar (eds.),
2023
ISBN 978-3-11-078572-2
e-ISBN (PDF) 978-3-11-078580-7

Elementary Linear Algebra with Applications.
MATLAB®, Mathematica®; and Maplesoft™
George Nakos, 2024
ISBN 978-3-11-133179-9
e-ISBN (PDF) 978-3-11-133185-0

Abstract Algebra.
With Applications to Galois Theory, Algebraic Geometry, Representation Theory
and Cryptography
3rd Edition
Gerhard Rosenberger, Annika Schürenberg and Leonard Wienke, 2024
ISBN 978-3-11-113951-7
e-ISBN (PDF) 978-3-11-114252-4

Peter D. Schumer

Number Theory

Multiplicative and Additive with Factorization
and Primality Testing

DE GRUYTER

Author
Prof. Peter D. Schumer
Middlebury College
Department of Mathematics and Statistics
303 College Street
Middlebury VT 05753, USA
schumer@middlebury.edu

ISBN 978-3-11-157867-5
e-ISBN (PDF) 978-3-11-157928-3
e-ISBN (EPUB) 978-3-11-157954-2

Library of Congress Control Number: 2025935975

Bibliographic information published by the Deutsche Nationalbibliothek
The Deutsche Nationalbibliothek lists this publication in the Deutsche Nationalbibliografie;
detailed bibliographic data are available on the internet at http://dnb.dnb.de.

www.degruyter.com
Questions about General Product Safety Regulation:
productsafety@degruyterbrill.com

Preface

The purpose of this book is to tell a story about the natural numbers and some of the key discoveries made about them over the centuries. Among the main characters are the primes, composites, triangular numbers, squares, quadratic residues, and Fermat and Mersenne primes. There are also many supporting characters such as the Carmichael numbers, pentagonal numbers, Fibonacci numbers, and pseudoprimes. Each has its own personality and unique and interesting relationship to the others. Despite my efforts, the theory of numbers is so ancient and so vast that there will always be much more than can be recounted here. Hopefully these pages will spark your interest to continue to go further and dig deeper. Numbers are immortal and their story will forever continue to grow.

Number theory has captured the imagination of many great scholars from a multitude of countries and civilizations over several millennia. Yet its treasures are more alive now than ever before. A.Y. Khinchin (1894–1959) declared number theory to be "the oldest, but forever youthful, branch of mathematics."

The study of number theory is a great journey into the recesses of one's mind. I have attempted to capture some of the essence of this vast historical adventure while at the same time pointing out routes yet uncharted. Like other adventures, it is a personal one. The choice of which theorems to contemplate, exercises to solve, sections to study, and unsolved problems to investigate is all up to you. The success of my book will depend on how coherent a story I have told and whether I have created a smooth and pleasant journey.

The book can serve as a textbook or source book in a wide variety of settings from an undergraduate course on number theory, as a primer for computer scientists needing some number theoretic background, to students working on senior theses or master's projects in mathematics and beyond. It contains a thoroughly up-to-date exposition of elementary number theory together with a host of advanced topics depending on the instructor's and student's needs. There are abundant examples throughout to illuminate how the principles actually apply in practice. Many examples are stated as problems with solutions. I hope this encourages students to take a more active role in solving the problems themselves. And I have included reference to many unsolved problems that might entice the interested student to investigate further.

There are many historical and biographical remarks interwoven throughout the text to highlight that much of mathematics is a very active human creation to which we can all play a part. Theorems and proofs include accurate attributions to the best of my ability, and I have given descriptive names to many results to aid in their retention. Give credit where it's due! The proofs are as direct as possible without requiring much beyond a solid background in calculus plus linear algebra. Only minimal familiarity with advanced analytic methods or abstract algebra is required.

The book itself is a thoroughly updated and vastly expanded version of a number theory textbook I wrote 30 years ago [200]. There is more material than can be covered

https://doi.org/10.1515/9783111579283-202

in any single course – enough for two semesters. Thus, the instructor has a great deal of freedom and flexibility. Although the ordering of the chapters and sections suits my preferences, there are a multitude of ways to reorder topics to fit various courses.

Over 800 exercises in all of varying depth and difficulty are included. Generally, the easier and more direct ones come first before those that require deeper analysis. Additionally, computational problems tend to precede more theoretical ones. I hope a good balance has been maintained. Rather than starring what I consider to be the difficult problems, let the students discover their own strengths and weaknesses. I expended as much care and effort in creating interesting problems as in writing the text itself.

Most of the computational exercises can be done by hand with a judicious choice of techniques and theorems. Some require a calculator (or smart phone), and others (especially in Chapters 7 and 8) would benefit from the use of a computer algebra system such as Maple or Mathematica. There are enough problems and enough variety to keep the best students happily occupied. Many problems are highly original, and I trust entertaining. The exercises are an essential part of the text, which include historical references as well as discussion of unsolved problems and open questions.

Chapter 1 serves as an introduction and is a bit chattier and less demanding than the others. It is both a primer for students with less mathematical background and a refresher for those with more. Much of the material is applicable to all of mathematics rather than just number theory. The instructor should decide how much time is appropriate (although I think it would be a mistake to skip it altogether).

Chapters 2–5 form the core of a complete standard number theory course with some enhancements for those who wish to include some nice extras. Normally, most of this material would be covered prior to discussing the somewhat more challenging topics in Chapters 6–10. However, actually almost any section of the book can be profitably studied once Section 3.3 is completed. The final chapters are largely independent of one another and can be saved for independent work, senior projects, or graduate level study.

Chapter 2 includes the fundamentals of congruence relations (such as the Euler-Fermat theorem and Lagrange's theorem on polynomial congruences) and an introduction to factorization (including Fermat's method). The major topics include the Euclidean algorithm, the solution of linear Diophantine equations with application to the Frobenius coin problem, the Chinese remainder theorem, the fundamental theorem of arithmetic, and Hensel's lemma. Most sections dig a bit deeper than is customary to lead students naturally toward modern results and the current state of affairs. For example, the Euclidean algorithm is succeeded by Lamé's result on its efficiency. After the discussion of the Chinese remainder theorem, there is is a full treatment of simultaneous linear congruences for moduli *not* relatively prime. Similarly, in addition to proving the infinitude of primes, analogous results for certain subclasses of primes are dealt with via examples or among the exercises once the appropriate tools are established. The treatment of Hensel's lemma includes the singular case.

In Chapter 3, the major number-theoretic functions are introduced and developed. The key concept presented is the Möbius inversion formula and its applicability in deriving closed formulas for many of the functions studied. The last section contains a fairly detailed treatment of perfect numbers and amicable pairs. In addition, the special topic of odd perfect numbers is included.

Chapter 4 contains a thorough characterization of integers possessing primitive roots. Our results are nicely applied to the mathematics of card shuffling. The rest of the chapter develops properties of the Legendre symbol and leads naturally up to Gauss's law of quadratic reciprocity including many illuminating examples. The reciprocity law is one of the highlights in any number theory course. In addition, the Jacobi symbol is introduced along with its reciprocity law.

Chapter 5 includes a detailed and lively investigation of the representation of integers as the sum of squares. The material begins with Pythagorean triplets and Fermat's last theorem for fourth powers followed by sums of 2, 3, and 4 or more squares in turn. In Section 5.3, Markov's equation is introduced and studied. There is a more complete exposition here than is found in most number theory books. The final section on Legendre's equation could serve as the capstone of a semester's course or as supplemental reading when the instructor wishes to jump ahead.

Chapter 6 contains a thorough introduction to finite and infinite continued fractions, including Lagrange's theorem on periodic continued fractions, the study of purely periodic continued fractions, and the revered Pell's equation. There is a full section on rational approximation to irrationals, including Hurwitz's theorem and a self-contained proof that π is irrational. An entire section is devoted to Farey fractions – a very appealing topic often overlooked. Farey fractions also serve as the quickest means to introduce the notion of rational approximations, which are studied in greater detail later. Euler's result giving the simple continued fraction for the number e is completely worked out. Finally, a brief discussion of algebraic numbers is included along with a proof that the algebraic numbers are denumerable as well as Liouville's construction of a set of transcendental numbers.

Chapter 7 covers the general theory of primality testing and factoring – two separate, but related topics. Several primality tests are presented, including Lucas's primality test, Pocklington's primality test, the Miller-Rabin-Jaeschke test. The dramatic AKS primality test is briefly discussed as well. Factoring techniques include the Continued Fraction Factorization Method, The Pollard Rho Factorization Technique, the Pollard p -1 Method, and the modern and highly applicable Quadratic sieve factorization method. A complete section on Fermat numbers offers much historical discussion and many up-to-date results. Similarly, a full state of the art section devoted to Mersenne primes complements previous material on perfect numbers.

Chapter 8 serves as a gateway to some important applications of number theory. Section 8.1 is a lively general introduction to cryptology with a bit of historical perspective. Section 8.2 presents the RSA encryption algorithm which is ever present in today's world. Applications to digital signatures and nuclear testing verification follows. Section 8.3 introduces many of the main ideas related to random number generation which has wide applicability.

Chapter 9 is a gentle introduction to analytic number theory. The material is by its nature a bit more advanced, so I have included a fair amount of historical and biographical exposition. Although the zeta function is introduced, no complex analysis is needed. The chapter begins with summation formulae and the sum of the reciprocals of the primes. Next, the average order of the lattice, divisor function, and the Euler phi function is presented as a fitting follow-up to the material in Chapter 3. The sum of the reciprocals of the squares is determined, along with some interesting applications. An introduction to the study of the distribution or primes follows, including Chebyshev's theorems and a discourse on the prime number theorem. Last, Bertrand's postulate is proven and several applications are presented. Included are such results as Richert's theorem that every positive integer beyond 6 is the sum of distinct primes and Mill's theorem that there exists a real A such that $[A^{3^n}]$ is prime for all $n \geq 1$.

Chapter 10 is made up of four self-contained sections dealing with additive number theory. They include a lengthy exposition on Waring's problem (including several variants), a section on the density of sets and Mann's $\alpha + \beta$ theorem, van der Waerden's theorem on arithmetic progressions, and an engaging section on the partition function including Euler's pentagonal number theorem.

In any book of limited length, to maintain coherence choices must be made of what to include and what to exclude. I have not included a detailed exposition of the general theory of binary quadratic forms, the geometry of numbers, algebraic geometry, or algebraic number theory. Though fully worthy of study, these topics are treated well elsewhere. I hope the reader will be encouraged to continue their study of number theory after learning a good deal here. As G. H. Hardy (1877–1947) commented,

The elementary theory of numbers [is} one of the very best subjects for early mathematical instruction. It demands very little previous knowledge; its subject matter is tangible and familiar; the processes of reasoning which it employs are simple, general, and few; and it is unique among the mathematical sciences in its appeal to natural human curiosity.

I hope I have made a modest contribution to this fascinating and most worthy subject.

I wish to thank Ranis Ibragimov and Scott Bentley for their initial encouragment and especially project manager Ambika Muthusamy and content editor Marie Hammerchmidt for their help and guidance throughout the writing process. I dedicate this book to Henry, Evelyn, and to all math enthusiasts present and future.

Contents

Chapter 1
Background

1.1 Brief historical introduction

Number theory deals with the fundamental properties of the integers and of their mathematical extensions. In this work, our main concern will be with the usual arithmetic operations on the natural numbers and on some special subsets of them such as the prime numbers and the sequence of squares.

The basic objects in number theory are simple ones. On the one hand, many of the definitions will already be familiar to you. As such, number theory is readily accessible and its beauty and appeal are quite immediate. On the other hand, many of the simplest observations are difficult to prove. In fact, some easily understood conjectures remain unproved despite the efforts of many of the greatest mathematicians of all time. Hence, the study of number theory can be a serious and difficult one.

Number theory is full of the sort of problems praised by David Hilbert (1862–1943) in his Paris address at the International Congress of Mathematicians (1900): ". . . a mathematical problem should be difficult to entice us, yet not be completely inaccessible, lest it mock at our efforts. It should be a guidepost on the mazy path to hidden truths, and ultimately a reminder of our pleasure in the successful solution." Carl Friedrich Gauss (1777–1855), arguably the greatest mathematician ever, expressed his view that "mathematics is the queen of the sciences, but number theory is the queen of mathematics."

The history of number theory is nearly as old as mathematics itself and on a rudimentary level probably as old as language and religion. Hence its origins have long been lost to antiquity. All cultures, no matter how primitive they appear to modern eyes, seem to have some ability to count. A wolf's tibia bone, approximately 30,000 years old, was found in what was then Czechoslovakia in 1937. Engraved in it are 55 notches organized in groups of 5. This is strong evidence that some Stone Age people had quantitative facilities. Surely tallying with stones or scratches in sand must be older yet.

Arithmetic and geometric sophistication developed over time as the need increased for keeping track of time, setting up a calendar, bartering for goods, surveying land, navigating, counting livestock, and so on. Significant mathematical achievements were made across the globe including Egypt, China, India, and among the Mayan culture in Central America. An intriguing archaeological find is a Babylonian cuneiform tablet dated approximately 1800 B.C.E., now tablet 322 of the Plimpton collection at Columbia University. On it are three columns of numbers written in the Babylonian sexagesimal system. The numbers across each row comprise solutions in integers to the equation $a^2 + b^2 = c^2$. These are now called Pythagorean triplets though this tablet predates Pythagoras by well over a millennium! It is important to note that these ancient people were

https://doi.org/10.1515/9783111579283-001

aware of the Pythagorean theorem concerning the sides of a right triangle. Beyond that they knew how to find integer solutions to the Pythagorean equation, which is more difficult and would certainly classify as a bona fide number-theoretic result.

The ancient Greeks are well-known for developing the axiomatic method and for their unparalleled achievements in geometry. Although not of the same magnitude, their number-theoretic studies were also significant. For the most part, these originate in the religious mysticism of Pythagoras (c.a. 572–497 B.C.E.) and his followers, the Pythagoreans. Indeed, the Pythagorean motto was "all is number" and the elementary properties of primes and composites as well as the distinction between odds and evens go back to them.

The study of polygonal numbers, the number of dots in diagrams comprised of ever-larger regular polygons, was a favorite of Pythagoreans. Examples include triangular numbers, squares, and pentagonal numbers (see Figure 1.1). For example, triangular numbers are those that can be drawn with an equilateral triangular pattern of dots (or with an appropriate number of bowling pins). The sequence of triangular numbers begins with 1, 3, 6, 10, 15, 21, 28, 36, and so on. See if you can rediscover the general formula for the nth triangular number. The Pythagoreans knew the formulas for all the polygonal numbers! And the study can be readily extended to figurative numbers consisting of other shapes and even further dimensions.

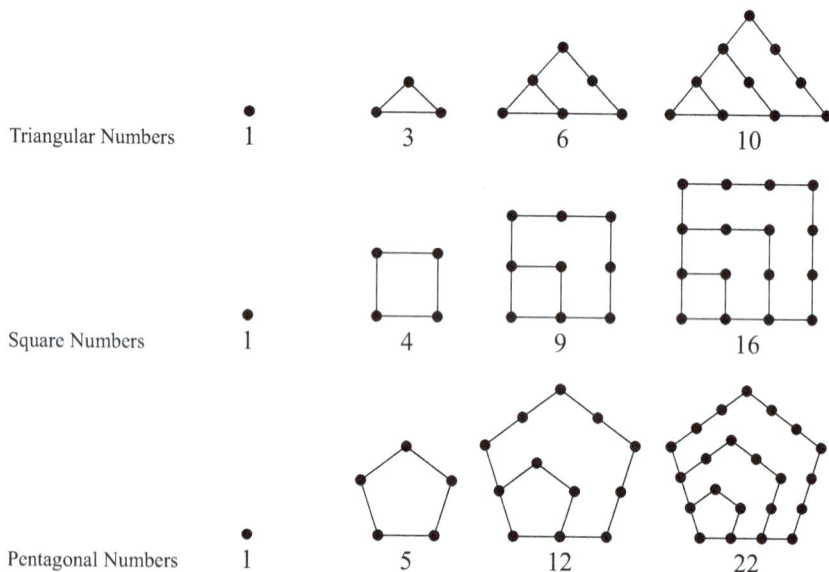

Figure 1.1: The first few polygonal numbers.

An excellent source for Greek number theory is Books VII through X of Euclid's *Elements* (c.a. 300 B.C.E.). Books VII through IX include 102 propositions dealing with the elementary number theory, the infinitude of primes, the unique factorization of integers, geometric and arithmetic progressions, the Euclidean algorithm for finding the greatest common divisor, and a discussion of perfect numbers. Perfect numbers are those like 6 and 28 which are equal to the sum of their proper divisors. Book X is comprised of 115 propositions dealing with "incommensurables" like $a \pm \sqrt{b}$ and $\sqrt{a \pm \sqrt{b}}$ as well as a discussion of Pythagorean triplets.

Greek algebra and number theory took a back seat to geometry for several centuries until the appearance of the *Arithmetica* written by Diophantus (fl. 250 C.E.). Though little is known with certainty about the life of Diophantus, the following riddle for his longevity dates from the fifth or sixth century:

> God granted him to be a boy from the sixth part of his life, and adding a twelfth part to this, he clothed his cheeks with down. He lit him the light of wedlock after a seventh part, and five years after his marriage He granted him a son. Alas! Late-born wretched child; after attaining the measure of half his father's life, chill fate took him. After consoling his grief by this science of numbers for four years he ended his life.

Six of the original 13 books of the *Arithmetica* are extant. They deal mainly with mixture, age, and other computational problems together with cookbook procedures of solution. Several points are worth noting, however. The first is Diophantus' interest in exact solutions rather than the approximate solutions considered perfectly appropriate by his Egyptian and Babylonian predecessors. The second is that Diophantus made several improvements in symbolism. He used abbreviations for commonly used phrases; hence, the *Arithmetica* is considered to be the first syncopated algebra text as opposed to a purely rhetorical one. The third and most significant for us is that only integer or rational solutions were allowed. For this reason, we now refer to such algebraic equations as Diophantine equations.

Although the Hindus and Arabs made significant advances in algebra during the Middle Ages, the European interest in number theory was not rekindled until Claude Gaspard de Bachet (1591–1639) published Greek and Latin versions of the *Arithmetica* in 1621. Bachet noticed that Diophantus tacitly assumed all positive integers are expressible as the sum of four squares. Bachet checked this assertion up to 325 and then asked his fellow scientists in Paris if any of them could prove it. The challenge was not met until Joseph Louis Lagrange (1736–1814) disposed of the problem in the affirmative in 1770.

Bachet's edition of the *Arithmetica* caught the attention of Pierre de Fermat (1601–1665), the "prince of amateur mathematicians." Fermat, by profession a legal counselor and jurist, was a highly capable and original mathematician. His mathematical discoveries were announced to friends through his voluminous correspondence, but unfortunately details were often rather scant and explanations somewhat cryptic. Independent of Descartes, Fermat discovered and developed what we now call analytic geometry. He also developed techniques for finding tangent and normal lines to various curves, studied various transcendental functions such as the spiral of Fermat $r^n = a\theta$, and worked on general methods of optimization. In addition, he contributed new methods for finding areas, volumes of revolution, and the rectification of plane curves. It is no wonder Laplace called Fermat the "discoverer of differential calculus".

Fermat's greatest passion, however, was the theory of numbers. He discovered many beautiful relationships among the integers and claimed to have proofs, though again he rarely disclosed their content. These include the following statements: (1) The area of an integral-sided triangle cannot be a perfect square. (2) Every odd prime can be expressed uniquely as the difference of two squares. (3) Every prime of the form $4n + 1$ can be expressed as the sum of two squares while no prime of the form $4n + 3$ can be so expressed. (4) If p is prime then p divides evenly into $n^p - n$ for any positive integer n. (5) All numbers in the sequence $2 + 1, 2^2 + 1, 2^4 + 1, 2^8 + 1, 2^{16} + 1$, etc. are prime. (6) The equation $x^3 = y^2 + 2$ has a unique integral solution. Attempts to prove many of these assertions were made through the concerted efforts of Leonhard Euler (1707–1783) and Lagrange in the eighteenth century. To Fermat's credit, most turned out to be true, but some were in error (e.g., Euler showed that conjecture (5) is false).

The most famous of all Fermat's assertions is that the equation $x^n + y^n = z^n$ has no non-trivial positive integer solutions for $n \geq 3$. In his copy of the *Arithmetica*, Fermat made the famous statement, "I have discovered a truly marvelous proof, but the margin is too small to contain it." Progress was made over the next couple of centuries by some of the world's greatest mathematicians including Leonard Euler (1707–1783), Sophie Germain (1776–1831), Adrien-Marie Legendre (1752–1833), Peter Gustav Lejeune Dirichlet (1805–1859), and Ernst Kummer (1810–1893). Other special cases of this assertion, commonly called Fermat's last theorem, were settled in the twentieth century, but a complete proof was lacking until the 1990s. For several centuries it appeared that the conjecture should more aptly be called Fermat's lost theorem. However, in what may be the most stunning mathematical achievement of the twentieth century, in June 1993, the British mathematician Andrew Wiles announced that he had proven Fermat's last theorem. Wiles actually claimed to prove a highly significant special case of the Taniyama-Shimura conjecture concerning semistable elliptic curves from which Fermat's last theorem follows as a corollary. The proof is long and difficult and requires deep results from the theory of modular forms and algebraic geometry. In fact, the proof had a small but significant gap. However, in October 1994, Wiles together with R.L. Taylor completed the proof by filling in the gap along somewhat different lines than originally envisioned. In any event, the mathematics that has been created in attempts to prove Fermat's last theorem have proven to be far more significant than the statement of the theorem itself. Wiles' contribution has had significant repercussions ever since. For example, the proof of the general case of the Taniyama-Shimura conjecture was finally completed in 2001 by Brian Conrad, Fred Diamond, Richard Taylor, and Christophe Breuil and is now called the modularity theorem.

Although we are just reaching the inception of number theory as a mature branch of mathematics, the vast number of people and depth of results prohibits us from doing justice to an historical discussion of it here. Hopefully, the discussion in this section will serve as motivation to read onward. We will discuss many other mathematicians and offer historical comments as it seems appropriate throughout the coming pages.

In case there are any lingering doubts about the importance or worthiness of studying number theory, we end with a quote from a letter from Carl Gustav Jacobi (1804–1851) to Adrien Marie Legendre (1752–1833), ". . . the only goal of Science is the honor of the human spirit, and that as such, a question of number theory is worth a question concerning the system of the world."

The exercises below give a little of the flavor of number theory. Some problems are easy while others are notoriously difficult. Hopefully, all will be appealing and will help serve as an incentive to see what's ahead!

Exercise 1.1

1. According to the riddle concerning Diophantus's life, how old did he live to be?
2. Solve the following ancient Babylonian problem: The total area of 2 squares is 10,000, and the side of 1 square is 10 less than 7/8 the side of the other. Find the lengths of the sides of the two squares.
3. The ancient Egyptians wrote fractions as sums of unit fractions, that is, with 1 in the numerator. For example, $\frac{2}{7} = \frac{1}{4} + \frac{1}{28}$.
   (a) Express $\frac{3}{7}$ as a sum of distinct unit fractions.
   (b) Express $\frac{11}{13}$ as a sum of distinct unit fractions.
   (c) Paul Erdös and *Ernst Straus* conjectured (1948) that the equation $\frac{4}{n} = \frac{1}{x} + \frac{1}{y} + \frac{1}{z}$ is solvable for all $n > 1$ (x, y, z need not all be distinct.) Verify the conjecture for $2 \le n \le 10$. (It has been verified for all $n \le 10^8$.)
   (d) Verify that the conjecture in part (c) is true for all n of the form $4k + 3$ via

   $$\frac{4}{4k+3} = \frac{1}{k+2} + \frac{1}{(k+1)(k+2)} + \frac{1}{(k+1)(4k+3)}.$$

4. *Nine Chapters on the Mathematical Arts* is a Chinese text well over 2000 years old. Solve the following problem from it: In the middle of a circular pond 10 ft in diameter there is a reed that extends to a height 1 ft out of the water. When it is drawn down it just touches the edge of the pond. How deep is the water?
5. Solve the following problem attributed to the ninth-century south Indian mathematician Mahavira: "Of a collection of mango fruits, the king took $\frac{1}{6}$, the queen $\frac{1}{5}$ of the remainder, and the three chief princes $\frac{1}{4}$, $\frac{1}{3}$, and $\frac{1}{2}$ of the successive remainders, and the youngest child took the remaining three mangoes. O you who are clever in miscellaneous problems on fractions, give out the measure of that collection of mangoes."

6. (a) What is the formula for the nth triangular number?
   (b) What is the formula for the nth pentagonal number?
   (c) A generalization of triangular numbers to three dimensions is the sequence of tetrahedral numbers (see Figure 1.2). The sequence begins with $1, 4, 10, 20, 35, ...$ What is the formula for the nth tetrahedral number?

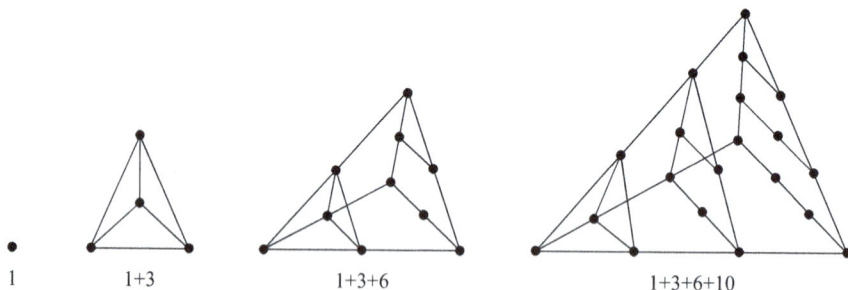

Figure 1.2: The first few tetrahedral numbers.

7. Show that the sum of two consecutive triangular numbers is a perfect square.
8. The Babylonian tablet Plimpton 322 contains the numbers 12,709 and 18,541. Find the third number for which they form a Pythagorean triplet.
9. Problem 8 of Book II of Diophantus's *Arithmetica* deals with dividing a given square into two rational squares. Express the number 16 as the sum of 2 rational squares.
10. Verify Fermat's claim that every odd prime number can be expressed uniquely as the difference of two squares. Is the assertion true if we eliminate the word "prime"?
11. Write out the numbers from 1 to 100 and express them as sums of squares using the least number of squares possible. Express them as sums of triangular numbers and finally as sums of cubes. Make some conjectures. Try to verify them.
12. Find an integral solution to Fermat's equation $x^3 = y^2 + 2$.
13. (a) Show that there are infinitely many positive integers which are not expressible by fewer than four squares.
    (b) Write out the numbers from 1 to 100 and express as many as you can as a sum of five nonzero squares. What is the largest number in your list not expressible as the sum of five nonzero squares?
    (c) Write out the numbers from 1 to 150 and express them as a sum of distinct squares. Can you posit a conjecture?
14. Why does the following trick work? Pick a number. Double it and add 6. Now double again and add 4 . Divide by 4 . Subtract your original number and then add 1 . You now have 5.

15. Explain the following: Let n be a 3-digit number with the first digit larger than the third digit. Let r be the number formed by reversing the digits of n. Let $N = n - r$ and let R be the number formed by reversing the digits of N. Then $N + R = 1089$.

16. Find two triangular numbers having sum and difference triangular.

17. What is the largest product a set of positive integers can have with a sum of 24 ? (It may be helpful to first try a smaller sum.)

18. (a) In a letter to Leonhard Euler dated June 7, 1742, Christian Goldbach (1690–1764) stated that every even integer greater than two was expressible as the sum of two primes. Goldbach's conjecture is still an outstanding problem (though it has been verified by Matti Sinisalo (1993) up to $4 \cdot 10^{11}$). Verify the assertion up to 100.

    (b) Count the number of ways each even integer less than 100 can be expressed as a sum of primes.

    (c) Try to express the even integers less than 100 as a sum of odd composites. Can you make any conjectures? Can you prove them?

19. Let $a_1 < b_1 < c_1 < d_1$ be four successive cubes. Let $a_2 = b_1 - a_1$, $b_2 = c_1 - b_1$, and $c_2 = d_1 - c_1$. Let $a_3 = b_2 - a_2$, $b_3 = c_2 - b_2$, and $a_4 = b_3 - a_3$. Show that $a_4 = 6$. Can you generalize this in any way?

20. The sequence 1, 25, 49 is a sequence of 3 squares in arithmetic progression. Are there other such sequences? What about longer sequences? This problem was stated by Fermat in 1656 but not fully addressed until Euler disposed of it in 1780. Due to Euler's tremendous output, the Saint Petersburg Academy did not publish the result until 1818.

21. Verify that $n^p - n$ is divisible by p for $1 \le n \le p$ for each of the primes $p = 2, 3, 5$, and 7. What is the situation for p prime and $n > p$? What happens if p is not prime?

22. Verify that if $p = 2, 3, 5$, or 7, then $(p - 1)! + 1$ is divisible by p. Does n divide $(n - 1)! + 1$ for any other $n \le 10$? Care to make a conjecture?

23. Find all values of $n \le 10$ for which $2^n - 1$ is prime. Care to make a conjecture? Check your conjecture on $n = 11$ and $n = 12$.

24. Show that every cube can be expressed as the difference of two squares.

25. On September 25, 1654, Fermat made the following assertion in a letter to Pascal: Every prime number of the form $3k + 1$ can be written as the sum of a square and three times a square. Verify this for all appropriate primes less than 100.

26. An integer is square-full (alternatively, powerful) if each of its prime factors appears to at least the second power.

    (a) Prove that there are infinitely many pairs of consecutive square-full numbers. Two such pairs are 8, 9 and 288, 289.

    (b) Can you find any consecutive triples of square-fulls? How about any pairs of consecutive cube-fulls? No one knows the answers to either of these questions. Good luck!

27. Collatz problem: A famous problem attributed to L. Collatz (1937) is the following (also called the Syracuse problem): Let m be a positive integer and let $a_1 = m$. Recursively define $a_{n+1} = a_n/2$ if a_n is even and $a_{n+1} = 3a_n + 1$ if a_n is odd. Verify that for all $m < 40$, there exists an n for which $a_n = 1$. The sequence will take over 100 steps for $m = 27$. (As of 2025, the conjecture has been verified up to $2^{71} \approx 2.36 \times 10^{21}$.)

1.2 Induction and the well-ordering principle

Two key concepts in number theory as well as in many other areas of mathematics are the well-ordering principle (WOP) and the principle of mathematical induction (PMI). No doubt you are already familiar with the latter principle, though you may be unaware of its wide applicability. The WOP, though perhaps not yet recognizable by name, will undoubtedly seem fairly intuitive. You may wonder why we even need to make such a big deal about so obvious an assertion. Interestingly, neither concept can be proved from other basic axioms of arithmetic. In fact, the PMI is the fifth and final of the so-called Peano axioms of arithmetic as delineated by Giuseppe Peano (1858–1932) in 1889. Even more intriguing, these two concepts are actually equivalent! That is, each is just a rewording of the other and hence really amounts to only one new axiom.

In this section we state the WOP and two variations of the induction principle. We prove the equivalence of them and then give a couple of examples demonstrating their utility. Keep in mind that these concepts will be of central importance throughout the rest of the book. As usual, let $\mathbf{N}$ represent the set of natural numbers $\{1, 2, 3, \ldots\}$.

WOP: Every non-empty subset of N has a least element.

We will see many important applications of the WOP in the pages ahead. The key to many problems is simply to identify the appropriate subset of $\mathbf{N}$. Here is a somewhat facetious example.

All natural numbers are interesting. The demonstration runs as follows: If the assertion were false, then there would be some natural numbers that were uninteresting. By the WOP there must be a least number, call it N, that was uninteresting. But then N would be the least uninteresting number. That's interesting! This is a contradiction and so the set of uninteresting numbers must be empty. That is, all natural numbers are interesting.

We now state our first version of the PMI.

PMI: If S is a subset of $\mathbf{N}$ such that $1 \in S$ and such that if $n \in S$ then $n + 1 \in S$, then $S = N$.

Another formulation of mathematical induction will prove quite useful in Chapter 2. It runs as follows:

Principle of complete induction (PCI): If S is a subset of $\mathbf{N}$ such that if $1,.., n \in S$ then $n+1 \in S$, then $S = N$.

PMI is sometimes referred to as weak induction. Consequently, PCI is then referred to as strong induction since it allows from a seemingly stronger hypothesis. Such distinctions are not fully accurate as we see in our next theorem.

Theorem 1.1: The WOP, PMI, and PCI are all equivalent.

Proof: We will prove Theorem 1.1 by showing that PCI implies WOP, WOP implies PMI, and finally PMI implies PCI. (Hopefully we're not overdoing the acronyms!) We utilize the fact that the contrapositive of a statement is equivalent to it.

(PCI $\Rightarrow$ WOP) Let T be a nonempty subset of $\mathbf{N}$ and suppose that T has no least element. Let $S = \mathbf{N} - T$, the complement of T in $\mathbf{N}$. Clearly $1 \notin T$ since 1 would be the least element of T. Hence $1 \in S$. Similarly, none of the numbers $2, 3,.., n$ are elements of T. So $1, 2,.., n \in S$. But then $n+1 \notin T$ since otherwise $n+1$ would be the least element of T. So $n+1 \in S$. By PCI, $S = N$. Hence T is empty contrary to our original assumption.

(WOP $\Rightarrow$ PMI) Let S be a subset of $\mathbf{N}$ with $1 \in S$ and such that if $n \in S$ then $n+1 \in S$. Let $T = \mathbf{N} - S$. If $S \neq \mathbf{N}$ then T is non-empty and WOP implies that T has a least element, call it t. Clearly $t > 1$ since $1 \in S$. Consider $n = t - 1$. Since n is less than t, n must be in S. But then $n+1 = t \in S$. This is a contradiction and hence $S = \mathbf{N}$.

(PMI $\Rightarrow$ PCI) Let S be a subset of $\mathbf{N}$ such that if $1,.., n \in S$ then $n+1 \in S$. It follows that $1 \in S$ since every natural number less than 1 is in S (i.e., $\varnothing \subset S$). Assume now that $1,.., n \in S$. In particular $n \in S$. By our condition above, $n+1 \in S$. But then $1 \in S$ and $n \in S$; hence $n+1 \in S$. By PMI, we have $S = N$.

This completes our proof. ■

Oftentimes we let $P(k)$ represent some proposition concerning the integer k. With this notation, PMI can be restated as follows: If $P(1)$ is true and $P(n)$ implies $P(n+1)$, then the proposition is true for all positive integers.

Example 1.1: Use induction (i.e., PMI) to prove that $\sum_{i=1}^{n} \frac{1}{i(i+1)} = \frac{n}{n+1}$.
Solution: Let $P(k)$ denote the above proposition for $n = k$.
Check $P(1)$: $\frac{1}{2} = \frac{1}{2}$.
Show $P(n) \Rightarrow P(n+1)$: Assume $P(n)$ is true; $\sum_{i=1}^{n} \frac{1}{i(i+1)} = \frac{n}{(n+1)}$.

Then $\sum_{i=1}^{n+1} \frac{1}{i(i+1)} = \sum_{i=1}^{n} \frac{1}{i(i+1)} + \frac{1}{(n+1)(n+2)} = \frac{n}{n+1} + \frac{1}{(n+1)(n+2)} = \frac{n+1}{n+2}$. This is $P(n+1)$ and we are done.

Often we wish to establish a number-theoretic property that holds for all n beyond a particular point. In such a case we need to modify the first step in PMI or PCI accordingly. Analogously we often modify the WOP to hold for all integers greater than a particular value. Our next example is of this type.

Example 1.2: Use induction to prove that $2^n \geq (n+1)^2$ for $n \geq 6$.
Solution: Let $P(k)$ denote the above proposition for $n = k$. We need to modify our first step appropriately.
Check $P(6)$: $2^6 = 64 \geq 49 = (6+1)^2$.
  Show $P(n) \Rightarrow P(n+1)$: Assume $P(n)$ is true for some n; $2^n \geq (n+1)^2$. Then

$$2^{n+1} = 2(2^n) \geq 2(n+1)^2 = 2n^2 + 4n + 2$$

$$= n^2 + (n^2 + 4n + 2) \geq 2 + (n^2 + 4n + 2) = (n+2)^2$$

So $P(n+1)$ is true. and by the PMI the assertion is proved.

There is some allowance for taste in the way in which an inductive proof is presented. However, there are also several common errors that some students make when they first learn to write proofs. To help avoid them, please note that there are always two steps involved in an inductive proof: First, an initial case must be checked. Second, the inductive hypothesis, $P(n)$, must be used to show that its successor, $P(n+1)$, necessarily follows.

   Now let us discuss an interesting sequence of numbers called the Fibonacci sequence. These are named after the Italian mathematician Leonardo of Pisa (ca. 1175–1250) who was also known as Fibonacci Pisano or simply Fibonacci (son of Bonaccio). In addition to promoting the Hindu-Arabic numeral system widely used today, his book *Liber Abaci* (1202) also contained the following problem: "How many pairs of rabbits can be produced from a single pair in a year if every month each pair begets a new pair which from the second month on becomes productive?". This leads to the following definition:

Definition 1.1: The *Fibonacci sequence* $\{F_i\}$ is defined recursively as $F_1 = 1$, $F_2 = 1$, and $F_{n+2} = F_n + F_{n+1}$ for $n \geq 1$.

The Fibonacci sequence begins 1, 1, 2, 3, 5, 8, 13, 21, 34, 55, 89, 144, 233, and so on. A natural question is whether there is a closed formula for the nth Fibonacci number. Indeed there is. It was discovered independently by A. DeMoivre (1718), L. Euler (1765), and others but is usually referred to as Binet's formula. The French mathematician J.P.M. Binet (1786–1856) published the following result in 1843.

Proposition 1.2: Let $\Phi = \frac{1+\sqrt{5}}{2}$ and $\Phi' = \frac{1-\sqrt{5}}{2}$. Then

$$F_n = \frac{1}{\sqrt{5}}\left(\Phi^n - \Phi'^n\right). \tag{1.1}$$

The number Φ is known as the *golden* ratio and appears often both in mathematics and in nature itself. The proof of Proposition 1.2 requires the following lemma:

Lemma 1.2.1: If $x^2 = x + 1$, then $x^n = F_n x + F_{n-1}$ for $n \geq 2$.

Proof of Lemma: We will use induction on n where $P(n)$ is the statement that $x^n = F_n x + F_{n-1}$.

$$P(2): x^2 = x + 1 = F_2 x + F_1.$$

$$P(n) \Rightarrow P(n+1): x^{n+1} = x^n(x) = (F_n x + F_{n-1})x = F_n x^2 + F_{n-1}x = F_n(x+1) + F_{n-1}x$$

$$= (F_n + F_{n-1})x + F_n = F_{n+1}x + F_n. \qquad \blacksquare$$

Proof of Proposition: The roots of $x^2 = x + 1$ are Φ and Φ'. By Lemma 1.2.1, $\Phi^n = F_n\Phi + F_{n-1}$ and $\Phi'^n = F_n\Phi' + F_{n-1}$. Thus $\Phi^n - \Phi'^n = F_n(\Phi - \Phi')$.
But $\Phi - \Phi' = \sqrt{5}$ and so $F_n = \frac{1}{\sqrt{5}}(\Phi^n - \Phi'^n)$. $\qquad \blacksquare$

Finally, we introduce an attractive result about Fibonacci numbers first proved by the Italian astronomer and mathematician J.D. Cassini in 1680 while serving as director of the Paris Observatory. We extend the Fibonacci sequence by defining $F_0 = 0$.

Proposition 1.3 (Cassini's Identity): For all $n \geq 1$, $F_{n+1}F_{n-1} - F_n^2 = (-1)^n$.

Proof: (Induction on n). For $n = 1$, $F_2F_0 - F_1^2 = -1 = (-1)^1$. Assume the proposition now holds for some positive integer n. Then $F_{n+2}F_n - F_{n+1}^2 = (F_n + F_{n+1})F_n - F_{n+1}^2 = F_n^2 + F_{n+1}F_n - F_{n+1}^2 = F_n^2 + F_{n+1}F_n - F_{n+1}(F_{n-1} + F_n) = F_n^2 - F_{n+1}F_{n-1} = -(F_{n+1}F_{n-1} - F_n^2) = -(-1)^n = (-1)^{n+1}$. $\qquad \blacksquare$

Exercise 1.2

1. (a) Let t_n be the nth triangular number. Use induction to prove that $t_n = \frac{n(n+1)}{2}$ for $n \geq 1$.
   (b) Let $S_n = 1^2 + 2^2 + 3^2 + \cdots + n^2$. Use induction to prove that $S_n = \frac{n(n+1)(2n+1)}{6}$.
   (c) Evaluate $\sum_{j=1}^{n} j(j+1)$.
2. (a) Let $C_n = 1^3 + 2^3 + \cdots + n^3$. Use induction to prove that $C_n = \frac{n^2(n+1)^2}{4}$.
   (b) State and prove a proposition suggested by the following pattern: $1 = 1^3$, $3 + 5 = 2^3, 7 + 9 + 11 = 3^3$, etc.

3. (a) Consider the general arithmetic progression $s_n = a + (a + d) + \cdots + (a + (n-1)d)$. Use induction to show that $s_n = \frac{n(2a + (n-1)d)}{2}$ for $n \geq 1$. You can remember this sum as the number (of terms) times the first (term) plus the last (term) all over two.

   (b) Consider the general geometric progression $g_n = a + ar + ar^2 + \cdots + ar^n$. Derive a formula for g_n and use induction to substantiate it.

4. (a) Let $f_k(n)$ be the nth k-gonal number defined by
   $$f_k(n) = 1 + (k-1) + (2k-3) + (3k-5) + \cdots + ((n-1)k + 3 - 2n).$$
   Prove $f_k(n) = \frac{n[(n-1)(k-2)+2]}{2}$.
   Check your result here with Exercise 1.1.6.

   (b) Show that $f_3(2n-1) = f_6(n)$, that is, the sequence of hexagonal numbers is the same as the sequence of every other triangular number.

5. Prove the following theorem of Nicomachus (c.a. 100 C.E.):
   $f_k(n) = f_{k-1}(n) + f_3(n-1)$ for $k \geq 4$ (e.g., the sum of the nth square and the $(n-1)$st triangular number equals the nth pentagonal number).

6. Prove that for all $n \geq 1$, the sum of the the first n odd positive integers equals n^2. (The Russian mathematician and scientist A.N. Kolmogorov (1903–1989) rediscovered this classic result at age 5.)

7. Into how many regions can we separate the plane using n straight lines?

8. Prove the following elementary facts about Fibonacci numbers:
   (a) $F_1 + F_2 + \cdots + F_n = F_{n+2} - 1$.
   (b) $F_2 + F_4 + \cdots + F_{2n} = F_{2n+1} - 1$.
   (c) $F_1 + F_3 + \cdots + F_{2n-1} = F_{2n}$.

9. (a) Prove that $F_1^2 + F_2^2 + \cdots + F_n^2 = F_n F_{n+1}$.
   (b) Prove that $F_n^2 + F_{n+1}^2 = F_{2n+1}$.

10. Give a matrix proof of Cassini's Identity (Proposition 1.3) as follows:
    (a) Let $M = \begin{bmatrix} F_2 & F_1 \\ F_1 & F_0 \end{bmatrix}$. Verify that $\det M = (-1)^1$.
    (b) Show by induction that $M^n = \begin{bmatrix} F_{n+1} & F_n \\ F_n & F_{n-1} \end{bmatrix}$.
    (c) Use the fact that the determinant function is multiplicative, i.e., $\det(AB) = (\det A)(\det B)$, to conclude the proof.

11. Use Binet's formula (Proposition 1.2) to calculate $\lim_{n \to \infty} \frac{F_{n+1}}{F_n}$

12. Define the Lucas sequence (Edouard Lucas 1842–1891) by $L_1 = 1$, $L_2 = 3$, and $L_{n+2} = L_n + L_{n+1}$ for $n \geq 1$.
    (a) Show $F_{n-1} + F_{n+1} = L_n$ for $n \geq 1$ where $F_0 = 0$.
    (b) Show $L_{n-1} + L_{n+1} = 5 F_n$ for $n \geq 1$ where $L_0 = 2$.

13. Let $\Phi = \frac{1+\sqrt{5}}{2}$ and $\Phi' = \frac{1-\sqrt{5}}{2}$. Prove that $L_n = \Phi^n + \Phi'^n$ for $n \geq 1$.

14. (a) Show that F_n is the integer closest to $\frac{\Phi^n}{\sqrt{5}}$ for $n \geq 1$.
    (b) Show that L_n is the integer closest to Φ^n for $n \geq 2$.

15. Let $a, b \in \mathbb{N}$. Let $x_1 = a$, $x_2 = b$, and $x_{n+2} = x_n + x_{n+1}$ for $n \geq 1$
    (a) Find a formula for x_n in terms of the Fibonacci numbers.

(b) Find $\lim_{n\to\infty} \frac{x_{n+1}}{x_n}$.

16. Prove the following generalization of Example 1.1:

$$\sum_{i=1}^{n} \frac{1}{((i-1)k+1)(ik+1)} = \frac{n}{nk+1}$$

17. (a) Show that if the Fibonacci numbers F_a, F_b, F_c are in arithmetic progression, then $a = b - 2$ and $c = b + 1$ for $b > 2$.

(b) Show that the Fibonacci sequence has no arithmetic progressions of length four.

18. (a) Find the smallest c for which $t_a \cdot t_b = t_c$ where $1 < a < b < c$ and t_n denotes the nth triangular number.

(b) Find the smallest $S = a + b + c$ for which S is triangular and $t_a + t_b = t_c$ (Problem 518, College Mathematics Journal, Jan. 1994 – due to K.R.S. Sastry).

19. (a) Show that 21, 2211, 222111, $\cdots$ are all triangular numbers.

(b) Show that 55, 5050, 500500, $\cdots$ are all triangular numbers.

20. (a) Verify that $t_{3n} + t_{4n+1} = t_{5n+1}$ for all $n \geq 1$.

(b) Verify that $t_{k-1} + t_n = t_k$ where $k = t_n$ for all $n \geq 1$.

21. Suppose you clear n Go stones off a Go board by collecting either one or two stones each time. If the number of stones remaining were noted throughout the clearing process, how many different ways are there to remove the stones?

22. Comment on our inductive "proof" of the following: All cars are the same. If there were just one car, then the assertion is trivially true. Assume the assertion is true for n cars. If there were $n+1$ cars, then the first n would be the same by our inductive hypothesis, as would the last n. But because of the overlap between the first n and the last n, all $n+1$ cars are the same!

23. What, if any, is the least element in the set $S = \left\{ 1, \frac{1}{2}, \frac{1}{4}, \cdots, \frac{1}{2^k}, \cdots \right\}$? Does this contradict the WOP?

24. Show that $\frac{n^3}{3} + \frac{n^2}{2} + \frac{n}{6}$ is an integer for all $n \geq 1$.

25. (a) Determine what type of polygonal number each of the following are: 1918, 1936, 1953, 1976, 2016, 2025, 2035.

(b) Verify that 40,755 is triangular, pentagonal, and hexagonal. (See Exercise 5.1.8 concerning triangular numbers that are squares.)

26. How many ways can a total of n adults and children be lined up so that no two children are standing next to each other?

27. (a) Show that if $k \geq 2$ and n is any positive integer, then n^k can be expressed as a sum of n consecutive odd integers.

(b) Show that if $k \geq 2$ and n is odd, then n^k can be expressed as a sum of n consecutive integers.

28. Use induction on the numerators to show that any reduced fraction between 0 and 1 can be written as a sum of distinct unit fractions (J.J. Sylvester – 1880).

1.3 Divisibility and congruences

The integers $\mathbb{Z}$ are comprised of the natural numbers $\mathbf{N}$ (or $\mathbb{Z}^+$) = {1, 2, 3, 4, . . .}, the number 0, and the negative integers which we will denote by $\mathbb{Z}^-$. From now on, the term "number" without qualifications will usually refer to natural number. Typically, we will use lower case Latin letters to denote integers.

One nice property of $\mathbb{Z}$ is that it is closed with respect to addition, subtraction, and multiplication. That is, if m and n are integers then so are $m+n, m-n$, and mn. Unfortunately, the same is not true for division in $\mathbb{Z}$ and so much greater care and study is required of it. In this section we will begin our study of divisibility as propounded by Gauss over two centuries ago. Of course, many of the basic notions are much older.

Definition 1.2: Let a and b be integers. We say that a *divides* b (written $a \mid b$) if there exists an integer c for which $b = ac$. In this case we say that a is *a factor* of b or that b is *divisible* by a or is *a multiple* of a. If b is not divisible by a we write $a \nmid b$.

For example, $5|35$, $12|-72$, and $a|0$ for all integers a. On the other hand, $3 \nmid 10$, $-12 \nmid 6$, and $0 \nmid a$ for all nonzero a. Below we establish some elementary properties of divisibility.

Proposition 1.4: Let a,b,c,x, and y be integers.
(a) (Reflexivity) $a \mid a$.
(b) (Transitivity) If $a \mid b$ and $b \mid c$ then $a \mid c$.
(c) (Linearity) If $a \mid b$ and $a \mid c$ then $a \mid (bx + cy)$.
(d) If $a \mid b$ then $a \mid bc$.
(e) Let $a > 0$ and $b > 0$. If $a \mid b$ then $a \leq b$.
(f) (Antisymmetry) If $a \mid b$ and $b \mid a$ then $a = \pm b$.

Proof:
(a) $a = 1a$.
(b) $a \mid b$ implies that there exists x for which $b = ax$. Since $b \mid c$, there exists y for which $c = by$. So $c = axy$ and hence $a \mid c$.
(c) $a \mid b$ and $a \mid c$ implies there are f and g for which $b = af$ and $c = ag$. So $bx + cy = afx + agy = a(fx + gy)$. Hence $a \mid (bx + cy)$.
(d) If $a \mid b$, then there exists x for which $b = ax$. So $bc = acx$ and hence $a \mid bc$.
(e) Since a and b are positive, $a \mid b$ implies there exists $c > 0$ such that $b = ac$. But since c is an integer, in fact $c \geq 1$. Hence $a \leq b$.
(f) If $a = 0$ then $b = 0$ and we are done. If $a \neq 0$, then $a \mid b$ ensures there exists x for which $b = ax$. Similarly, $b \mid a$ implies there exists y for which $a = by$. So $a = axy$ and so $1 = xy$. Since x and y are integers, either $x = y = 1$ or $x = y = -1$. Hence $a = \pm b$.

Theorem 1.5 (Division Algorithm): Let a and b be integers with $b > 0$. Then there exist unique integers q and r such that $a = qb + r$ where $0 \leq r < b$.

The division algorithm simply formalizes what you've known since grade school. When you divide an integer (the dividend) by another integer (the divisor), you get a unique quotient (q) and remainder (r). For example, if $a = 100$ and $b = 13$ then $q = 7$ and $r = 9$. If a is positive, we can think of the division algorithm in terms of successive subtractions. The number r is the last nonnegative integer obtained after repeatedly subtracting b from a (in fact, q times).

Proof: Given a and b, consider the set $S = \{a - nb\colon n \in \mathbb{Z}\}$. If $a \geq 0$, then $a \in S$ (just let $n = 0$) and so S contains a nonnegative integer. If $a < 0$, then let $n = a$. In this case $a - nb = -a(b - 1)$ which is nonnegative. By the WOP (suitably modified to include the number zero), there is a least nonnegative element of S; call it r. Now define q by $a = qb + r$. To see that $0 \leq r < b$, first note that r is defined to be nonnegative. If $r \geq b$, then $a - (q + 1)b \geq 0$ and r would not be the least nonnegative element of S. We only need to establish uniqueness.

Suppose there exist r_1, r_2, q_1, and q_2 such that $a = q_1 b + r_1 = q_2 b + r_2$. Then $b(q_1 - q_2) = r_2 - r_1$. Hence $b \mid (r_2 - r_1)$. But $0 \leq r_1, r_2 < b$ and so $|r_2 - r_1| < b$. By Proposition 1.3(c), $r_2 - r_1 = 0$ and $r_1 = r_2$. But then $q_1 b = q_2 b$ and so $q_1 = q_2$. ∎

It is useful to note that q is the greatest integer less than or equal to a/b. We denote this by $q = [a/b]$.

Definition 1.3: *Greatest integer function:* If r is a real number then $[r]$ denotes the greatest integer less than or equal to r.

For example, $[2.9] = 2, [\sqrt{10}] = 3, [5] = 5$, and $[-1.3] = -2$. The greatest integer function is often called the *floor function*. It can also be thought of as the age function since one usually gives one's age as being that at his/her last birthday (if not earlier).

It can be a bit tricky to deal with sums involving the greatest integer function. But one nice exception is the following result due to the great French analyst and number theorist Charles Hermite (1822–1901).

Proposition 1.6: Let $x \in \mathbf{R}$ and n be a positive integer, then

$$[nx] = [x] + \left[x + \frac{1}{n}\right] + \left[x + \frac{2}{n}\right] + \cdots + \left[x + \frac{n-1}{n}\right].$$

Proof: Let $f(x) = [x] + \left[x + \frac{1}{n}\right] + \cdots + \left[x + \frac{n-1}{n}\right] - [nx]$.

Then

$$f\left(x + \frac{1}{n}\right) = \left[x + \frac{1}{n}\right] + \left[x + \frac{2}{n}\right] + \cdots + [x + 1] - [nx + 1].$$

Thus,

$$f\left(x + \frac{1}{n}\right) - f(x) = ([x + 1] - [x]) + ([nx] - [nx + 1]) = 1 - 1 = 0.$$

Hence, f has period $1/n$ (or some number that divides $1/n$). But if $0 \le x < 1/n$, then each term of $f(x)$ is 0 since $0 \le x < x + \frac{n-1}{n} < 1$ and $0 \le nx < 1$. Hence, $f(x) = 0$ for $0 \le x < 1/n$. By the periodicity of f, $f(x) = 0$ for all x and the proposition is proven. ∎

We now move on to a brief discussion of congruences. The following definition, due to Gauss in his *Disquisitiones Arithmeticae* (1801), should seem straightforward. Gauss's careful choice of notation was itself a significant milestone and will have far-reaching consequences in the pages ahead.

Definition 1.4: Let a, b, and n be integers with $n > 0$. We say that a is *congruent to b modulo n* if $n \mid (a - b)$. This is denoted $a \equiv b \pmod{n}$. If $n \nmid (a - b)$ then we say that a is *incongruent* to b modulo n and write $a \not\equiv b \pmod{n}$.

For example, $7 \equiv 37 \pmod{10}, 18 \equiv -15 \pmod{11}$, and $8 \not\equiv 4 \pmod{6}$. In other words, a is congruent to b modulo n if and only if a and b have the same remainder on division by n. Below we collect some basic but important properties of congruences. Proposition 1.5 establishes that congruence to a fixed modulus is an equivalence relation.

Proposition 1.7: Let $a, b, c \in \mathbb{Z}$ and $n \in \mathbb{N}$.
(a) (Reflexivity) $a \equiv a \pmod{n}$.
(b) (Symmetry) If $a \equiv b \pmod{n}$ then $b \equiv a \pmod{n}$.
(c) (Transitivity) If $a \equiv b \pmod{n}$ and $b \equiv c \pmod{n}$ then $a \equiv c \pmod{n}$.

Proof:
(a) $n \mid (a - a)$.
(b) $a \equiv b \pmod{n}$ implies $n \mid (a - b)$ and so there exists a d such that $nd = a - b$. But then $n(-d) = b - a$ and so $b \equiv a \pmod{n}$.
(c) If $a \equiv b \pmod{n}$, then there exists an integer d such that $nd = a - b$. $b \equiv c \pmod{n}$ means there exists an f such that $nf = b - c$. But then $n(d + f) = a - c$ and $a \equiv c \pmod{n}$. ∎

The fact that congruence modulo n is an equivalence relation means that the set of integers can be partitioned into n equivalence classes. Each integer is placed into a class dependent upon its remainder or residue modulo n. This observation leads to our next proposition.

Proposition 1.8: Let $n \in \mathbf{N}$. Every integer is congruent to exactly one of $0, 1, \cdots$, or $n - 1$ modulo n.

Proof: Let $a \in \mathbf{Z}$. By the division algorithm (Theorem 1.4), there exist unique integers q and r such that $a = qn + r$ where $r \in \{0, 1,.., n-1\}$. So $a \equiv r \pmod n$ as desired. Uniqueness also follows directly from the division algorithm. ∎

Definition 1.5: A *complete set of residues modulo n* is a set of n integers $r_1, r_2,..., r_n$ for which every integer is congruent to exactly one of $r_1, r_2,..., r_n \pmod n$.

For example, by Proposition 1.8, $\{0, 1, 2, 3, 4, 5, 6\}$ is a complete set of residues modulo 7. This is the canonical complete residue set. Other complete residue sets modulo 7 are $\{-3, -2, -1, 0, 1, 2, 3\}$ and $\{12, 24, 43, 76, 88, 93, 112\}$.

Proposition 1.9: Let $a, b, c, d \in \mathbf{Z}$ and $n \in \mathbf{N}$. If $a \equiv b \pmod n$ and $c \equiv d \pmod n$, then
(a) $a + c \equiv b + d \pmod n$.
(b) $a - b \equiv c - d \pmod n$.
(c) $ac \equiv bd \pmod n$.

Proof: The proof is left as Exercise 1.3.2(a).
From an abstract algebra perspective, the set of integers modulo n form an additive group $\mathbf{Z}/n\mathbf{Z}$, often denoted simply by $\mathbf{Z}_n$, consisting of the equivalence classes $[1], \ldots$ $[n]$ where the equivalence class $[j]$ includes all numbers congruent to j (mod n). We denote the equivalence class $[j]$ simply by j, though any element of $[j]$ can be chosen to be its representative. The group $\mathbf{Z}_n$ is a cyclic group generated by 1.

Notice that by part (c) of Proposition 1.9, if $a \equiv b \pmod n$, then $a^2 \equiv b^2 \pmod n$. It follows by induction that if $a \equiv b \pmod n$, then $a^r \equiv b^r \pmod n$ for any positive integer r (Exercise 1.3.2(b)). In addition, by parts (a) and (b), we can add and subtract congruent expressions without disturbing the congruence relation. An important consequence is that if $f(x)$ is a polynomial with integer coefficients and $a \equiv b \pmod n$, then $f(a) \equiv f(b) \pmod n$.
For example, if $f(x) = 3x^8 + 12x^2 - 5x + 7$, then
$f(12) \equiv f(25) \pmod{13}$ since $12 \equiv 25 \pmod{13}$.

Example 1.3: Let us find the remainder when we divide 230^{61} by 11. $230 \equiv -1 \pmod{11}$ and so $230^{61} \equiv (-1)^{61} = -1 \equiv 10 \pmod{11}$. The remainder when we divide 230^{61} by 11 is 10.

Example 1.4: We now discuss an interesting math puzzle involving hats and prisoners. To begin with, suppose there are two prisoners. The warden places either a black or white hat on each of them. They only see the other prisoner and cannot communicate with each other in any way. At the count of three, they must simultaneously yell out the color of the hat on their head. If exactly one of them is correct, they will be released. Otherwise, they remain imprisoned. What strategy guarantees success?

Solution: Prisoner A yells out the color of the hat he sees while prisoner B yells out the opposite color of the hat he sees. Either both hats are the same color or both hats are different. Either way, exactly one of them will be correct!

Now for a generalization: There are 12 prisoners. The warden has a large number of hats of 12 different colors all of which are known to the prisoners. He seats the prisoners in a large circle and places a hat on each prisoner so that they can see all the hats but their own. He uses as many hats of any given color as he wishes. The prisoners cannot communicate with one another. At the count of three, all must announce the color of the hat on their own head. If exactly one of them is correct, then they are all set free. What strategy guarantees success?

Solution: The prisoners label themselves 1, 2, 3, . . ., 12 perhaps depending on their position around the circle (hours of the clock, for example). They also label the different colors by the numbers 1, 2, . . ., 12 which they memorize. Each of them adds up the numbers corresponding to the 11 hats that they see. They then subtract that number from their own prisoner number and reduce (modulo 12) to obtain one of the numbers 1, 2, 3, 4, 5, 6, 7, 8, 9, 10, 11, or 12. That number corresponds to the hat color that they announce.

For example, the following table lists the 12 prisoners followed by their hat color numbers:

$$1 - 5, 2 - 3, 3 - 3, 4 - 10, 5 - 1, 6 - 3, 7 - 9, 8 - 12, 9 - 10, 9 - 5, 10 - 11, 11 - 5, 12 - 6$$

Prisoner 1 adds up the values of the hats he sees getting 78 and $1 - 78 = -77$ which is 7 (mod 12). He says color 7 (which is incorrect since he wears hat color 5). The other prisoners do a similar calculation. Prisoner 2 gets 80 and $2 - 80 = -78$ which is 6 (mod 12). He is incorrect. Prisoner 3 gets 80 and $3 - 80 = -77$ which is 7 (mod 12); this is wrong. Prisoner 4 gets 73 and $4 - 73 = -69$ which is 3 (mod 12); this is incorrect. Prisoner 5 gets a sum of 82 and $5 - 82 = -77$ which is 7 (mod 12). His answer is wrong. Prisoner 6 gets 80 and $6 - 80 = -74$ which is 10 (mod 12); this leads to an incorrect guess. Prisoner 7 gets a sum of 74 and $7 - 74 = -67$ which is 5 (mod 12); this is wrong. Prisoner 8 gets 71 and $8 - 71 = -63$ which is 9 (mod 12); this is wrong. Prisoner 9 gets a sum of 73 and $9 - 73 = -64$ which is 8 (mod 12); this is incorrect. Prisoner 10 gets a sum of 72 and $10 - 72 = -62$ which is 10 (mod 12); this is wrong. Prisoner 11 gets a sum of 78 and $11 - 78 = -67$ which is 5 (mod 12); this is correct! Prisoner 12 gets a sum of 77 and $12 - 77 = -65$ which is 7 (mod 12); this leads to a wrong guess. Exactly one prisoner (in this case prisoner 11) got the right answer. The 12 prisoners are set free.

Explanation: The total sum of the 12 hat numbers is some number h. Prisoner number n gets a sum of $h - x_n$ where x_n is the color number of his hat. That prisoner announces

$n-(h-x_n) = n-h + x_n$ (mod 12). But $n-h$ runs through a complete residue system mod 12 as the prisoners run through n from 1 to 12. So, for exactly one prisoner $n-h$ will be congruent to 12 which is also congruent to 0 modulo 12. That's the one prisoner who announces x_n which is the correct color of his hat.

This concludes our introduction to divisibility. Chapter 2 is devoted to a deeper study of these concepts.

Exercise 1.3

1. Find q and r in the division algorithm for given values of a *and* b:

    $$(a, b) = (112, 9), (2167, 13), (-45, 7), (-176, 11), (1234, 2345).$$

2. (a) Prove Proposition 1.9.
    (b) Show that if $a \equiv b \pmod{n}$, then $a^r \equiv b^r \pmod{n}$ for any positive integer r.
3. Let n be a natural number.
    (a) Show that n has the same remainder upon division by 9 as does the sum of the digits of n.
    (b) Show that n has the same remainder upon division by 11 as does the alternating sum of the digits of n. For example, $11 \mid 1342$ and $11 \mid (1-3+4-2)$.
4. (a) Show that all squares are either congruent to 0 or 1 modulo 4 .
    (b) Show that no number in the sequence $2, 22, 222, 2222,...$ is a perfect square.
    (c) Show that no number in the sequence $5, 105, 205, 305, 405, 505,...$ is a perfect square.
5. Which of the following sets form a complete set of residues modulo 9 ?

    $$S_1 = \{-4, -3, -2, -1, 0, 1, 2, 3, 4\}, S_2 = \{9, 18, 27, 36, 45, 54, 63, 72, 81\},$$

    $$S_3 = \{5, 18, 31, 44, 57, 70, 83, 96, 109\}, S_4 = \{1, 2, 3, 4, 5, 6, 7, 8\}.$$

6. (a) What is the remainder when 100100 is divided by 11 ?
    (b) What is the remainder when 702^{10} is divided by 7 ?
7. (a) Show that two-thirds of all triangular numbers are divisible by 3. (b) Show that exactly one of $n-2, n$, and $n+2$ is divisible by 3 for any integer n.
8. Show that the product of any 4 consecutive integers is divisible by 24.
9. If $ax \equiv ay \pmod{n}$, does it follow that $x \equiv y \pmod{n}$? Investigate under what conditions it does follow.
10. How many consecutive zeros appear at the end of 1,000!?
11. Consider the sequence $S = \{1, 12, 123, 1234,..., N\}$ where the nth entry is the previous entry with the number n appended at the end. If N is the 300th entry, how many elements of S are divisible by 3? How many are divisible by 5?
12. (a) Show that there is only one solution in positive integers x, y, and z for which $x! + y! = z!$.

(b) Show that there are infinitely many solutions in positive integers $x < y < z$ for which $x!y! = z!$

13. Show there are no integral solutions to the Diophantine equation $3x - 6y + 12z = 4000$.

14. Let n be a positive integer with digital representation $d_t d_{t-1}...d_3 d_2 d_1$. Starting from the right, let a_1 be the integer with digits $d_3 d_2 d_1$, a_2 be the integer with digits $d_6 d_5 d_4$, and in general $a_{(c)}$ is the integer with digits $d_{3i} d_{3i-1} d_{3i-2}$ (if necessary, append zeros to the left of d_t so the last $a_{(c)}$ is a three-digit number). Form the alternating sum $a_1 - a_2 + a_3 - ...$ and then repeat this process to the resulting number until you obtain a three-digit number m. Show that m is divisible by p if and only if n is divisible by p for $p = 7$, 11, and 13.

15. (a) Show that $10a + b$ is divisible by 17 if and only if $a - 5b$ is divisible by 17.
    (b) Use the result in part (a) to determine which of the following are divisible by 17: 221, 357, 459, 2142, and 56,100.

16. Without direct calculation, fill in the missing digits.
    (a) $15^{10} = 576\_039062\_$
    (b) $2^2 \cdot 3^7 \cdot 5^4 \cdot 7^5 \cdot 11^3 = 1223086146\_75\_\_$

17. Show that 30 divides $n^5 - n$ for all integers n.

18. (a) What can we conclude if $a|b, b|c, c \mid d$, and $d \mid a$?
    (b) Could a, b, c, and d all be distinct if $a \mid b, b|c, c \mid d$, and $d \mid 2a$? Explain.

19. Show that the following assertions concerning the greatest integer function are false: Let r and s be reals and n an integer:
    (a) $[r] + [s] = [r + s]$
    (b) $[rs] = [r] \cdot [s]$
    (c) $[nr] = n[r]$
    (d) $[r_n] = [r]_n$.

20. (Variation on a problem from the Soviet "Tournament of Towns")
    Suppose that an American history course has an enrollment of 9, a British history course has 10, and Chinese history 11. No student is enrolled in more than one history course. Whenever two students from different courses speak to each other, they decide to drop their current courses and both add the third course. Is it ever possible that all 30 students could be in the same history course?

21. (a) In the game of Last Draw, two players alternately draw from a pile of counters. The person to draw the last counter wins. If there are initially s counters and each player may draw from 1 to n counters each time, show that the first player has a winning strategy as long as $s \equiv 0 \pmod{m+1}$.
    (b) If the rules are such that the player to draw the last counter loses, show that the first player has a winning strategy as long as $s \equiv 1 \pmod{m+1}$.

22. (a) What proportion of triangular numbers are divisible by 2^k for $k \geq 1$?
    (b) What proportion of triangular numbers are divisible by p for p an odd prime?

23. Variation on a Green Chicken problem (2013): Let n be a positive integer. Show that there are infinitely many Fibonacci numbers divisible by n. Hint: There are only finitely many pairs (F_i, F_{i+1}) (mod n). Show there is an m for which $F_0 = 0 \equiv F_m \equiv F_{2m} \equiv F_{3m} \equiv \cdots$ (mod n).

1.4 Basic combinatorics

Much of number theory involves combinatorial analysis. In this section we begin by briefly introducing two related concepts: permutations and combinations. Next, we define binomial coefficients and give some examples of their utility. Finally, we discuss the pigeonhole principle and discuss some further examples.

Definition 1.6: Let S be a set of n elements. Any arrangement of r elements from the set with $1 \leq r \leq n$ is called a *permutation* of S.

For example, let $S = \{$red, white, blue$\}$. Then the two-element permutations of S consist of red-white, white-red, red-blue, blue-red, white-blue, and blue-white. Notice that the order counts. In general, we will denote the number of r-element permutations from a set of n elements as $P(n, r)$.

It is easy to determine a formula for $P(n, r)$. Notice that in order to arrange r elements from an n-element set, we have n choices for the first element, $n - 1$ choices for the second element, . . ., and $n - r + 1$ for the last element. Hence

$$P(n, r) = n(n - 1) \cdots (n - r + 1) = \frac{n!}{(n-r)!} \tag{1.2}$$

For example, the number of three-letter "words" with no letter repeated from the set $\{a, b, c, d, e, f\}$ is $P(6, 3) = 120$.

Definition 1.7: Any r-element subset of an n-element set S is called a *combination* of S.

If $S = \{$red, white, blue$\}$, then the two-element combinations consist of $\{$red, white$\}$, $\{$red, blue$\}$, and $\{$white, blue$\}$. Notice that the order does not matter with combinations. Let $\binom{n}{r}$ represent the number of combinations of r-elements from a set of n elements. We read $\binom{n}{r}$ as "n choose r."

The calculation of $\binom{n}{r}$ is easily accomplished. The number of permutations of r-elements out of n was $P(n, r) = \frac{n!}{(n-r)!}$. However, rearrangements do not alter the set chosen, so we need to divide by $r!$ to get an accurate count of the number of r-element subsets of an n-element set. Hence $\binom{n}{r} = \frac{n!}{r!(n-r)!}$. You no doubt recognize this as the binomial coefficient from calculus. It is noteworthy that $\binom{n}{r}$ is an integer. It follows that the product of r successive integers is divisible by $r!$. This leads to the following definition.

Definition 1.8: The *binomial coefficient* $\binom{n}{r}$ is defined as

$$\binom{n}{r} = \frac{n!}{r!(n-r)!} \tag{1.3}$$

For integers $0 \le r \le n$ and $n \ge 1$.

For $r < 0$ or $r > n$ we define $\binom{n}{r} = 0$.

The binomial coefficient $\binom{n}{r}$ represents the number of combinations of r objects taken out of a set of n distinguishable objects, while $P(n, r)$ is the number of permutations of r objects from a set of n. For example, there are $\binom{6}{3} = 20$ ways to choose a 3-person committee from the group {Amy, Bill, Christopher, Doug, Evelyn, and Frank}. However, there are $P(6, 3) = 120$ ways to choose a president, treasurer, and secretary for the group.

The next result is fundamental in many areas of mathematics.

Theorem 1.10 (Binomial Theorem): Let $a, b \in \mathbb{Z}$ and $n \in \mathbb{N}$. Then

$$(a+b)^n = \sum_{k=0}^{n} \binom{n}{k} a^{n-k} b^k. \tag{1.4}$$

The result seems intuitive. When multiplying $(a+b)$ times itself n times, the coefficient of $a^{n-k}b^k$ for $0 \le k \le n$ comes from choosing k factors of b out of n possible choices. Hence the coefficient of $a^{n-k}b^k$ is $\binom{n}{k}$. We now proceed more formally.

Proof: We use induction. Let $P(n)$ denote formula (1.4). It can be readily checked that $P(1)$ is true. Assume $P(n)$ is true. Next we deduce $P(n+1)$:

$$(a+b)^{n+1} = (a+b)(a+b)^n = (a+b)\sum_{k=0}^{n} \binom{n}{k} a^{n-k} b^k$$

$$= \left[a^{n+1} + \sum_{k=1}^{n} \binom{n}{k} a^{n+1-k} b^k \right] + \left[\sum_{k=0}^{n-1} \binom{n}{k} a^{n-k} b^{k+1} + b^{n+1} \right]$$

But $\sum_{k=0}^{n-1} \binom{n}{k} a^{n-k} b^{k+1} = \sum_{k=1}^{n} \binom{n}{k-1} a^{n+1-k} b^k$ by a change of index.

Furthermore,

$$\binom{n}{k} + \binom{n}{k-1} = \binom{n+1}{k} \quad \text{(Exercise 1.4.10(a))}$$

Hence

$$(a+b)^n = a^{n+1} + \sum_{k=1}^{n} \binom{n+1}{k} a^{n+1-k}b^k + b^{n+1} = \sum_{k=0}^{n+1} \binom{n+1}{k} a^{n+1-k}b^k$$

Thus $P(n+1)$ follows and the theorem is established. ∎

In our discussion prior to Theorem 1.10, we assumed that all elements of a set are distinct, so there is no difficulty in discerning one element from another. The situation is slightly different if some elements are identical.

Proposition 1.11: The number of n-element permutations from an n-element set where a_1 of the elements are alike and of one type, a_2 of them are alike and of another type, . . ., and a_r are alike and of a final type is

$$\frac{n!}{a_1 ! a_2 ! \cdots a_r !} \tag{1.5}$$

where $a_1 + a_2 + \cdots + a_r = n$.

Proof: Let N be the number sought. If the a_1 elements of the first type were distinguishable, then there would be $N \cdot a_1!$ Permutations in all. If in addition the a_2 elements of the second type were distinguishable, then there would be $N \cdot a_1! \cdot a_2!$ permutations in all. Continuing in this way, we have $N \cdot a_1! \cdot a_2! \cdots a_r !$ n-element permutations from an n-element set. But

$$P(n, n) = n! \text{ And so } N = \tfrac{n!}{a_1! a_2! \cdots a_r!}$$ ∎

For example, the number of "words" formed from jumbling the word bookkeeper is $10!/2!2!3! = 151200$. Numbers of the form (1.5) are called *multinomial* coefficients.

Example 1.5: How many ways can m integers be chosen from the set $\{1, 2, .., n\}$ if no two consecutive integers can be chosen? So that the answer will be non-trivial, assume $n \geq 2m - 1$. (It is instructive to attempt to discover the answer before reading ahead.)

Solution: Consider n markers of which m are white and the remaining $n - m$ are black. Set aside $m - 1$ black markers and place the remaining $n - (m-1)$ markers in a row in any order we wish. By Proposition 1.10 there are $\frac{(n-m+1)!}{m!(n-2m+1)!} = \binom{n-m+1}{m}$ ways to do this. Now intersperse the remaining $m - 1$ black markers so that each is placed between consecutive white markers. There is only one distinguishable way to do this.

The white markers represent the chosen integers and the black ones those not chosen. There is a one-to-one relationship between marker patterns and desired integer sets. Hence the number of ways m integers can be chosen from $\{1, 2, .., n\}$ with no two of them consecutive is $\binom{n-m+1}{m}$.

We now discuss the pigeonhole principle (also called the Dirichlet box principle). This simple principle often has far-reaching consequences.

Pigeonhole Principle: If n sets contain more than n distinct elements, then at least one of the sets contains more than one element.

The pigeonhole principle can be proved by using the WOP, but we will accept its veracity as being self-evident. Here is an application of it.

Example 1.6: The Burj Khalifa in Dubai is the tallest building in the world with 163 floors which we number 1 to 163. Suppose that an elevator stops 82 times as it descends from the top floor. Show that it stops at two floors whose sum is 163.

Solution: Suppose the elevator stops at floors $f_1, f_2, ..,\ f_{82}$ where

$$1 \le f_1 < f_2 < \cdots < f_{82} < 163 \tag{1.6}$$

Now consider the numbers $f_1, f_2, .., f_{82}, 163 - f_1, 163 - f_2, .., 163 - f_{82}$.

By (1.6), all 164 numbers above are between 1 and 163 inclusive.

By the pigeonhole principle two of them are equal. But none of the $f_i's$ are equal and hence none of the $163-f_i's$ are equal. Thus, there exists i, j such that $f_i = 163 - f_j$. But then $f_i + f_j = 163$ as claimed. Furthermore, $i \ne j$ since 163 is odd.

Example 1.7: Given a set of 2000 natural numbers, show there is a subset whose sum is divisible by 2000.

Solution: Let the set consist of the numbers $a_1, a_2, .., a_{2000}$. Define $s_1 = a_1$, $s_2 = a_1 + a_2, .., s_{2000} = a_1 + a_2 + \cdots + a_{2000}$. Consider the set $\{0, s_1, s_2, .., s_{2000}\}$. Since there are only 2000 different congruence classes modulo 2000, two of the numbers above must be in the same class. But then 2000 divides their difference, which is a sum of numbers from the set.

We complete this chapter by discussing the enumeration of all possible ways to partition a set of n elements. We begin with a definition.

Definition 1.9: The nth *Bell number* B_n counts the number of different ways to partition a set of n elements.

Equivalently, B_n represents the number of possible equivalence class relations among n objects. The sequence begins $B_0 = 1$, $B_1 = 1$, $B_2 = 2$, $B_3 = 5$, and $B_4 = 15$. For example, we can express the five possible partitions of the numbers 1, 2, 3 by $\{(1, 2, 3)\}$, $\{(1, 2), (3)\}$, $\{(1, 3), (2)\}$, $\{(1), (2, 3)\}$, and $\{(1), (2), (3)\}$. The Bell numbers are named after the Scottish-American mathematician Eric Temple Bell (1883–1960) who wrote about them in the 1930s. Bell was a prolific writer best known in the mathematical world for his highly influential and inspiring (if somewhat fictional) historical biographies in *Men of Mathematics*. However, the study of Bell numbers themselves actually predates Bell himself by several centuries.

In Edo period Japan (1603–1868), the art of incense appreciation was widely practiced, known as kodo (the way of fragrance). One popular game involved the hostess carefully choosing and then burning five incense sticks to see if the guests could identify which ones were the same and which were different. The number of possibilities is precisely $B_5 = 52$. For example, having the first, third, and fourth incense sticks all the same with the second and fifth distinct corresponds to the partition {(1, 3, 4), (2), (5)}. The participants even had a specially developed stick figure notation to record their guesses. In time, the 52 possible partitions were each named after a chapter title in the ancient Heian period book *The Tale of Genji* (ca. 1,000) which fortunately consists of 54 chapters with 2 chapter titles to spare. This particular incense guessing game was known as Genji-ko.

Bell numbers also arise in identifying rhyming schemes in poetry. The number of possible schemes for a quatrain is B_4 while B_5 is the number of conceivable quintains. For example, a limerick has rhyming scheme usually denoted by aabba where the first, second, and fifth lines rhyme as do the third and fourth. We represent this numerically as the partition {(1, 2, 5), (3, 4)}.

The Bell numbers increase rather rapidly and can become quite difficult to calculate directly. However, there is a formula for calculating the next Bell number from previous ones which we now state and prove.

Proposition 1.12: For $n \geq 0$, $B_{n+1} = \sum_{k=0}^{n} \binom{n}{k} B_k$, where we define $\binom{0}{0} = B_0 = 1$.

Proof: For any partition of the numbers $1, 2, \ldots, n + 1$, we remove the set containing 1 and all numbers in its equivalence class. What remains is a partition of a set containing k elements for some k where $0 \leq k \leq n$. For each k, there are $\binom{n}{n-k} = \binom{n}{k}$ ways to choose a set of $n-k$ elements that join 1 in its equivalence class. And for the remaining k numbers, there are B_k ways to partition them. Hence the total number of ways to partition $1, 2, \ldots, n + 1$ is the sum

$$B_{n+1} = \sum_{k=0}^{n} \binom{n}{k} B_k \qquad \blacksquare$$

For example, for $n = 4$, $B_5 = \sum_{k=0}^{4} \binom{4}{k} B_k = 1 \cdot 1 + 4 \cdot 1 + 6 \cdot 2 + 4 \cdot 5 + 1 \cdot 15 = 52$.

Exercise 1.4

1. (a) Calculate $P(n, r)$ and $\binom{n}{r}$ for $n = 6$ and $r = 1, 2, 3, 4, 5$, and 6.
   (b) How many 11-letter permutations are there of Tallahassee? How many 10-letter permutations are there of Cincinnati (treat the c's as being equivalent)?

2. How many different numbers larger than one trillion can be created by permuting the digits of 3141592653589 ? How many are larger than 4 trillion?

3. (a) Explain why $P(n, n) = P(n, n-1)$ for all $n \geq 1$.
   (b) Explain why $\binom{n}{r} = \binom{n}{n-r}$ for $n \geq 1$ and $0 \leq r \leq n$.

4. (a) How many n-digit natural numbers are there?
   (b) How many n-digit numbers are divisible by 9 ?

5. (a) Show that there are 2^n subsets of a set with n elements.
   (b) Show that for any set, the number of subsets with an odd number of elements equals the number of subsets with an even number of elements (including the empty subset).

6. (a) How many ways can seven different colored balls be placed in a row?
   (b) A juggler can juggle seven discernible balls in a circular pattern. How many different possible patterns are there?

7. No matter which 1001 distinct integers are chosen from $\{1, 2, ..., 1991\}$, prove that 2 must have difference 9.

8. Given any $m + 1$ integers, prove that two can be selected having difference divisible by m.

9. (a) Of the numbers from 1 to 1000, how many are divisible by 2? By 3? By 5?
   (b) How many of the numbers from 1 to 1000 are divisible by 2 and 3 but not by 5?

10. (a) Show that $\binom{n}{k} + \binom{n}{k-1} = \binom{n+1}{k}$.
    (b) Show that $\sum_{k=0}^{n} \binom{n}{k} = 2^n$. (See Exercise 1.4.5(a).)
    (c) Show that $\sum_{k=0}^{n} (-1)^k \binom{n}{k} = 0$. (See Exercise 1.4.5(b).)
    (d) Show that $\sum_{k=0}^{n} \binom{n}{k} m^k = (m+1)^n$.

11. Let there be given nine lattice points in three-dimensional Euclidean space. Show that there is a lattice point on the interior of one of the line segments joining two of these points (Putnam Exam – 1971, A-1).

12. (a) Show that $\sum_{j=1}^{n} (2j-1)^2 = \binom{2n+1}{3}$ (F. Mariares identities – 1913).
    (b) Show that $\sum_{j=1}^{n} (2j)^2 = \binom{2n+2}{3}$.

13. (a) How many non-empty subsets are there of the set $\{1, 2, ..., n\}$?
    (b) Show that the number of non-empty subsets of the set $\{1, 2, ..., n\}$ containing no two consecutive terms is $F_{n+2} - 1$. Hint: Apply the result of Example 1.4.

14. Verify that $\sum_{k=0}^{n} \binom{n}{k}^{2} = \binom{2n}{n}$.

15. Human twins can be either fraternal or identical. Triplets can be all identical, two identical and one fraternal, or all three fraternal. How many possible genotypes are there for quadruplets, quintuplets, sextuplets? (For n children see Section 9.4.)

16. (a) Show that $\sum_{k=0}^{r} \binom{n}{k}\binom{m}{r-k} = \binom{n+m}{r}$ for $r \leq n + m$.

    (b) Show that $\sum_{k=0}^{N} \binom{N}{k}^{2} = \binom{2N}{N}$.

    (The identity in (a) dates back at least to Chu Chi-kie (1303), but is often called Vandermonde's Identity (1772) and has many significant applications.)

17. Prove that p is prime if and only if all binomial coefficients $\binom{p}{k}$ for $1 \leq k \leq p-1$ are divisible by p. This observation forms the basis of some important primality tests.

18. (a) How many ways can a coin be flipped 8 times with outcome 4 heads and 4 tails and such that the number of heads at any point is always at least as large as the number of tails ("heads ahead, tails trail")?

    (b) Let C_n denote the number of ways a coin can be flipped $2n$ times with n heads and n tails with the number of heads always at least the number of tails. Try to find a closed formula for C_n. The number C_n is called the nth Catalan number, named after the Belgian mathematician Eugene Catalan (1814–1894) who wrote about it in a paper of 1838.

19. Show for $n \geq 1$, $\sum_{k=0}^{n} \frac{(-1)^{k}}{2k+1} \binom{n}{k} = \prod_{k=1}^{n} \frac{2k}{2k+1}$.

20. Inclusion-Exclusion Principle: Let $|S|$ denote the number of elements in a finite set S.

    (a) Show that $|A \cup B| = |A| + |B| - |A \cap B|$.

    (b) Show that

    $$|A \cup B \cup C| = |A| + |B| + |C| - |A \cap B| - |A \cap C| - |B \cap C| + |A \cap B \cap C|.$$

    (c) Generalize part (b) for the case of n sets ($n \geq 2$) (D.A. Da Silva – 1854).

    (d) How many integers from 1 to 1,000 are not divisible by 2 or 7?

    (e) How many integers from 1 to 1,000 are not divisible by 3, 4, or 5?

21. Consider the (n-element) permutations of the numbers $1, 2, \ldots, n$. A derangement is a permutation where none of the numbers are in their original position. Let d_n be the total number of such derangements.

    (a) Determine d_n for $n = 1, 2, 3, 4$, and 5.

    (b) Show that $d_n = n! \sum_{k=0}^{r} (-1)^{k} \frac{1}{k!}$.

    (c) Show that $d_{n+1} = (n+1)d_n + (-1)^{n+1}$.

    (d) Show that $\lim_{n \to \infty} d_n/n! = 1/e = 0.36787944\ldots$. (Hence the probability that a permutation is a derangement approaches $1/e$ as n gets large).

22. (a) Derive the formula $\sum_{k=1}^{n} k \binom{n}{k} = n \cdot 2^{n-1}$ by letting $a = x$ and $b = 1$ in Theorem 1.9, differentiating with respect to x, and then letting $x = 1$.

    (b) By a similar method, use part (a) to derive the formula $\sum_{k=1}^{n} k^2 \binom{n}{k} = n(n+1)2^{n-2}$.

23. A group of friends play a round-robin singles tennis tournament. Of course, their total number of wins and losses were equal. Show that the sum of the squares of their number of wins equals the sum of squares of their number of losses (due to Paul Vaderlind).

24. (a) Thirty-two players sign up for a singles tennis tournament. How many possible ways can players be paired in the first round?

    (b) Thirty-two players sign up individually for a doubles tennis tournament. How many ways can teams be created and then paired in the first round?

25. (a) P. Erdös and J. Selfridge (1975) proved that $\binom{n}{k}$ is never a perfect power for $4 \le k \le \frac{n}{2}$.

    However, show that $\binom{n}{2}$ is a perfect square for infinitely many n by showing that the equation

    $x^2 - 2y^2 - x = 0$ is solvable for $(x_1, y_1) = (2, 1)$ and if (x_n, y_n) solves the equation, then so does (x_{n+1}, y_{n+1}) with $x_{n+1} = 3x_n + 4y_n - 1$ and $y_{n+1} = 2x_n + 3y_n - 1$.

    (b) Verify that $\binom{50}{3}$ is a perfect square.

26. (a) Calculate the Bell numbers B_4 and B_5 directly by explicitly listing all appropriate partitions.

    (b) Calculate B_6 and B_7 by using Proposition 1.11.

27. The *Stirling numbers of the second kind* $\left\{ {n \atop k} \right\}$ are defined as the number of ways to partition a set of n distinguishable objects into k nonempty subsets with $\left\{ {n \atop n} \right\} = 1$

    for $n \ge 0$ and $\left\{ {n \atop 1} \right\} = 1$ for $n \ge 1$. It follows that $B_n = \sum_{k=0}^{n} \left\{ {n \atop k} \right\}$. Show the following:

    (a) $\left\{ {n \atop n-1} \right\} = \binom{n}{2}$ (b) $\left\{ {n \atop 2} \right\} = 2^{n-1} - 1$ (c) $\left\{ {n+1 \atop k} \right\} = k \left\{ {n \atop k} \right\} + \left\{ {n \atop k-1} \right\}$.

28. *Josephus Problem*: A group of n children sit in a circle numbered 1, . . ., n. To eliminate children from the game, the teacher taps every other student on the shoulder in turn until just one child remains. Denote the winner by $J(n)$.

    (a) Determine $J(n)$ for all $2 \le n \le 12$.

    (b) Show that $J(2k) = 2J(k)-1$ and that $J(2k + 1) = 2J(k) + 1$.

    (c) Prove that $J(2^a + t) = 2t + 1$ for $0 \le t < 2^a$.

    (d) Evaluate $J(n)$ for $n = 50, 64, 100, 1{,}000$.

Chapter 2
Congruences and prime factorization

2.1 The Euclidean algorithm and some consequences

In this section we define and develop the greatest common divisor (gcd) concept. Next, we state and prove the Euclidean algorithm for determining the gcd. The Euclidean algorithm is one of the most ancient algorithms in number theory and yet remains today as one of the most useful theories. In particular, we apply it to the solution of linear Diophantine equations. Finally, we prove the result of Gabriel Lamé (1795–1871) concerning the efficiency of the Euclidean algorithm.

Definition 2.1: Let a and b be integers. We call d the gcd of a and b if
(a) $d > 0$.
(b) $d \mid a$ and $d \mid b$.
(c) If $f \mid a$ and $f \mid b$, then $f \mid d$.

Below we will prove that the gcd of two integers always exists and is unique. Hence it is legitimate to define the gcd as we have done above.

Let us write gcd (a, b) for the gcd of a and b.

Occasionally we will simply write (a, b) for gcd (a, b) when there is little chance of notational confusion.

Definition 2.2: If $\gcd(a, b) = 1$, then we say that a and b are relatively prime. More generally, a set of integers is pairwise relatively prime if all pairs of distinct integers are relatively prime.

For example, $\gcd(6, 15) = 3$, $\gcd(-100, -30) = 10$, and the three numbers 6, 11, and 35 are pairwise relatively prime. (The notation $a \perp b$ is becoming fashionable to denote that a and b are relatively prime).

In the examples above, the gcd of two numbers could be determined by factoring each and listing all common prime factors. This technique is useful and perfectly valid. However, for larger numbers, carrying out the factorization might be very difficult. The theory of factorization is itself a deep and very active area of research today. It also has important applications in cryptology and hence is of national security and international commerce importance. You may be surprised how pure mathematics often has such practical applications!

Fortunately, there is a constructive algorithm for finding the gcd of two integers, which does not depend on their factorization. It's called the Euclidean algorithm since it appears as Proposition 2 of Book VII in *Euclid's Elements*.

https://doi.org/10.1515/9783111579283-002

Proposition 2.1: Let a and b be positive integers. Then $\gcd(a, b)$ exists and is unique.

Proof (Euclidean algorithm): By the division algorithm, we can write
$a = q_1 b + r_1$ where $0 \le r_1 < b$.

$b = q_2 r_1 + r_2$ where $0 \le r_2 < r_1$.
$r_1 = q_3 r_2 + r_3$ where $0 \le r_3 < r_2$.
$r_{n-3} = q_{n-1} r_{n-2} + r_{n-1}$ where $0 \le r_{n-1} < r_{n-2}$.
$r_{n-2} = q_n r_{n-1}$. (So $r_n = 0$.)

Here r_n is defined as the first zero remainder. This process must eventually terminate in n steps for some $n \ge 1$ since the remainders are strictly decreasing nonnegative integers. In fact, clearly $n \le min\{a, b\}$.

We claim that r_{n-1} is a gcd of a and b. To verify this the three conditions in Definition 2.1 must be checked:
(i) By definition $r_{n-1} > 0$.
(ii) By our last equation, $r_{n-1} \mid r_{n-2}$. But then r_{n-1} divides both terms on the right in the penultimate equation. By Proposition 1.3 (c), $r_{n-1} \mid r_{n-3}$.
     Similarly, r_{n-1} divides $r_{n-4}, \ldots, r_1$, etc, so $r_{n-1} \mid b$ and $r_{n-1} \mid a$.
(iii) If $f \mid a$ and $f \mid b$, then, since $r_1 = a - q_1 b$, we have that $f \mid r_1$. But $r_2 = b - q_2 r_1$. So $f \mid r_2$. Similarly, f divides $r_3, \ldots, r_{n-2}$ and hence $f \mid r_{n-1}$.

Now let d be a gcd of a and b. Since r_{n-1} is a gcd, we have that $d \mid r_{n-1}$ and $r_{n-1} \mid d$. By Proposition 1.3(f), $d = \pm r_{n-1}$. But d and r_{n-1} are positive and thus $d = r_{n-1}$. Thus, the gcd of a and b exists and is unique. ■

Example 2.1: Find $\gcd(54, 231)$.
Solution: Let us apply the Euclidean algorithm:

$$231 = 4(54) + 15$$
$$54 = 3(15) + 9$$
$$15 = 1(9) + 6$$
$$9 = 1(6) + 3$$
$$6 = 2(3)$$

So $\gcd(54, 231) = 3$.
It is useful to note that the Euclidean algorithm process can be reversed. This leads to the following *porisms* to Proposition 2.1, that is, results following from the proof of Proposition 2.1 rather than from the proposition itself. In any event, rather than splitting hairs, we label it a corollary.

Corollary 2.1.1: Let $d = \gcd(a, b)$.

(a) There exist integers x and y such that $d = ax + by$.
(b) If $d = 1$ and $a \mid bc$, then $a \mid c$.
(c) Let $d = 1$. If $a \mid c$ and $b \mid c$, then $ab \mid c$.
(d) If there are x and y for which $ax + by = 1$, then $d = 1$.
(e) If $ax + by = c$, then $d \mid c$.
(f) If $\gcd(a, b) = 1$ and $\gcd(a, c) = 1$, then $\gcd(a, bc) = 1$.
(g) $\gcd(a/d, b/d) = 1$.

Proof:
(a) (Induction): We adopt the notation from our proof of Proposition 2.1. Note that $r_1 = 1(a) - q_1(b)$ and that

$$r_2 = b - q_2(r_1) = -q_2(a) + (1 + q_1 q_2)b$$

Now we make the inductive assumption that we can write all of $r_1, r_2, \dots, r_{n-2}$ as a linear combination of a and b. In particular, we assume that $r_{n-3} = ax_{n-3} + by_{n-3}$ and $r_{n-2} = ax_{n-2} + b_{n-2}$. But then $d = r_{n-1} = r_{n-3} - q_{n-1}r_{n-2} = (x_{n-3} - q_{n-1}x_{n-2})a + (y_{n-3} - q_{n-1}y_{n-2})b$.
  Let $x = x_{n-3} - q_{n-1}x_{n-2}$ and $y = y_{n-3} - q_{n-1}y_{n-2}$.
(b) Since $d = 1$ there exist x and y for which $1 = ax + by$. Multiplying both sides of the equation by c, we obtain $c = acx + bcy$. But $a \mid a$ and $a \mid bc$ and so by Proposition 1.3 (c), $a \mid c$.
(c) Since $a \mid c$ and $b \mid c$, there exist r and s for which $ar = c$ and $bs = c$. But then $b \mid ar$. Since $d = 1$, part (b) implies $b \mid r$. So there exists a t such that $bt = r$. Thus $c = ar = abt$ and $ab \mid c$.
(d) Suppose $d > 1$. $d \mid a$ and $d \mid b$ implies that $d \mid (ax + by)$ for all x and y by Proposition 1.3(c). But $d \nmid 1$. So for any choice of x and y, we have $ax + by \neq 1$.
(e) Since $d \mid a$ and $d \mid b$, $d \mid ax + by$ for any x and y. Hence $d \mid c$.
(f) Since $\gcd(a, b) = \gcd(a, c) = 1$, there exist x_1, x_2, y_1, and y_2 such that $ax_1 + b_1 = 1$ and $ax_2 + cy_2 = 1$. Multiplying together, $(ax_1 + b_1)(a_2 + cy_2) = 1$. Expanding, $a(a_1x_2 + bx_2y_1 + cx_1y_2) + bc(y_1y_2) = 1$. Hence $\gcd(a, bc) = 1$ by part (d).
(g) By part (a) there are integers x and y such that $d = ax + by$. Since all terms are divisible by d, $1 = (a/d)x + (b/d)y$. But by part (d), it follows that $\gcd(a/d, b/d) = 1$. ∎

Notice that in our proof of Corollary 2.1.1(a) we needed the PCI version of mathematical induction.

Example 2.2: Write 3 as a linear combination of 54 and 231.
Solution: Since $3 = \gcd(54, 231)$, we simply reverse the steps in Example 2.1:

$$3 = 9 - 1(6)$$
$$= 9 - 1[15 - 1(9)] = -1(15) + 2(9)$$
$$= -1(15) + 2[54 - 3(15)] = 2(54) - 7(15)$$
$$= 2(54) - 7[231 - 4(54)] = -7(231) + 30(54)$$

In Section 6.1 we use continued fractions to derive another method to express the gcd of two integers as a linear combination of them. In fact, there are infinitely many such representations. We prove a slightly more general result presently.

Corollary 2.1.2 (Linear Diophantine Equation Theorem): Let a and b be nonzero integers and let $d = \gcd(a, b)$. Consider the linear Diophantine equation

$$ax + by = c \tag{2.1}$$

If $d \mid c$, then eq. (2.1) has infinitely many integer solutions.
    If $d \nmid c$, then eq. (2.1) has no solution.
    In the former case, if $x = x_0, y = y_0$ is a particular solution, then all solutions are given by

$$x = x_0 + (b/d)n, y = y_0 - (a/d)n, n \text{ any integer.} \tag{2.2}$$

Proof: Let (x, y) be a solution to (2.1). Since $d \mid a$ and $d \mid b$, it follows that $d \mid c$. So if $d \nmid c$, then eq. (2.1) has no solution.
    Now assume that $d \mid c$. From Corollary 2.1.1(a), there are integers s and t such that $as + bt = d$. Since $d \mid c$, there is an integer f for which $c = df$. So $c = df = (as + bt)f = a(sf) + b(tf)$. Thus eq. (2.1) is solvable with $x = sf, y = tf$.
    Next we show there are infinitely many solutions in this case. Let x_0 and y_0 be a particular solution of eq. (2.1) and let x and y be as in eq. (2.2) for some n. Then $ax + by = a[x_0 + (b/d)n] + b[y_0 - (a/d)n] = a_0 + b_0 = c$, as desired. Since a and b are nonzero, we get infinitely many solutions as n ranges over all integers.
    Finally, we show that every solution of eq. (2.1) is of the form prescribed. Notice that $x = x_0$ and $y = y_0$ is of the form (2.2) with $n = 0$. Now let (x, y) be any solution of eq. (2.1). Then $ax + by = ax_0 + by_0$ and $a(x - x_0) = b(y_0 - y)$. Dividing both sides by d, $(a/d)(x - x_0) = (b/d)(y_0 - y)$. By Corollary 2.1.1(g), $\gcd((a/d), (b/d)) = 1$. By Corollary 2.1.1(b), $(a/d) \mid (y_0 - y)$. Hence there exists an n with $(a/d)n = y_0 - y$. So $y = y_0 - (a/d)n$. Substituting into the equation $(a/d)(x - x_0) = (b/d)(y_0 - y)$ and solving for x yields $x = x_0 + (b/d)n$. ∎

Example 2.3: At a used bookstore all paperbacks cost 3 dollars apiece and all hardbacks cost 7 dollars apiece. Describe what can be purchased for precisely 100 dollars.
Solution: In this case $a = 3$, $b = 7$, $c = 100$, and $d = \gcd(a, b) = 1$. Since $d \mid c$, the linear Diophantine equation $3x + 7y = 100$ is solvable. However, we must find solutions with $x \geq 0, y \geq 0$. It is readily apparent that $3x_0 + 7y_0 = 1$ is solvable with $x_0 = -2$ and $y_0 = 1$ (the Euclidean algorithm is hardly necessary when a and b are so small).

By Corollary 2.1.2, all solutions of $3x + 7y = 100$ are given by $x = -200 + 7n, y = 100 - 3n, n$ an integer.

Since x and y are nonnegative integers, it follows that $29 \leq n \leq 33$. Hence there are five possibilities depending on the choice of n. In particular, $(x, y) = (3, 13), (10, 10), (17, 7), (24, 4)$, or $(31, 1)$.

When a and b are relatively prime, the linear Diophantine equation theorem guarantees solutions to $ax + by = c$ for any value of c. A closely related but more practical problem is to determine solutions where x and y are nonnegative integers. In particular, we wish to determine the largest integer c which cannot be expressed as a nonnegative linear combination of a and b. This has many conflicting attributions in the literature but is often called the *Frobenius coin problem* although the first published investigations seem to be due to J.J. Sylvester. Apparently, it then became a favorite topic of the German algebraist F.G. Frobenius (1849–1917) who often lectured on it.

Given a set of positive integers $a_1, a_2, \ldots, a_n$, let $g(a_1, a_2, \ldots, a_n)$ denote the largest integer which cannot be expressed as a nonnegative linear combination of them. We call this number the *Frobenius number* of $a_1, a_2, \ldots, a_n$. If the numbers are not relatively prime, then it makes no sense to talk of its Frobenius number since there will be infinitely many positive integers that are not multiples of their gcd and hence cannot be written as a linear combination of them. We narrow our attention to explicitly determining $g(a, b)$ with a and b relatively prime. Interestingly, no general formula for $g(a_1, a_2, \ldots, a_n)$ for any $n \geq 3$ has been discovered although many special cases and some upper and lower bounds have been established.

Proposition 2.2 (J.J. Sylvester, 1884): Let a and b be relatively prime, then g(a, b) = $ab - a - b$.

Proof: We must show that there is no solution to $ax + by = ab - a - b$ with $x, y \geq 0$, but all larger numbers have such a solution. For argument's sake, suppose there exists x, $y \geq 0$ such that $ax + by = ab - a - b$. In this case, $-b \equiv by \pmod{a}$ implying $y \equiv -1 \pmod{a}$ since $\gcd(a, b) = 1$. Similarly, $-a \equiv ax \pmod{b}$ and hence $x \equiv -1 \pmod{b}$.

Thus, $ab - a - b = ax + by \geq a(b - 1) + b(a - 1) = 2ab - a - b > ab - a - b$ is a contradiction.
Hence, $g(a, b) \geq ab - a - b$. We next find solutions for all larger integers.
Let $N > ab - a - b$. By Bezout's identity there exist integers x' and y' such that $ax' + by' = 1$. It follows that $aNx' + bNy' = N$.

Let $x_0 = Nx'$ and $y_0 = Ny'$. Then

$$ax_0 + by_0 = N$$

Of course, x_0 or y_0 might be negative.

By Corollary 2.1.2, all solutions to the previous equation are given by

$$x = x_0 + bn, y = y_0 - an, n \in \mathbf{Z}$$

Now choose n such that $0 \le x \le b - 1$. For this value of n,

$$ax + by = N > ab - a - b$$

This implies

$$b(y + 1) > a(b - 1 - x)$$

But $a, b > 0$ and $b - 1 \ge x$ implies $y + 1 > 0$ or $y \ge 0$.

This gives nonnegative integers x and y for which $ax + by = N$.

Therefore, $g(a, b) = ab - a - b$.   ■

For example, the largest amount of change that cannot be made with coins of denominations 5 and 11 is $g(5, 11) = 55 - 5 - 11 = 39$.

Linear Diophantine equations can also be used to solve some standard measuring puzzles where someone needs to pour out an exact amount of liquid but doesn't have a properly sized vessel.

Example 2.4: Suppose a recipe requires exactly 6 oz of wine, but Gabby has only a 5-oz and a 9-oz cup both unmarked. She has a full bottle of wine. What can she do?

Solution: Find a reasonable solution to the linear Diophantine equation $5x + 9y = 6$ which will then determine what needs to be done allowing only filling, emptying, and transferring of the liquid. For example, since 5 and 9 are relatively prime, Gabby could first solve $5x + 9y = 1$. One such solution is $x = 2$ and $y = -1$. Multiplying both sides of the equation by 6 gives $5(12) + 9(-6) = 6$. All solutions to $5x + 9y = 6$ are given by $(x, y) = (12 - 9t, -6 + 5t)$, $t \in \mathbf{Z}$. It is preferable to have small absolute values of x and y. Gabby chooses $t = 1$ giving $(x, y) = (3, -1)$. Hence it suffices to fill the 5-oz container three times and empty the 9-oz container once. In particular, if we let (a, b) represent the amount of wine in the 5-oz and 9-oz containers, respectively, then the solution can be represented as $(0, 0) \to (5, 0) \to (0, 5) \to (5, 5) \to (1, 9) \to (1, 0) \to (0, 1) \to (5, 1) \to (0, 6)$.

One measure of the utility of an algorithm is its efficiency. In particular, given two integers, can we obtain an upper bound for the number of steps required to find their gcd using the Euclidean algorithm? The answer is yes as follows from a beautiful result (published in 1844) by the French mathematician Gabriel Lamé (1795–1870).

Lamé is best known for the introduction of curvilinear coordinates in handling partial differential equations. He also did significant work in characterizing the elastic properties of an isotropic body. Outside of mathematics, Lamé served as the chief engineer of mines in France and helped plan and build the first railroads from Paris to Versailles and Paris to St. Germain. In number theory, Lamé was the first to prove Fermat's last lheorem for the case $n = 7$ (1840).

Theorem 2.3: The number of steps required in the Euclidean algorithm is never more than five times the number of digits in the smaller number.

Let's first see another example of the Euclidean algorithm which shows that the number 5 in Theorem 2.3 cannot be replaced by a smaller integer.

Example 2.5: Use the Euclidean algorithm to find $\gcd(55, 89)$.
Solution:
$$89 = 1(55) + 34$$
$$55 = 1(34) + 21$$
$$34 = 1(21) + 13$$
$$21 = 1(13) + 8$$
$$13 = 1(8) + 5$$
$$8 = 1(5) + 3$$
$$5 = 1(3) + 2$$
$$3 = 1(2) + 1$$
$$2 = 2(1). \ \gcd(55, 89) = 1$$

Notice that the remainders in the example above were all Fibonacci numbers. The Euclidean algorithm took nine steps here with the smaller number comprising only two digits. It would have taken 10 steps if we had reversed 55 and 89 in our first step.

Lemma 2.3.1: Let F_r denote the rth Fibonacci number. Then $F_{r+5} \geq 10 \cdot F_r$ for all $r \geq 2$.

Proof: The assertion is true for $r = 2, 3$, and 4 as can be seen by checking directly. For $r > 4$ we have $F_r = F_{r-1} + F_{r-2} = (F_{r-3} + F_{r-2}) + F_{r-2}$. Therefore

$$F_r = 2 F_{r-2} + F_{r-3} \tag{2.3}$$

Similarly,

$$F_{r+5} = 2\ F_{r+3} + F_{r+2} = 2(F_{r+2} + F_{r+1}) + F_{r+2} = 3\ F_{r+2} + 2\ F_{r+1}$$
$$= 3(F_{r+1} + F_r) + 2\ F_{r+1} = 5\ F_{r+1} + 3\ F_r = 5(F_r + F_{r-1}) + 3\ F_r$$
$$= 8\ F_r + 5\ F_{r-1} = 8(F_{r-1} + F_{r-2}) + 5\ F_{r-1} = 13\ F_{r-1} + 8\ F_{r-2}$$
$$= 13(F_{r-2} + F_{r-3}) + 8\ F_{r-2} = 21\ F_{r-2} + 13\ F_{r-3}$$
$$> 20\ F_{r-2} + 10\ F_{r-3} = 10(2\ F_{r-2} + F_{r-3}) = 10\ F_r\ \textit{by formula}\ (2.3)$$

For the sake of our next proof, it will be convenient to define $r' =: r + 1$. So $F_{1'} = 1$, $F_{2'} = 2, F_{3'} = 3, F_{4'} = 5$, etc. Thus Lemma 2.2.1 can be summarized as

$$F_{r+5'} \geq 10 \cdot F_{r'} \quad \text{for all } r' \geq 1 \tag{2.4}$$

Hence $F_{r+5'}$ has at least 1 more digit than $F_{r'}$. ∎

Proof of Theorem 2.3: Without loss of generality, let a and b be arbitrary positive integers for which $0 < b \leq a$. Let $r_{-1} = a$ and $r_0 = b$. Then we can list the steps in the Euclidean algorithm as follows:

$$r_{-1} = q_1 r_0 + r_1, 0 < r_1 < r_0.$$
$$R_0 = q_2 r_1 + r_2, 0 < r_2 < r_1.$$
$$r_{n-3} = q_{n-1} r_{n-2} + r_{n-1}, 0 < r_{n-2} < r_{n-1}.$$

$r_{n-2} = q_n r_{n-1}$. So $r_{n-1} = \gcd(r_{-1}, r_0)$ and there are n steps altogether.

Substitute $c_i =: r_{n-i}$ for $i = 1, \ldots, n+1$ and $d_i =: q_{n+1-i}$ for $i = 1, \ldots, n$. Note that $c_n = r_0 = b$ and $c_{n+1} = r_{-1} = a$. The above now becomes:

$$c_{n+1} = d_n c_n + c_{n-1}, 0 < c_{n-1} < c_n$$
$$c_n = d_{n-1} c_{n-1} + c_{n-2}, 0 < c_{n-2} < c_{n-1}$$
$$C_3 = d_2 c_2 + c_1, 0 < c_1 < c_2$$
$$C_2 = d_1 c_1.\ \text{So } c_1 = \gcd(c_{n+1}, c_n).$$

All the c_i's and d_i's are at least 1 since all are positive integers. But $d_1 \neq 1$ since otherwise $c_1 = c_2$ contrary to $0 < c_1 < c_2$. Hence $d_1 \geq 2$.

Working backward, we get $c_1 \geq 1 = F_{1'}; c_2 \geq 2 \cdot 1 = 2 = F_{2'}; c_3 \geq 1 \cdot 2 + 1 = 3 = F_{3'}$. By induction, it can be seen readily that

$$c_i \geq F_{i'} \text{ for all } i = 1, \ldots, n \tag{2.5}$$

Notice now that $F_{r'}$ has 1 digit for $0 \leq r \leq 5$. By eq. (2.4) we see

$F_{r'}$ has at least 2 digits for $1.5 < r \leq 2.5$, $F_{r'}$ has at least 3 digits for $2 \cdot 5 < r \leq 3 \cdot 5$, ...,
and in general $F_{r'}$ has at least $k+1$ digits for $k \cdot 5 < r \leq (k+1) \cdot 5$.

Recall that n was the number of steps in the Euclidean algorithm for a and b. There exists a k such that

$$k \cdot 5 < n \leq (k+1) \cdot 5 \tag{2.6}$$

In fact, $k = \left[\frac{n-1}{5}\right]$.

So $F_{n'}$ has at least $k+1$ digits. But by eq. (2.5), $c_n \geq F_{n'}$ and so c_n has at least $k+1$ digits, that is, $5 \cdot \#$ digits in $c_n \geq 5(k+1)$. By eq. (2.6), $n \leq (k+1)5 \leq 5 \cdot \#$ digits in c_n.

Thus, the number of steps in the Euclidean algorithm is at most 5 times the number of digits in the smaller number. ∎

A complementary concept to that of gcd is that of least common multiple. We define it below and study some of its basic properties in the exercises.

Definition 2.3: Let a and b be nonzero integers. We call m the *least common multiple* of a and b if

(a) $m > 0$.
(b) $a \mid m$ and $b \mid m$.
(c) If $a \mid n$ and $b \mid n$ then $m \mid n$.

By Exercise 2.1.10(a), the least common multiple of a and b equals $ab/\gcd(a,b)$. Since the gcd of a and b is unique, so is the least common multiple. Hence it is legitimate to define the least common multiple of a and b, heretofore denoted as $lcm[a,b]$. When there is no cause for confusion, the abbreviated $[a,b]$ is sometimes used. For example, $lcm[10,15] = 30$ and $lcm[36,150] = 900$.

In many situations, extensions of Definitions 2.1 and 2.3 are required.

In particular, let $x_1, \ldots, x_n$ be a set of integers. Define $\gcd(x_1, \ldots, x_n)$ as the largest positive integer dividing all elements of the set. Similarly, $lcm[x_1, \ldots, x_n]$ is the smallest positive multiple of all elements. For example, $\gcd(12, 18, 36) = 6$ and $lcm[2, 3, 4, 5, 6, 7] = 420$.

Exercise 2.1

1. (a) Use the Euclidean algorithm to find $\gcd(495, 4900)$.
   (b) Find x and y such that $495x + 4900y = \gcd(495, 4900)$.
2. (a) Use the Euclidean algorithm to find $\gcd(462, 2002)$.
   (b) Find x and y such that $462x + 2002y = \gcd(462, 2002)$.

3. (a) Use the Euclidean algorithm to find $\gcd(1234, 5678)$.
    (b) Find x and y such that $1234x + 5678y = \gcd(1234, 5678)$.
    (c) Verify Theorem 2.3 in this case.

4. (a) Use the Euclidean algorithm to show that 143 and 343 are relatively prime.
    (b) Find x and y such that $143x + 343y = -10$.

5. (a) Use the Euclidean algorithm to find $\gcd(2002, 2600)$.
    (b) Describe all the solutions to $2002x + 2600y = \gcd(2002, 2600)$.

6. (a) Show that any two consecutive squares are relatively prime.
    (b) Show that any two consecutive Fibonacci numbers are relatively prime.
    (c) Let L_n denote the nth Lucas number (see Exercise 1.2.12). Show that any two consecutive Lucas numbers are relatively prime.

7. Prove that if F_n and F_{n+1} are consecutive Fibonacci numbers, then for any integer d there are integers x and y such that $xF_n + yF_{n+1} = d$.

8. (a) Show that F_n and F_{n+2} are relatively prime for all $n \geq 1$.
    (b) Show that if F_n is even, then $\gcd(F_n, F_{n+3}) = 2$.
    (c) Show that if F_n is odd, then $\gcd(F_n, F_{n+3}) = 1$.
    (d) Investigate $\gcd(F_n, F_{n+k})$ for $k \geq 4$.

9. (a) Find $\operatorname{lcm}[495, 4900]$.
    (b) Find $\operatorname{lcm}[F_n, F_{n+1}]$.
    (c) Find $\operatorname{lcm}[1234, 5678]$.

10. (a) Prove that $\gcd(a, b) \cdot \operatorname{lcm}[a, b] = ab$.
    (b) Is it true that $\gcd(a, b, c) \cdot \operatorname{lcm}[a, b, c] = abc$? Explain fully.

11. (a) Find $\gcd(21, 81, 120)$.
    (b) Find $\operatorname{lcm}[21, 81, 120]$.

12. Show that $\gcd(a, b, c) = \gcd(\gcd(a, b), c)$. Extend this inductively.

13. If $\gcd(a, b, c) = 1$, must it be the case that a, b, and c are pairwise relatively prime? Explain.

14. Let $l = \operatorname{lcm}[a, b]$. Show that eq. (2.2) may be rewritten as

$$x = x_0 + (1/a)n, y = y_0 - (1/b)n$$

15. Describe $\gcd(t_n, t_{n+1})$ where t_n is the nth triangular number.

16. Let $a = r_{-1}$ and $b = r_0$ in the Euclidean algorithm (proof of Proposition 2.1). Show that $\sum_{i=1}^{n} q_i r_{i-1} = a + b - \gcd(a, b)$.

17. Let S be any set of $n+1$ integers chosen from $1, 2, 3, \ldots, 2n$. Prove that there are two relatively prime integers in S. (According to Paul Erdös, this problem was solved by the 11-year-old Hungarian prodigy Louis Posa in half a minute.)

18. Generalize Corollary 2.1.1 (parts b, c, e, f) for a product of n integers.

19. Determine whether the following linear Diophantine equations are solvable. If so, find all solutions:
    (a) $24x + 16y = 200$
    (b) $36x - 162y = 3600$

(c) $33x + 121y = 1000$

(d) $105x + 286y = -3$

20. Determine whether the following linear Diophantine equations are solvable. If so, find all solutions:

(a) $14x + 35y = 106$

(b) $51x - 153y = 34$

(c) $135x + 57y = 1000$

(d) $5x + 55y = 125$

21. Chocolate candies come with or without nuts. Those with nuts weigh 4 oz each while those without nuts weigh 3 oz each. List all possible assortments for a 6 lb bag.

22. A restaurant is full with 160 patrons. All tables have either 4 or 8 people at them. If there are 28 tables altogether, what is the composition of tables?

23. Let $g(a, b)$ be the Frobenius number for a and b. Find

(a) $g(3, 11)$, (b) $g(10, 21)$, and (c) $g(3, 5, 7)$ with an extended definition of the function g.

24. Let a, b, and c be nonzero integers. Show that the equation $ax + by + cz = d$ is solvable if and only if $\gcd(a, b, c)$ divides d.

25. Green, red, and yellow peppers cost 40, 75, and 90 cents apiece, respectively. How many different assortments of peppers can be purchased for 10 dollars?

26. (a) Chicken nuggets come in boxes of 6, 9 or 20. (1984 Middlebury/Williams Green Chicken Contest) What is the smallest integer N for which one could order exactly n chicken nuggets for any $n > N$? (N is the largest number which cannot be ordered exactly.)

(b) If we make the (admittedly artificial) assumption that the restaurant is willing to both buy and sell in boxes of 6, 9, or 20, show that a customer can end up with any predetermined number of chicken nuggets.

(c) If 6-nugget boxes cost $1.80, 9-nugget boxes cost $2.25, and 20-nugget boxes cost $4.00, what is the least expensive way to order 96 chicken nuggets? What is the least expensive way to order 96 chicken nuggets if a customer wants at least one box of each size?

27. One refinement of the Euclidean algorithm involves modifying the division process in the proof of Proposition 2.1 so that the remainders are integers (not necessarily positive) chosen to be as small as possible in absolute value. (Modified Euclidean Algorithm) In particular, let $r_i = q_{i+2} r_{i+1} + r_{i+2}$ where $|r_{i+2}| \leq \frac{r_{i+1}}{2}$. (For the sake of definiteness, define $r_{i+2} = \frac{r_{i+1}}{2}$ in the case of equality.)

(a) Show that $|r_{n-1}| = \gcd(a, b)$ where r_{n-1} is the last nonzero remainder.

(b) Show that the modified Euclidean algorithm is at least as fast as the standard Euclidean algorithm.

(c) Compare the two algorithms in finding $\gcd(141, 36)$.

(d) Compare the two algorithms in finding $\gcd(89, 144)$.

(e) Show that if there is an i for which $r_{i+2} = \frac{r_{i+1}}{2}$, then $\gcd(a, b) = r_{i+2}$.

28. Show that among any ten consecutive positive integers, there is always at least one integer relatively prime to the other nine. (For n consecutive integers, the proposition is true for all $n \leq 16$ and false for all $n \geq 17$ as shown by Pillai (1940) and Brauer (1941).)

29. What is the largest value that cannot be made with coin denominations of 3 and 13? How about 7 and 22?

30. (a) Let a and b be relatively prime. Show that the set of positive integers that are not a nonnegative linear combination of a and b are the positive numbers n of the form $ab - a - b - ka - la$ where $k, l \geq 0$.

    (b) Investigate the set of numbers $N(a, b)$ which are not nonnegative linear combinations of a and b for various values of a and b. Can you prove Sylvester's result that $N(a, b) = (a - 1)(b - 1)/2$?

31. (a) Use induction to establish the formula: $\sum_{k=1}^{n} \binom{n}{k} F_{k+r} = F_{2n+r}$ where F_n is the nth Fibonacci number. Recall that $\binom{0}{0} = 1$ and $\binom{n}{k} = 0$ for $k > n$ or $k < 0$.

    (b) Conclude that $\sum_{k=1}^{n} \binom{n}{k} F_k = F_{2n}$. (Compare with Proposition 1.11.)

32. Show that for any integers a, b, c, $[a, b, c]^2 / [a, b] [b, c][c, a] = (a, b, c)^2 / (a, b)(b, c)(c, a)$. (USA Olympiad – 1972 – Problem 1)

2.2 Congruence equations and the Chinese remainder theorem

The Chinese mathematician Sun-Tzi (ca. 300 C.E.) proposed the following problem: "There are things of an unknown number which when divided by 3 leave 2, by 5 leave 3, and by 7 leave 2. What is the number?" The method of finding a solution to such problems will be discussed in this section. That there is a solution is guaranteed by the so-called Chinese remainder theorem (CRT).

We begin with an important preliminary proposition.

Proposition 2.4: Suppose that $\gcd(a, n) = 1$. Then there is an a^* such that $aa^* \equiv 1 \pmod{n}$. Furthermore, a^* is unique modulo n. Conversely, if there exists an a^* such that $aa^* \equiv 1 \pmod{n}$, then $\gcd(a, n) = 1$.

Proof: (Existence) Let $\gcd(a, n) = 1$. Then there exist x and y such that $ax + ny = 1$ by Corollary 2.1.1(a). But then $ax \equiv 1 \pmod{n}$ and we may choose $a^* = x$.

(Uniqueness) Let a^* and b^* be such that $aa^* \equiv ab^* \equiv 1 \pmod{n}$.
   Then $n \mid a(a^* - b^*)$. But $\gcd(a, n) = 1$ and so by Corollary 2.1.1(b), $n \mid (a^* - b^*)$. Hence a^* is unique modulo n.
   (Converse) If $aa^* \equiv 1 \pmod{n}$, then there is an r such that $rn = aa^* - 1$. But then $aa^* + (-r)n = 1$ and $\gcd(a, n) = 1$ by Corollary 2.1.1(d). ■

Definition 2.4: Let $a \in \mathbb{Z}$. An integer $a*$ for which $aa* \equiv 1 \pmod{n}$ is called an *arithmetic inverse* of a modulo n.

For example, both 9 and 20 are arithmetic inverses of 5 modulo 11. If the numbers are large, then the Euclidean algorithm can be used to find arithmetic inverses.

Example 2.6: Find an arithmetic inverse $a*$ of $a = 1271 \pmod{1996}$ with $1 \le a* \le 1996$.
Solution: Apply the Euclidean algorithm to verify that $\gcd(1271, 1996) = 1$.

$$1996 = 1(1271) + 725$$
$$1271 = 1(725) + 546$$
$$725 = 1(546) + 179$$
$$546 = 3(179) + 9$$
$$179 = 19(9) + 8$$
$$9 = 1(8) + 1$$
$$8 = 8(1). \text{ Therefore } \gcd(1271, 1996) = 1$$

Now work backward to find x and y such that $1271x + 1996y = 1$. Then $x \pmod{1996}$ is the answer:

$$1 = 9 - 1(8)$$
$$= 9 - 1[179 - 19(9)] = -1(179) + 20(9)$$
$$= -1(179) + 20[546 - 3(179)] = 20(546) - 61(179)$$
$$= 20(546) - 61[725 - 1(546)] = -61(725) + 81(546)$$
$$= -61(725) + 81[1271 - 1(725)] = 81(1271) - 142(725)$$
$$= 81(1271) - 142[1996 - 1(1271)] = 223(1271) - 142(1996)$$

Thus if $x = 223$ and $y = -142$, then $1271x + 1996y = 1$. Consequently $a* = 223$.
    The next result can be viewed as a follow-up to Proposition 1.7.

Proposition 2.5: Let $d = \gcd(a, n)$ and suppose $ax \equiv ay \pmod{n}$.
(a) If $d = 1$ then $x \equiv y \pmod{n}$.
(b) $x \equiv y \pmod{n/d}$.

Proof: Although (a) is a special case of (b), it is helpful to prove (a) first.
(a) Since $ax \equiv ay \pmod{n}$ we have that $n \mid a(x-y)$. But $\gcd(a, n) = 1$ and so $x \equiv y \pmod{n}$ by Corollary 2.1.1(b).

(b) $ax \equiv ay \pmod{n}$ implies that $n \mid a(x-y)$ as above. So there exists r such that $nr = a(x-y)$. But $d \mid a$ and $d \mid n$. Hence $(n/d)r = (a/d)(x-y)$. Thus $\left(\frac{a}{d}\right)x \equiv \left(\frac{a}{d}\right)y \pmod{\frac{n}{d}}$. But $\gcd\left(\frac{a}{d}, \frac{n}{d}\right) = 1$ by Corollary 2.1.1(g). By part (a), $x \equiv y \pmod{n/d}$. ■

For example, if $15x \equiv 15y \pmod{35}$, then $x \equiv y \pmod 7$ since $\gcd(15, 35) = 5$. Similarly, if $14x \equiv 14y \pmod{45}$, then $x \equiv y \pmod{45}$ since 14 and 45 are relatively prime.

Let us consider the general linear congruential equation

$$ax + b \equiv 0 \pmod n \tag{2.7}$$

The solution to eq. (2.7) is quite straight and will be worked out below.

We might expect the general quadratic congruential equation

$$ax^2 + bx + c \equiv 0 \pmod n$$

to be only somewhat more difficult. Surprisingly, its solution is substantially more involved and was only first worked out by Gauss. We will return to a full study of the quadratic case in Chapter 4.

Proposition 2.6 (Linear Congruence Theorem): Let $d = \gcd(a, n)$.
(a) If $d = 1$ then eq. (2.7) has a unique solution modulo n.
(b) If $d \mid b$ then eq. (2.7) has d incongruent solutions modulo n.

If $d \nmid b$ then eq. (2.7) has no solutions.

Proof: Although (b) subsumes (a), part (a) is an important special case worthy of separate note.
(a) (Existence) Since $d = 1$, there exist x_0 and y_0 such that $ax_0 + ny_0 = 1$ by Corollary 2.1.1(a). Let $x = -bx_0$.
   Then $ax + b = -b(ax_0 - 1) = nby_0 \equiv 0 \pmod n$. Thus $x = -bx_0$ is a solution to eq. (2.7).
   (Uniqueness) Let $ax_1 + b \equiv 0 \pmod n$ and $ax_2 + b \equiv 0 \pmod n$.
   Then $a(x_1 - x_2) \equiv 0 \pmod n$. But $\gcd(a, n) = 1$ and so $x_1 \equiv x_2 \pmod n$ by Corollary 2.1.1(b).
(b) Equation (2.7) is solvable if and only if there exists y such that $ax - ny = -b$. By Corollary 2.1.1(c), if eq. (2.7) is solvable then $d \mid b$. So if $d \nmid b$ then eq. (2.7) has no solutions.
   Now let x be a solution to eq. (2.7). By Proposition 2.4(b),

$$\left(\frac{a}{d}\right)x + \frac{b}{d} \equiv 0 \pmod{n/d} \tag{2.8}$$

Furthermore, $\gcd(a/d, n/d) = 1$. By part (a) above, eq. (92.8) has a unique solution x_0 with $0 \le x_0 < n/d$. Hence $x = x_0 + (n/d)t$ are all solutions to eq. (2.7) for $0 \le t \le d - 1$. ■

Example 2.7: Find all distinct solutions of the linear congruential equation

$$30x + 18 \equiv 0 \pmod{42}$$

Solution: Since $\gcd(30, 42) = 6$ and $6 \mid 18$, there are 6 incongruent solutions (mod 42). The congruence $30x + 18 \equiv 0 \pmod{42}$ is equivalent to the congruence $5x + 3 \equiv 0 \pmod 7$. The latter congruence has the unique solution $x_0 = 5$ with $0 \le x_0 < 7$. So $x = 5 + 7t$ for $0 \le t < 6$ gives the distinct solutions $x \equiv 5, 12, 19, 26, 33, 40 \pmod{42}$.

Now we state and prove the CRT. The key point to note is that our proof is completely constructive, thus enabling us to actually solve a given system of linear congruence equations. Our proof is similar to that given by Gauss (*Disquisitiones Arithmeticae*, Art. 36) which to the author's knowledge is the first careful proof of the CRT.

Theorem 2.7 (CRT): Suppose that $m_1, m_2, \ldots, m_n$ are positive pairwise relatively prime integers. Let $b_1, b_2, \ldots, b_n$ be integers (not necessarily distinct). Then the system of congruences

$$x \equiv b_1 \pmod{m_1}$$

$$x \equiv b_2 \pmod{m_2}$$

$$x \equiv b_n \pmod{m_n}$$

has a simultaneous solution.

Furthermore, the solution is unique modulo m where $m = m_1\, m_2 \cdots m_n$.

Proof: (Existence) We will write x in the form

$$x = y_1\, b_1 + \cdots + y_n b_n$$

where, for all $i, y_i \equiv 1 \pmod{m_i}$ and $y_i \equiv 0 \pmod{m_j}$ for $1 \le j \le n$ except $j = i$. Now set $m_i' =: m/m_i$ for $1 \le i \le n$.

Since the moduli are pairwise relatively prime, $\gcd(m_i, m_i') = 1$ by Corollary 2.1.1(f). So m_i' has an arithmetic inverse $m_i'^*$ (mod m_i), that is

$$m_i'^*\, m_i' \equiv 1 \pmod{m_i} \tag{2.9}$$

Set

$$x = m_1'^*\, m_1'\, b_1 + \cdots + m_n'^*\, m_n' b_n \ (\text{i.e., } y_i = m_i'^*\, m_i')$$

We claim that x is a simultaneous solution of the system of congruences.

Fix i $(1 \le i \le n)$. For $j \ne i, m_i \mid m'_j$ and so

$$m^*_j m'_j \equiv 0 \ (\text{mod } m_i) \tag{2.10}$$

By eqs. (2.9) and (2.10), $x \equiv b_i (\text{mod } m_i)$. Since i is arbitrary, $x \equiv b_i (\text{mod } m_i)$ for all i, $1 \le i \le n$.

(Uniqueness) Let x and x' be two solutions. Then $x \equiv x' (\text{mod } m_i)$ for $1 \le i \le n$.

i.e. $m_i \mid (x - x')$ for all i. But $\gcd(m_i, m_j) = 1$ for $i \ne j$.

By Corollary 2.1.1(c), $m \mid (x - x')$. Thus $x \equiv x' (\text{mod } m)$. ∎

Example 2.8: Find a number between 1 and 100 which is divisible by 3, leaves a remainder of 2 when divided by 5, and a remainder of 3 when divided by 7.

Solution: The problem asks for an integer x with $1 \le x \le 100$ for which $x \equiv 0 \ (\text{mod } 3)$, $x \equiv 2 \ (\text{mod } 5)$, and $x \equiv 3 \ (\text{mod } 7)$.

We set $x \equiv 0y_1 + 2y_2 + 3y_3 \ (\text{mod } 105)$ where

$$y_1 \equiv 1 \ (\text{mod } 3), 0 \ (\text{mod } 5), \text{ and } 0 \ (\text{mod } 7)$$
$$y_2 \equiv 0 \ (\text{mod } 3), 1 \ (\text{mod } 5), \text{ and } 0 \ (\text{mod } 7)$$
$$y_3 \equiv 0 \ (\text{mod } 3), 0 \ (\text{mod } 5), \text{ and } 1 \ (\text{mod } 7)$$

By Corollary 2.1.1(c), $y_1 \equiv 0 \ (\text{mod } 35), y_2 \equiv 0 \ (\text{mod } 21)$, and $y_3 \equiv 0 \ (\text{mod } 15)$.

We could follow the technique of our proof, but with small moduli it is just as easy to do the following: y_1 is a multiple of 35 which is $\equiv 1 (\text{mod } 3)$. Add 35 to itself enough times until we have just such a multiple. We find $y_1 = 70$ works. (Of course, in this example we didn't have to find y_1 since it will be multiplied by 0.) Similarly, y_2 is a multiple of 21 which is $\equiv 1 (\text{mod } 5)$. So $y_2 = 21$ suffices. Finally, $y_3 = 15$.

So $x \equiv 0 \cdot 70 + 2 \cdot 21 + 3 \cdot 15 = 87 \ (\text{mod } 105)$. Our answer is 87. Luckily, there was no need to reduce modulo 105.

If you want to try this number trick on four good-natured friends, there is no need to repeat our work above. Have one choose a number and the others calculate the remainders upon division by 3, 5, and 7. Simply multiply the respective remainders by 70, 21, and 15 and sum the results. Then reduce modulo 105. (Of course, the number chosen can range from 1 to 105, but saying, "Pick a number from 1 to 100" sounds more natural.) The mental arithmetic may be simpler when 70 is replaced by −35.

Notice that an important condition in the CRT is that the moduli be pairwise relatively prime. In fact, without that assumption there might be no solution at all. For example, the systems $x \equiv 1 \ (\text{mod } 3)$ and $x \equiv 2 \ (\text{mod } 6)$ have no solutions since the second condition implies that $x \equiv 2 \ (\text{mod } 3)$ which is at variance with the first condition.

There are plenty of situations, however, where the moduli are not all relatively prime and yet solutions do exist. Such problems go back as far as the Chinese priest Yih-hing (717 C.E.). The following proposition and its corollary are generalizations of the CRT.

Proposition 2.8: Let $m = \text{lcm}\,[m_1, m_2]$. The systems

$$x \equiv a_1 (\text{mod}\, m_1)$$

$$x \equiv a_2 (\text{mod}\, m_2)$$

has a solution if and only if $\gcd(m_1, m_2) \mid (a_1 - a_2)$. In this case, the solution is unique modulo m.

Proof: Let $d = \gcd(m_1, m_2)$.
$\Rightarrow$ If there exists a solution to the simultaneous congruences, then $x \equiv a_1 (\text{mod}\, d)$ and $x \equiv a_2 \pmod{d}$. Hence $a_1 \equiv a_2 \pmod{d}$ and $d \mid (a_1 - a_2)$.

($\Leftarrow$) If $d \mid (a_1 - a_2)$, then the solutions to the first congruence $x \equiv a_1 (\text{mod}\, m_1)$ are given by $x = a_1 + m_1 y$ for integral y. Substituting this into the second congruence gives $m_1\, y + (a_1 - a_2) \equiv 0 \pmod{m_2}$. This must be solvable in order for the system to have a solution. But by the proof of Proposition 2.5(b), this has a unique solution y modulo m_2/d. Therefore, by the CRT the simultaneous congruences have a unique solution modulo $m_1 \cdot (m_2/d)$. The result follows by noting that $m = m_1 \cdot (m_2/d)$ by Exercise 2.1.10(a). ∎

For example, the systems $x \equiv 7 \pmod{15}$ and $x \equiv 3 \pmod{21}$ have no solution since $\gcd(15, 21) = 3 \nmid (7 - 3)$. On the other hand, the systems $x \equiv 10 \pmod{15}$ and $x \equiv 4 \pmod{21}$ have a unique solution $\pmod{105}$ since $3 \mid (10 - 4)$. (What is the solution?)

Corollary 2.8.1: Let $m = \text{lcm}[m_1, \dots, m_n]$. The linear system $x_i \equiv a_i (\text{mod}\, m_{(c)})$ for $1 \le i \le n$ has a solution if and only if $\gcd(m_i, m_j) \mid (a_i - a_j)$ for $1 \le i < j \le n$.
In this case, the solution is unique modulo m.

Proof: Use induction on n. ∎

By combining the linear congruence theorem with the CRT or its generalization (Proposition 2.8), we are now in a position to determine whether or not a system of linear congruences is solvable, and if so, determine its solution. One final example may be instructive.

Example 2.9: Solve the simultaneous system $3x + 8 \equiv 12 \pmod{14}$ and $6x + 7 \equiv 11 \pmod{20}$.
Solution: $3x + 8 \equiv 12 \pmod{14}$ implies that $3x \equiv 4 \pmod{14}$. Since 5 is an arithmetic inverse of 3 modulo $14, x \equiv 4 \cdot 5 \equiv 6 \pmod{14}$.

If $6x + 7 \equiv 11 \pmod{20}$, then $6x \equiv 4 \pmod{20}$ and $3x \equiv 2 \pmod{10}$. Since 7 is an arithmetic inverse of 3 modulo $10, x \equiv 2 \cdot 7 \equiv 4 \pmod{10}$.

Since $\gcd(10,14) = 2$ and $2 \mid (6-4)$, Proposition 2.8 guarantees a solution unique modulo $lcm[10,14] = 70$. In fact, $x \equiv 34 \pmod{70}$ which is readily found by adding multiples of 14 to 6 until we reach a number congruent to 4 modulo 10.

Exercise 2.2

1. Find an arithmetic inverse of a modulo n for the following:
    (a) $a = 5$ and $n = 23$
    (b) $a = 4$ and $n = 17$
    (c) $a = 39$ and $n = 11$.
2. Use the Euclidean algorithm to find an arithmetic inverse of 1001 $(\bmod\, 2048)$.
3. Use the Euclidean algorithm to find an arithmetic inverse of 4821 $(\bmod\, 10000)$ between -15000 and -5000.
4. (a) Find all solutions to the linear congruence $3x + 7 \equiv 0 \pmod{11}$.
    (b) Find all solutions to the linear congruence $7x + 25 \equiv 0 \pmod{16}$.
5. (a) Find all solutions to the linear congruence $4x + 22 \equiv 0 \pmod{12}$.
    (b) Find all solutions to the linear congruence $3x + 15 \equiv 0 \pmod{33}$.
6. Find all solutions to the linear congruence $283x + 121 \equiv 0 \pmod{563}$.
7. (a) Show that $3x + 3100 \equiv 0 \pmod{120}$ is not solvable.
    (b) Show that $55x - 3000 \equiv 0 \pmod{121}$ is not solvable.
8. Answer the question of Sun-Tzi posed at the beginning of this section by finding the least positive solution.
9. (a) What number between 1 and 1000 is $\equiv 2 \pmod{7}$, $\equiv 3 \pmod{11}$, and $\equiv 8 \pmod{13}$?
    (b) What number between 1 and 1000 is $\equiv 14 \pmod{7}$, $\equiv 33 \pmod{11}$, and $\equiv 28 \pmod{13}$?
10. How many numbers between 3000 and 3990 are $\equiv 1 \pmod{5}$, $\equiv 2 \pmod{9}$, and $\equiv 3 \pmod{11}$?
11. Find a multiple of 7 having remainders of $1, 2, 3, 4,$ and 5 when divided by $2, 3, 4, 5,$ and 6, respectively (Fibonacci – 1202).
12. Explain why the systems $x \equiv 1 \pmod{12}$, $x \equiv 3 \pmod{50}$, and $x \equiv 5 \pmod{81}$ are unsolvable.
13. Prove Corollary 2.8.1.
14. Show that $(ab)^* \equiv a^* \, b^* \pmod{n}$.
15. A positive integer is square-free if it is not divisible by any square larger than 1.
    (a) How many consecutive square-free numbers can there be?
    (b) Show that there are arbitrarily long strings of non-square-free integers.
16. (a) Show that there are infinitely many triples of consecutive integers which are pairwise relatively prime.
    (b) Could there be four consecutive integers that are pairwise relatively prime?

17. Determine all solutions to the simultaneous system $x \equiv 10 \pmod{12}, x \equiv 4 \pmod{21}$, and $x \equiv 11 \pmod{35}$.

18. Determine all solutions to the simultaneous system $4x + 11 \equiv 3 \pmod{18}$ and $6x + 9 \equiv 33 \pmod{45}$.

19. Find all (x, y) with $1 \le x, y \le 35$ such that $3x + 2 \equiv 0 \pmod 5$, $4y + 5 \equiv 0 \pmod 7$, and $x + y \equiv 0 \pmod 9$.

20. A system of congruences $a_i \pmod{n_i}$ is a *covering system* if every integer y satisfies $y \equiv a_i \pmod{n_i}$ for at least one value of $i (1 \le i \le r)$.

    (a) Show that $0 \pmod 2, 1 \pmod 3, 2 \pmod 3$, and $3 \pmod 6$ forms a covering system.

    (b) A covering system is proper if $n_1 < n_2 < \cdots < n_r$. Show that $0 \pmod 2$, $0 \pmod 3$ $1 \pmod 4$, $5 \pmod 6$, 7 and $\pmod{12}$ are a proper covering system.

    (c) Try to find a proper covering system with $n_1 = 3$. (Erdös offered \$500 for the proof or disproof of proper covering systems with n_1 arbitrarily large.)

21. Explain why the following algorithm gives the correct day of the week for any day of the twentieth century (Perpetual).

    Let $S = \left[\frac{5Y}{4}\right] + M + D \pmod 7$, where Y is the last two digits of the year, M is the month described below, and D is the day of the month. M equals 0 for July or April, 1 for January or October, 2 for May, 3 for August, 4 for February, March, or November, 5 for June, and 6 for September or December. (In leap years, January is 0 and February is 3.) S equals 1 for Sunday, 2 for Monday, 3 for Tuesday, 4 for Wednesday, 5 for Thursday, 6 for Friday, and 0 for Saturday.

    For any day in the twenty-first century, simply calculate S − 1.

22. Use the perpetual calendar algorithm above to find the day of the week of

    (a) November 11, 1918; (b) December 7, 1941; (c) August 28, 1963; (d) July 20, 1969; (b) July 4, 1976; (f) September 11, 2001; (g) March 11, 2011; and (h) your birthday.

2.3 Primes and the fundamental theorem of arithmetic

In this section we begin our study of prime numbers. In particular, we show that they are the basic multiplicative building blocks for all natural numbers. We shall state and prove the fundamental theorem of arithmetic which asserts that the prime factorization of any positive integer is unique. Next, we show that there are infinitely many primes, and finally we discuss primes of various forms.

Definition 2.5: A *prime* is a positive integer p for which

(a) $p > 1$

(b) p is divisible by only ±1 and ±p.

The list of primes begins $2, 3, 5, 7, 11, 13, 17, 19, 23, 29, 31, 37$, and so on. The distribution of primes is full of great subtlety and mystery. To better understand the primes it is necessary to also study numbers which are not primes.

Definition 2.6 A positive integer greater than 1 which is not prime is called a *composite*.

The list of composites begins $4, 6, 8, 9, 10, 12, 14, 15, 16$, and so on. Our next proposition lays the groundwork for the fundamental theorem of arithmetic.

Proposition 2.9: Every integer greater than 1 is expressible as a product of primes.

Proof: We reason indirectly. Suppose that there are some positive integers greater than 1 which are not expressible as a product of primes. By the WOP there is a smallest such integer n. Then n must be a composite and thus there is an integer a with $1 < a < n$ such that $a \mid n$. By definition of divisibility there is a number b with $ab = n$ and consequently $1 < b < n$. But a and b must be expressible as products of primes by the minimality of n. So

$$a = p_1 p_2 \cdots p_s \text{ and } b = q_1 q_2 \cdots q_t$$

for some primes p_i and q_j where $s \geq 1$ and $t \geq 1$. Hence $n = p_1 p_2 \cdots p_s \cdot q_1 q_2 \cdots q_t$ contrary to our assumption about n. Therefore, every integer greater than 1 is expressible as a product of primes. ∎

At this point, it is useful to make the following observation (Elements VII, Prop. 32):

Proposition 2.10 (Euclid's Lemma): Let p be a prime and suppose $p \mid ab$.
Then either $p \mid a$ or $p \mid b$.

Proof: For argument's sake suppose $p \nmid a$. Then $\gcd(p, a) = 1$ and by Corollary 2.1.1(b), $p \mid b$. ∎

Notice that the primality of p is necessary in Proposition 2.9. For example, $15 \mid 6 \cdot 10$ and yet $15 \nmid 6$ and $15 \nmid 10$.

Corollary 2.10.1: Let p be a prime and suppose $p \mid a_1.a_2 \ldots a_s$. Then there exists an i with $1 \leq i \leq s$ for which $p \mid a_i$.

Proof: The proof is left as an exercise.

We next prove that the prime factorization of a given integer is unique. For example, $91 = 7.13$, and there is no other way to represent 91 as a product of primes save for $13 \cdot 7$. We see now why 1 is not considered prime. If it were, then the prime factorization of an integer would no longer be unique. For example, $91 = 7 \cdot 13 = 1 \cdot 7 \cdot 13 = 1 \cdot 1 \cdot 7 \cdot 13$, etc.

The fundamental theorem of arithmetic is stated and essentially proved by Euclid as Proposition 14 of Book IX of the *Elements*. However, the degree of generality in the *Elements* was not up to the high standards of Gauss. Euclid's proof only applies to integers having at most three prime factors. This lacuna in the Euclidean proof is due to the Greek geometric view of number where three dimensions seemed quite adequate. Due to this Gauss included it as Article 16 of Section 2 of his *Disquisitiones Arithmeticae*.

Theorem 2.11 (Fundamental Theorem of Arithmetic): Every natural number greater than 1 is expressible as a product of primes uniquely up to rearrangement of the factors.

Proof: By Proposition 2.9, every natural number greater than 1 is expressible as a product of primes. Suppose that there exist natural numbers having more than one prime factorization. By WOP suppose n is the smallest such number. Then $n = p_1 p_2 \cdots p_s = q_1 q_2 \cdots q_t$ for some primes p and q_j. Hence $p_1 \mid q_1 q_2 \cdots q_t$ and by Corollary 2.10.1, p_1 divides one of the $q_j's$. Without loss of generality, we may assume $p_1 \mid q_1$. But q_1 is prime and as such is only divisible by 1 and itself. Hence $p_1 = q_1$.

Let $m = p_2 \cdots p_s = q_2 \cdots q_t$. Our assumptions on n imply that m has a unique factorization. Hence, after suitable rearrangement, $s = t$ and $p_2 = q_2, \ldots, p_s = q_t$. So $n = p_1 m$ has a unique prime factorization and the assertion is proved. ■

By Theorem 2.11, we can now refer to the prime factorization of an integer. Let $n = \prod_{i=1}^{t} p_i^{a_i}$. This is the canonical prime factorization of n if the $p_i's$ are distinct and written in ascending order.

Unique factorization is a central property of the integers and should not be taken for granted. The following variation on an example due to David Hilbert helps illustrate this point.

Example 2.10: Let S denote the set of all positive integers of the form $3k + 1$ (i.e. all positive integers congruent to $1 \, modulo \, 3$). Like **N**, S is closed with respect to multiplication. Now define an S-prime as an element of S larger than 1 which is only divisible in S by 1 and itself. The first few S-primes are $4, 7, 10, 13, 19$, and so on. The first S-composite is 16.

As with **N**, all elements of S are representable as products of S-primes. However, the factorization is not unique. For example, $220 = 10 \cdot 22 = 4 \cdot 55$. Notice that 4, 10, 22, and 55 are all S-primes. It is worth a moment's reflection to understand why the structure of S is different from **N**.

How many primes are there? One might reason that as the integers get larger, there are more natural numbers less than a given integer and hence a greater chance to find some divisors of it. Thus, it seems conceivable that there might be a largest prime number beyond which all integers are composite. In fact, this is not the case. There is no largest prime as the following beautiful theorem demonstrates.

Theorem 2.12 (*Elements* – Book IX, Prop. 20): There are infinitely many primes.

Proof: Let $p_1, p_2, \ldots, p_n$ be a non-empty set of primes (not necessarily distinct). Consider the integer $N = p_1 p_2 \cdots p_n + 1$. By Proposition 2.9, N is expressible as a product of primes. Let p be a prime such that $p \mid N$. Clearly $p \neq p_i$ for all i since $p_i \nmid N$ for $1 \leq i \leq n$. Hence p is not part of our set of primes. Similarly, no finite list of primes is complete and hence the set of primes is infinite. ∎

Although there is no largest prime, finding the largest yet discovered has some appeal. For the record, at the present time the largest known prime number is the behemoth $2^{136,279,841} - 1$ which is a special type of prime recently discovered (2024). We will give a fuller discussion of such primes in Chapter 7. The number above has 41,024,320 decimal digits!

The distribution of the primes has been of keen interest to mathematicians for centuries and continues to be a fruitful area of research right up to the present. For example, 2 and 3 are the only consecutive integers that are both prime. With no gap between them, 2 and 3 are sometimes called "Siamese" twins. The next smallest gap possible is that of just one intervening integer. Numbers like 3 and 5, 5 and 7, 11 and 13, 17 and 19 are called twin primes. No one has been able to prove that there are infinitely many pairs of twin primes! Yet some enjoy continuing to search for ever larger ones. The largest known pair of twin primes (found in 2016) is the pair 2, 996, 863, 034, 895 $\cdot 2^{1,290,000} \pm 1$.

The following proposition shows that gaps between consecutive primes can be arbitrarily large.

Proposition 2.13: There are arbitrarily long strings of consecutive composite numbers.

Proof: Let n be a positive integer. Consider the n consecutive integers

$$(n+1)! + 2, (n+1)! + 3, \ldots, (n+1)! + (n+1)$$

For each i with $2 \leq i \leq n+1$, we have $(n+i) \mid (n+i)! + i$ and yet $(n+i)! + i > n+i$. So all the consecutive integers are composite. Since n, is arbitrary, the result follows. ∎

It is instructive to realize that although there are arbitrarily long strings of consecutive composites, there are no infinite strings. The distinction between "arbitrarily many" and "infinitely many" is an important one.

Next we make a slight refinement to Theorem 2.12.

Proposition 2.14: There are infinitely many primes of the form $4k + 3$.

Proof: Let $p_1, p_2, \ldots, p_s$ be a non-empty set of primes of the form $4k + 3$. (Since 3 is prime, such a set can be constructed.) Consider $N = 4p_1p_2 \cdots p_s - 1$. Note that $N \equiv 3 \pmod 4$. Since products of primes congruent to 1 (mod 4) are congruent to 1(mod 4), there is a prime $p \equiv 3 \pmod 4$ with $p \mid N$. But $p_i \nmid N$ for $1 \leq i \leq s$ and so no finite list of primes of the form $4k + 3$ can be complete. ∎

In 1837, G.L. Dirichlet (1805–1859), Gauss' successor at the University of Göttingen, used sophisticated analytical methods to prove a remarkable generalization concerning primes in an arithmetic progression.

Dirichlet's Theorem: If $\gcd(a, b) = 1$ then there are infinitely many primes $p \equiv a \pmod b$.

For example, there are infinitely many primes that end with the digits 123 since they are of the form $p = 1000k + 123$. And mathematicians have continued to make further improvements. In 1920, S. Chowla conjectured that if $q \geq 3$ and $\gcd(a, q) = 1$, then there exist infinitely many pairs (p_{n+1}, p_{n+2}) such that $p_{n+1} \equiv p_{n+2} \equiv a \pmod q$ where p_n is the nth prime. (See Exercise 2.3.12 for comparison.) In 2020, Daniel Shiu proved Chowla's conjecture and established the following amazing theorem.

Shiu's String of Congruent Primes Theorem: Given $q \geq 3$ and $\gcd(a, q) = 1$, for every $k > 1$ there exist strings of consecutive primes $p_{n+1}, \ldots, p_{n+k}$ such that $p_{n+1} \equiv \cdots \equiv p_{n+k} \equiv a \pmod q$.

For example, there are arbitrarily long strings of consecutive primes $\equiv 23$ (mod 1,001).

Though we leave the proof of Dirichlet's theorem and Shiu's theorem to other treatises, we close with an interesting extension of Proposition 2.13. Not only are there arbitrarily long sequences of composites, but there are also primes that are isolated, being arbitrarily distant from other primes.

Proposition 2.15 (W. Sierpinski): For any $n \geq 2$, there is a prime for which all n integers preceding it and n integers following it are composite.

Proof: Let $n \geq 2$ and suppose $p > n$ is prime. Let

$$N = (p - n)(p - n + 1) \cdots (p - 1)(p + 1) \cdots (p + n - 1)(p + n)$$

be the product of the n numbers to the left of p and the n numbers to the right of p. Since p doesn't divide any of the factors of N, N and p are relatively prime. Dirichlet's Theorem implies that there are an infinite number of primes $\equiv p \pmod N$. Let $q = mN + p$ be one such prime with $m \geq 1$ (and hence $q > p$).

Now consider the n numbers on either side of q:

$$q-n, \ q-n+1, \ \ldots, \ q-1, \ q+1, \ \ldots q+n-1, \ q+n$$

For all $i = 1, \ldots, n$, the number $p \pm i$ divides N and so it divides mN and $mN + (p \pm i)$. Hence, $p + i$ divides $(mN + p) \pm i = q \pm i$. But $p \pm i < mN + (p \pm i)$.

So $q \pm i$ is composite for all $i = 1, \ldots, n$. ∎

Exercise 2.3

1. Explain why if n is a perfect square and a perfect cube, then n must be a perfect sixth power.

2. (a) Let $p_1 = 2, p_2 = 3, \ldots$, and in general let p_r be the rth prime number. Verify that $\left(\prod_{i=1}^{n} p_i\right) + 1$ is prime for $n = 1, 2, 3, 4$, and 5.

   (b) Verify that the product above is not prime for $n = 6$ by checking that $30031 = 59 \cdot 509$. Does this contradict our proof of Theorem 2.11?

3. (a) Prove that there are infinitely many primes of the form $3k + 2$.

   (b) Prove that there are infinitely many primes of the form $6k + 5$.

   (c) Prove that there are infinitely many primes of the form $2k + 101$.

4. Prove that there are infinitely many primes by filling in the necessary details: Assume there are only finitely many primes $p_1, \ldots, p_n$. Let $ab = p_1 p_2 \cdots p_n$ where $a > 1$ and $b > 1$. Show that $a + b$ is divisible by a new prime (T.L. Stieltjes – 1890).

5. Prove Corollary 2.10.1.

6. A magician asks person A to write down a three-digit number. Person B multiplies it by 7, person C multiplies the result by 13, and person D multiplies that by 11. Upon seeing the final product, the magician immediately calls out the original number. How is the trick done? (Cf. Exercise 1.3.14.)

7. Verify that all positive integers n with $1 < n < 30$ that are relatively prime to 30 are in fact prime (30 is the largest number with this property).

8. If p and q are twin primes with $3 < p < q$, then show that $9 \mid pq + 1$ and that $pq + 1$ is a perfect square.

9. Let a and b be positive integers and let $a = p_1^{a_1} \cdots p_t^{a_t}$ and $b = p_1^{b_1} \cdots p_t^{b_t}$ where some of the a_i's and b_i's may be zero. Let $m_i = \min\{a_i, b_i\}$ and let $M_i = \max\{a_i, b_i\}$.

   (a) Show that $\gcd(a, b) = p_1^{m_1} \cdots p_t^{m_t}$.

   (b) Show that $\text{lcm}[a, b] = p_1^{M_1} \cdots p_t^{M_t}$.

10. (a) Show that if a and b are relatively prime and ab is a square, then a and b are each square.

    (b) Show that if $a_1, \ldots, a_n$ are pairwise relatively prime and $a_1 a_2 \cdots a_n$ is a square, then $a_1, \ldots, a_n$ are all squares.

11. Use Dirichlet's theorem on primes in an arithmetic progression to deduce that if $\gcd(a, b) = 1$ then there are infinitely many integers $n \equiv a \pmod{b}$, which are the product of exactly k distinct primes for any $k \geq 1$.

12. Let p, $p+r$, $p+2r$ be three primes in arithmetic progression. Show that $6 \mid r$ if and only if $p > 3$. Can you generalize this result? (In 1944, S. Chowla proved there were infinitely many triples of primes in arithmetic progression.)

13. (a) Show that there are infinitely long arithmetic progressions of difference d consisting solely of composites for any d.

    (b) Show that there are no infinitely long arithmetic progressions consisting solely of primes.

14. Show that 3 is the only prime of the form $k^4 + k^2 + 1$.

15. The real numbers $r_1, \ldots, r_n$ are linearly independent if the only solution in integers to $k_1 r_1 + \cdots + k_n r_n = 0$ is $k_1 = \cdots = k_n = 0$. Show that if $p_1, \ldots, p_n$ are n distinct primes, then $\log p_1, \log p_2, \ldots, \log p_n$ are linearly independent.

16. Carry out Example 2.9 with S the set of all positive integers of the form $4k + 1$. Is the factorization in S now unique?

17. (a) Use Theorem 2.11 to show that $\sqrt{2}$ is irrational.

    (b) Show that $\log_3 10$ is irrational.

    (c) Show that $m^{1/n}$ is irrational for all positive integers m that are not perfect nth powers.

18. (a) Given that $71 \mid (7! + 1)$, show that $71 \mid (9! + 1)$.

    (b) Given that $61 \mid (16! + 1)$, show that $61 \mid (18! + 1)$.

19. (a) Use Proposition 1.2 and Exercise 1.2.13 to show that if $k \mid n$, then $F_k \mid F_n$.

    (b) Show that there are arbitrarily long strings of consecutive composite Fibonacci numbers.

20. What is the largest even integer not expressible as the sum of two odd composite integers? Prove it (E. Just and N. Schaumburger – 1973). (In Exercise 1.1.18, you were asked to think about this question.)

21. (a) Use the CRT to show that for any n and k there is a string of n consecutive integers each divisible by at least k distinct primes.

    (b) Let $p_1, \ldots, p_n$ be the first n primes. Given any n and k, show that there are n consecutive integers $r+1, \ldots, r+n$ for which $p_i^k \mid (r+i)$ for $1 \le i \le n$.

    (c) Given $d \ge 1$, show that the statements in (a) and (b) remain true for n integers in arithmetic progression with constant difference d.

22. Explicitly work through the proof of Proposition 2.15 with $n = 6$ and $p = 7$.

23. A prime has the "multiplicative property" if the product of its digits (in base 10) equals its position among the primes. Find two such primes less than 100. C. Pomerance and C. Spicer [105] have found three such examples and have shown there are only finitely many in all.

24. Use Shiu's string of congruent primes theorem to show that there are arbitrarily many strings of 1 million consecutive primes ending with the digits 9973. (Let $a = 9{,}973$, $q = 10000$.)

2.4 Introduction to primality testing and factoring

Mathematicians have long sought methods of distinguishing between primes and composites. There is no known simple formula which gives all the primes in succession; nor is there even a practical algorithm which will readily identify a given huge integer (say 2000 digits) as being prime or not. Thus, in some sense, progress toward prime specification is not wholly adequate. Great progress has been made in the last couple of decades, however, in determining whether a given integer of say 100 or less digits is prime. Though considerably more difficult, if it is not prime, modern techniques combined with computing power can often be used to factor it. In addition, if the given integer is of a special form – say $2^p - 1$ or $2^{2^n} + 1$ – then there are specialized results that enable us to consider much larger integers.

In this section we describe some of the rudiments of primality testing and techniques used in factoring composites. We then go on to show that there is no non-constant polynomial having range only primes. Chapter 7 is devoted to a more detailed study of primality testing and factoring.

We begin with a description of the sieve of Eratosthenes (276–190 B.C.E.). Eratosthenes achieved prominence in mathematics, poetry, geography, astronomy, philosophy, history, and even athletics. He was a close friend of Archimedes, a royal tutor to the son of Ptolemy III, and the chief librarian at Alexandria. In addition to writing on means and loci and creating a detailed star catalog, his crowning scientific achieve-

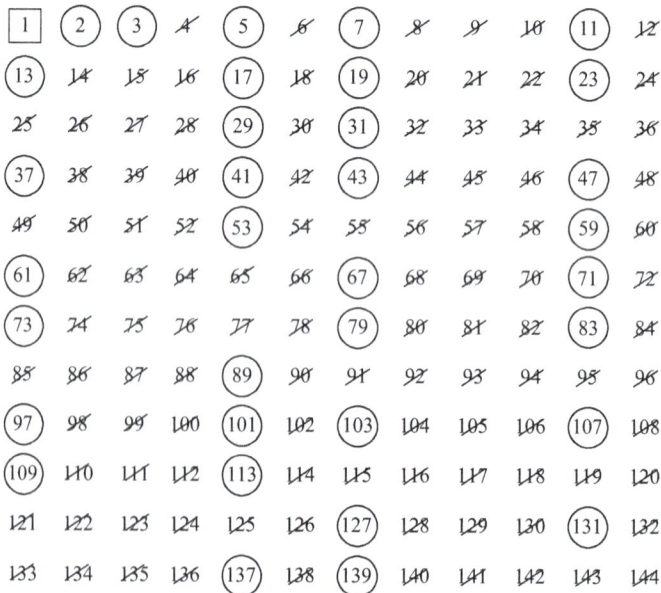

Figure 2.1: Sieve of Eratosthenes up to 144.

ment was the accurate determination of the circumference of the Earth. His sieve can be described as follows:

List all the integers from 1 to n (for convenience we let $n = 144$). See Figure 2.1.

We put a box around 1 since it has special standing as the multiplicative identity. The next number, 2, is circled since it must be prime. Next, we put a slash through all multiples of 2 since they are necessarily composite. The first number not crossed off is 3. We circle it since it must be prime. Next, we put a slash through all multiples of 3 not already crossed off. Then we circle 5 and continue similarly. In the diagram above we stop after eliminating all multiples of 11 (as follows from Proposition 2.14). The remaining numbers that haven't been sieved out are circled and are exactly the primes between 1 and 144.

Proposition 2.16: If n is composite, then n is divisible by a prime $p \le \sqrt{n}$.

Proof: If n is composite, then there exist integers a and b with $1 < a \le b < n$ such that $n = ab$. If $a > \sqrt{n}$, then $ab \ge a^2 > n$ (a contradiction). So $a \le \sqrt{n}$. By Proposition 2.9, there exists a prime p for which $p \mid a$; hence $p \mid n$. By Proposition 1.3(e), $p \le a \le \sqrt{n}$. ∎

In our example, we need only eliminate multiples of primes less than $\sqrt{144} < 13$. The largest such prime is 11.

We now turn to a factorization technique called Fermat's factorization method (from an undated letter of his from about 1643) which is often helpful when an odd composite is not divisible by any small primes. If n is an odd composite, then $n = ab$ with $1 < a \le b$. Let $x = (b + a)/2$ and $y = (b - a)/2$. Since a and b are odd, x and y are nonnegative integers. We can now write $n = x^2 - y^2$.

Conversely, to factor n we search for x and y such that $n = x^2 - y^2$ to obtain $n = ab$ where $a = x - y$ and $b = x + y$. In theory, we could then apply this technique to a and b, etc. until we had the prime factorization of n. This can be a fairly practical factorization technique when a and b are roughly the same size (approximately $\sqrt{n}$).

Example 2.12: Let us factor $n = 12319$. We see that $110 < \sqrt{12319} < 111$. Next, we compute $f(c) = c^2 - 12319$ for $c = 111, 112, \ldots$ successively until $f(c)$ is a perfect square. Much effort is saved by noting that $f(c + 1) = f(c) + c + (c + 1)$. In our example, $f(112) = 225 = 15^2$. Hence $12319 = (112 - 15)(112 + 15) = 97 \cdot 127$. It is easily checked that 97 and 127 are prime.

If there is no simple formula for all the primes, is there at least one that gives only primes? For example, Euler noticed that the polynomial $f(n) = n^2 - n + 41$ was prime for $n = 1, \ldots, 40$. Goldbach made the observation that no polynomial could represent primes exclusively. Euler's proof (1762) of this assertion is given below. We will return to several related questions in subsequent chapters.

Proposition 2.17: Let f be a non-constant polynomial with integer coefficients. Then $f(n)$ is composite for infinitely many n.

Proof: Let $f(n) = a_k n^k + \cdots + a_1 n + a_0$ where $a_i \in \mathbb{Z}$ for $0 \le i \le k$. In fact, we may assume $a_k \ge 1$. It is readily seen that $\lim_{n \to \infty} f(n) = \infty$. So there is an m for which $f(m) > 1$. Let $f(m) = t$.

Consider $f(m + rt) = a_k (m + rt)^k + \cdots + a_1 (m + rt) + a_0$ for $r \ge 1$. By the binomial theorem, $f(m + rt) = f(m) + t \cdot g(r) = t[1 + g(r)]$ where g is a polynomial with integer coefficients.

But the leading term of $g(r)$ is $a_k t^{k-1} r^k$. Hence $\lim_{r \to \infty} g(r) = \infty$. So there is an R such that $1 + g(r) > 1$ for all $r > R$.

But then $f(m + rt) = t \cdot [1 + g(r)]$ is composite for all $r > R$. ∎

Exercise 2.4

1. Are the following prime? If not, factor them completely:
   (a) 127
   (b) 289
   (c) 307
   (d) 343
   (e) 409
   (f) 899

2. With the aid of Fermat's factorization method, completely factor the composite numbers:
   (a) 8137
   (b) 9919
   (c) 20711
   (d) 24708
   (e) 86989
   (f) 8999999

3. Show that infinitely many odd integers of the form $n^2 + 1$ are composite. (It is conjectured that there are infinitely many primes of the form $n^2 + 1$.)

4. Carry out the sieve of Eratosthenes algorithm to determine all primes less than 300. When eliminating multiples of a prime p, explain why it suffices to begin with numbers greater than or equal to p^2.

5. Show if $n = a_1 \cdot a_2 \cdot \cdots \cdot a_r$ with $a_i > 1$ for all i then n has a prime divisor less than $\sqrt[r]{n}$.

6. Comment on the following "proof" of Proposition 2.16:
   $f(a_0) = a_k a_0 k + \cdots + a_1 a_0 + a_0$ is divisible by a_0 and hence is not prime. Similarly, $a_0 \mid f(r \cdot a_0)$ for all $r \in \mathbb{Z}$. So $f(n)$ is composite for infinitely many n.

7. Verify that if n is an odd prime less than 19, then the Fibonacci number F_n is prime. Show that F_{19} is composite. (It is unknown if there are infinitely many prime Fibonacci numbers.)

8. Prove the following assertion due to Fermat: Every odd prime can be written uniquely as the difference of two squares. What about the square of an odd prime? What about the cube of an odd prime?

9. (a) Verify Euler's assertion that $f(n) = n^2 - n + 41$ is prime for $n = 1, \dots, 40$.
   (b) Verify that $f(n) = n^2 - n + 41$ is prime for $n = -39, \dots, 0$ too.
   (c) Use parts (a) and (b) to show that $f(n) = n^2 - 81n + 1601$ is prime for $n = 1, \dots, 80$.
   (d) Find a quadratic polynomial that is prime for $161 \le n \le 240$.
   (e) (Euler) Verify that $f(n) = 2n^2 + 29$ is prime for $n = 0, \dots, 28$.

10. Show that if n is expressible as $a^3 + b^3$ where $1 \le a < b$, then n is composite. What can be said of n if it has two distinct representations as the sum of two cubes?

11. P. Fletcher, W. Lindgren, and C. Pomerance call a pair of odd primes $p > q$ a symmetric pair if $\gcd(p-1, q-1) = p - q$. (Equivalently, in Figure 4.1, the number of lattice points in S_1 and S_2 are equal.)
    (a) Show that p, q form a symmetric pair if and only if there is an $r \ge 1$ for which $p - q = 2r$ and $p \equiv 1 \pmod{2r}, q \equiv 1 \pmod{2r}$.
    (b) Determine all primes less than 100 belonging to some symmetric pair. (It is known that there are infinitely many primes not belonging to any symmetric pair. Recently Banks, W., Pollack, P., and Pomerance, C. have proven infinitely many primes belong to some symmetric pair as well.)

12. (Lagrange Interpolation Formula): Let n be a positive integer and $p_1, \dots, p_n$ be primes. Let $f(x) = \sum_{j=1}^{n} P_j \cdot P_j(x)$ where $P_j(x) = \prod_{i=1}^{n} \frac{x-i}{j-i}$ (where the product excludes $i = j$) for $1 \le j \le n$.

    Show that $f(j) = p_j$ for $1 \le j \le n$. Hence there are polynomials that give primes for arbitrarily many consecutive arguments (cf. Proposition 2.16).

2.5 Some important congruence relations

In this section we introduce several theorems of historical importance that deal with congruence relations. The first of these is Fermat's little theorem and its extension, the Euler-Fermat theorem. Fermat's little theorem is extremely useful in helping to simplify congruences involving large numbers.

The history of these theorems is long and convoluted. There is a Chinese reference dating from the fifth century B.C.E. where it is reported that if p is prime then p divides $2^p - 2$. Unfortunately, some scholars later erroneously asserted that that the converse is true, namely that if n divides $2^n - 2$ then n is prime. The first such counterexample to this assertion is $n = 341 = 11 \times 31$. Notice that $2^{10} \equiv 1 \pmod{341}$ and hence $2^{341} = 2^{10 \times 34 + 1} \equiv 1^{34} \times 2 = 2 \pmod{341}$.

Fermat stated the generalization given below in a letter of 1640 to Frenicle de Bessy (1605–1675) but omitted the demonstration. Euler took it upon himself to attempt to prove and extend many of Fermat's assertions. In 1736, he gave the first of several proofs of Theorem 2.18.

Theorem 2.18 (Fermat's Little Theorem): Let p be prime and suppose $p \nmid a$, then $a^{p-1} \equiv 1 \pmod p$.

In some books the following corollary is referred to as Fermat's little theorem:

Corollary 2.18.1: Let p be prime. Then $a^p \equiv a \pmod p$ for any positive integer a.

Proof of Corollary: If $p \nmid a$, then Fermat's little theorem implies $a^{p-1} \equiv 1 \pmod p$. Multiplication by a gives the desired result. If $p \mid a$ then $a^p \equiv 0 \equiv a \pmod p$. The corollary follows. ∎

We will temporarily defer the proof of Fermat's little theorem since the result follows directly from Theorem 2.19. For now, we give an example of its computational utility.

Example 2.13: Let us calculate 25^{8000} (modulo 7).
Note that $8000 = 6 \cdot 1333 + 2$. So $25^{8000} = 25^{6 \cdot 1333 + 2} = (25^6)^{1333} \cdot 25^2$.

Since $7 \nmid 25$, Fermat's little theorem implies $25^6 \equiv 1 \pmod 7$. Hence $25^{8000} \equiv 1^{1333} \cdot 25^2 \equiv 25^2 \pmod 7$. But $25^2 \equiv 4^2 = 16 \equiv 2 \pmod 7$. Therefore, $25^{8000} \equiv 2 \pmod 7$.

The method used above is extremely efficient. In general, to calculate $a^b \pmod p$, first check if $p \mid a$. If so, then $a^b \equiv 0 \pmod p$. If not, then calculate the remainder r when a is divided by $p - 1$ with $0 \le r \le p - 1$. Then $a^b \equiv a^r \pmod p$. If r is small, then the remaining calculation is not difficult. If p is large and $r < p$ is large, then Fermat's little theorem is often used in conjunction with other algorithms. Later, in this section. we describe one such algorithm, an efficient exponentiation algorithm well suited for computer calculation. Mastering some computational techniques will prove extremely useful throughout your study of number theory (especially in Chapter 7).

Another application of Fermat's little theorem is the calculation of an arithmetic inverse of $a \pmod p$ for p not too large. The number $a^* \equiv a^{p-2} \pmod p$ is the arithmetic inverse of $a \pmod p$ since $aa^* \equiv a^{p-1} \equiv 1 \pmod p$.

Fermat's little theorem is a crucial result often utilized in primality testing. Given a large odd integer n, it is fairly easy to check $2^{n-1} \pmod n$. If it happens that $2^{n-1} \not\equiv 1 \pmod n$, then by Fermat's little theorem, n is composite. However, we have no knowledge of its factors. In general, factoring is much more difficult than primality testing. The fact that factorization is difficult can be advantageous in some applications (see Section 8.1).

Before we proceed, we need to introduce two important definitions.

Definition 2.7 (*Euler phi function*): If n is a positive integer, then $\phi(n)$ denotes the number of positive integers less than or equal to n that are relatively prime to n.

For example, $\phi(1) = 1, \phi(6) = 2, \phi(8) = 4$, and $\phi(p) = p - 1$ for p prime. The Euler phi function was formerly called the Euler totient function. The nomenclature was due to the fact that integers smaller than n relatively prime to n were referred to as totitives of n by Euler. The Euler phi function is one of the most essential functions in all of number theory. We will discuss it in greater detail in Chapter 3.

Notice that if $a \equiv b \pmod{n}$, then $\gcd(a, n) = \gcd(b, n)$. This follows from writing $a = An + r, b = Bn + r$. Let $d_1 = \gcd(a, n)$ and $d_2 = \gcd(b, n)$. Since $r = a - An, d_1 \mid r$ and consequently $d_1 \mid b$. So $d_1 \mid d_2$. Analogously, $r = b - Bn$ and so $d_2 \mid r$ and $d_2 \mid a$.

So $d_2 \mid d_1$. By Proposition 1.3(*f*) $d_1 = d_2$. In particular, if $\gcd(a, n) = 1$ and $a \equiv b \pmod{n}$, then $\gcd(b, n) = 1$. Since there are $\phi(n)$ integers between 0 and n relatively prime to n, we can make the next definition.

Definition 2.8: Let n be a positive integer. A *reduced set of residues modulo n* is a set of $\phi(n)$ integers $r_1, \ldots, r_{\phi(n)}$ for which every integer relatively prime to n is congruent to exactly one of $r_1, \ldots, r_{\phi(n)} \pmod{n}$.

For example, the numbers 1, 3, 7, 9 form a reduced set of residues modulo 10 (as do 13, 27, 31, 49.) For another example, the numbers 1, 2, 3, 4, 5, 6 form a reduced set of residues modulo 7. Now choose an integer relatively prime to 7, say 12, and multiply each element of the reduced residue set above by 12. We get the numbers 12, 24, 36, 48, 60, 72, which are congruent to 5, 3, 1, 6, 4, 2 (mod 7), respectively. This is simply a rearrangement of our original reduced set of residues. The observation that this holds in general is the basis for our next important theorem.

Theorem 2.19 (Euler–Fermat Theorem): If $\gcd(a, n) = 1$, then $a^{\phi(n)} \equiv 1 \pmod{n}$.

Proof: Let $r_1, \ldots, r_{\phi(n)}$ be a reduced set of residues modulo n. Then each of $ar_1, \ldots, ar_{\phi(n)}$ is relatively prime to n. Furthermore, if $ar_i \equiv ar_j \pmod{n}$ for some $i \neq j$, then $r_i \equiv r_j \pmod{n}$ by Proposition 2.4(a), contrary to our original assumption. Hence $ar_1, \ldots, ar_{\phi(n)}$ is a reduced set of residues modulo n. So $r_1, \ldots, r_{\phi(n)}$ and $ar_1, \ldots, ar_{\phi(n)}$ are simply reorderings of one another modulo n. Thus,

$$(ar_1) \cdots \left(ar_{\phi(n)}\right) \equiv (r_1) \cdots (r_n)(\bmod\, n).$$

$$\text{Hence, } a^{\phi(n)}(r_1) \cdots (r_n) \equiv (r_1) \cdots (r_n)(\bmod\, n)$$

But $(r_1) \cdots (r_n)$ is relatively prime to n as follows by induction from Corollary 2.1.1(f).

Proposition 2.5(a) implies $a^{\phi(n)} \equiv 1 \pmod{n}$. ∎

Example 2.14: Show that if n is a number relatively prime to 10 then n divides infinitely many numbers of the set $\{9, 99, 999, 9999, 99999, \ldots\}$.

Solution: Often the crux of the matter for problems of this type is to show that n divides one of the numbers in the set. Since $\gcd(10, n) = 1$, we can apply the Euler-Fermat theorem obtaining $10^{\phi(n)} \equiv 1 \pmod{n}$. But then n divides $10^{\phi(n)} - 1 = 99\ldots9$, which is a number made up of $\phi(n)$ nines. It follows that n divides $10^{k\phi(n)-1}$ for any $k \geq 1$. The result follows.

The Euler-Fermat theorem is a great aid in calculation. It is an extremely useful tool whenever we wish to compute $a^r \pmod{n}$ for $r \geq \phi(n)$. For example, let us compute $7^{179} \pmod{20}$. Since $\phi(20) = 8$, it follows that $7^8 \equiv 1 \pmod{20}$. Hence

$$7^{179} = 7^{8(22)+3} \equiv 7^3 \pmod{20} \equiv 3 \pmod{20}$$

What can we do to lessen the difficulty of calculating $a^r \pmod{n}$ for r large but with $r < \phi(n)$? In this case the Euler-Fermat theorem does not apply. However, there is a useful technique which utilizes the binary representation of the exponent r.

Suppose we wish to compute $3^{41} \pmod{79}$. In this case 79 is prime and $41 < \phi(79) = 78$. Start by writing 41 in binary: $(41)_{10} = (101001)_2$. Next, we calculate 3^k for k appropriate powers of 2 corresponding to the 1's in the binary representation of 41. We then multiply the partial results to obtain the final answer. By so doing, we reduce the number of multiplications from $r - 1$ to the order of magnitude of $\log r$. In this instance,

$$3^1 \equiv 3 \pmod{79}$$

$$3^2 \equiv 9 \pmod{79}$$

$$3^4 = \left(3^2\right)^2 \equiv 9^2 \equiv 2 \pmod{79}$$

$$3^8 = \left(3^4\right)^2 \equiv 2^2 = 4 \pmod{79}$$

$$3^{16} = \left(3^8\right)^2 \equiv 4^2 = 16 \pmod{79}$$

$$3^{32} = \left(3^{16}\right)^2 \equiv 16^2 \equiv 19 \pmod{79}$$

Hence $3^{41} = 3^{32} \cdot 3^8 \cdot 3^1 \equiv 19 \cdot 4 \cdot 3 = 228 \equiv 70 \pmod{79}$.

In fact, there is a method, the binary exponentiation algorithm, which further codifies our work. Here are the details.

2.5.1 Binary exponentiation algorithm

To calculate $a^r \pmod{n}$:

(i) Write r in binary as $r = r_1 r_2 \ldots r_t$ where $r_i = 0$ or 1 for all i.

(ii) Let $m_0 = 1$.

(iii) For $i = 1, \ldots,$ let $\begin{cases} m_i = (m_{i-1})^2 \pmod{n} & \text{if } r_i = 0 \\ m_i = a \cdot (m_{i-1})^2 \pmod{n} & \text{if } r_i = 1 \end{cases}$ and

(iv) $m_t \equiv a^r \pmod{n}$.

For example, to calculate $3^{41} \pmod{79}$:

$41 = 32 + 8 + 1 = (101001)_2$. So the appropriate sequence $\pmod{79}$ is

$$1 \xrightarrow{1} 3 \cdot 1^2 = 3 \xrightarrow{0} 3^2 = 9 \xrightarrow{1} 3 \cdot 9^2 \equiv 6 \xrightarrow{0} 6^2 = 36 \xrightarrow{0} 36^2 = 1296 \equiv 32 \xrightarrow{1} 3 \cdot 32^2 = 3072 \equiv 70 \pmod{79}.$$

One final example should help to summarize what we have accomplished thus far.

Example 2.15: Let us calculate $6^{10,021} \pmod{41}$.

The number 41 is prime and hence $\phi(41) = 40$. Now $10021 = 25 \cdot 40 + 21$. Hence $6^{10021} = 6^{25 \cdot 40 + 21} \equiv 1^{25} \cdot 6^{21} \equiv 6^{21} \pmod{41}$ by the Euler-Fermat theorem. But $(21)_{10} = (10101)_2$. The binary exponentiation algorithm gives the sequence

$$1 \xrightarrow{1} 6 \cdot 1^2 = 6 \xrightarrow{0} 6^2 = 36 \equiv -5 \xrightarrow{1} 6 \cdot (-5)^2 = 150 \equiv -14 \xrightarrow{0} (-14)^2 = 196 \equiv -9 \xrightarrow{1} 6 \cdot (-9)^2$$

$$\equiv 6(-1) \equiv -6 \equiv 35 \pmod{41}.$$

Hence $6^{10.021} \equiv 35 \pmod{41}$.

We now turn to a stunning result due to Joseph Louis Lagrange. Lagrange made significant contributions to all areas of mathematics of his day and is ranked along with Euler as the greatest mathematician of the eighteenth century. His first academic post was at the artillery school in Turin when he was just 19 years old! When Euler left the Berlin Academy in 1766 for St. Petersburg, Frederick the Great invited Lagrange to fill the vacancy saying, "it is necessary that the greatest geometer of Europe should live near the greatest of kings."

Let us consider an example first. The congruence equation $x^5 - 5x^3 + 4x \equiv 0 \pmod{p}$ obviously has at most p solutions modulo p. What may be less apparent is that there are actually just five solutions for any prime $p \geq 5$. The solutions are $x = 0, 1, 2, p - 2,$ and $p - 1$. That there can be no more is guaranteed by our next result.

Theorem 2.20 (Lagrange's Theorem): If p is prime and $f(x) = \sum_{i=0}^{n} a_i x^i$ is a polynomial of degree $n \geq 1$ with integral coefficients and $a_n \not\equiv 0 \pmod{p}$, then the congruence equation $f(x) \equiv 0 \pmod{p}$ has at most n incongruent solutions modulo p.

Proof: We prove the result by induction on n. For $n=1$ we have $f(x) = a_1 x + a_0$ and the result follows from the linear congruence theorem. Assume the theorem is valid for polynomials of degree n. Now let $f(x)$ be a polynomial of degree $n+1$. Suppose, contrary to what we wish to demonstrate, that $f(x)$ has at least $n+2$ incongruent solutions modulo p. Let s be one of these solutions. Then $f(x) = (x-s)q(x) + r$ where degree of q is n and a_{n+1} is the leading coefficient of q as well as f. Since $f(s) \equiv 0 \pmod{p}$, it follows that $r \equiv 0 \pmod{p}$. Hence $f(x) \equiv (x-s)q(x) \pmod{p}$. Let t be any solution to $f(x) \equiv 0 \pmod{p}$ with $t \equiv s \pmod{p}$. Then

$$f(t) \equiv (t-s)q(t) \equiv 0 \pmod{p}$$

But $t - s \equiv 0 \pmod{p}$ and so $q(t) \equiv 0 \pmod{p}$ by Euclid's lemma.

The fact that there are at least $n+1$ choices for t contradicts our inductive assumption that $q(x) \equiv 0 \pmod{p}$ has at most n solutions. This establishes the result. ∎

Note that the primality of p is necessary in Lagrange's theorem. For example, $x^2 - 1 \equiv 0 \pmod 8$ has solutions $1, 3, 5$, and $7 \pmod 8$.

The next result is universally referred to as Wilson's theorem since it was incorrectly ascribed to John Wilson (1741–1793) by his teacher Edward Waring (1736–1798) in Waring's influential treatise *Meditationes Arithmeticae* published in 1770. In fact, a special case of the theorem was discussed by ibn al-Haythan (ca. 965–1040). The *Meditationes* is well known for two conjectures it contains (each given as assertions without proof). One is Goldbach's conjecture that every even integer greater than 2 can be expressed as the sum of two primes. The other is Waring's problem to prove that every natural number is the sum of at most 4 squares, 9 cubes, 19 fourth powers; and for general n, there is an N dependent only on n such that every natural number is the sum of at most Nth powers. We take up a closer analysis of Waring's problem in Sections 5.4 and 9.1.

John Wilson was Senior Wrangler at Cambridge University and was mathematically inclined. He spent some time tabulating primes but early on switched to a successful career in law and was eventually knighted. One accomplishment that he probably lacked was the demonstration of Wilson's theorem itself. The first published proof was due to Lagrange in 1771 to which Euler added several of his own.

Corollary 2.20.1 (Wilson's Theorem): If p is a prime, then $(p-1)! \equiv -1 \pmod{p}$.

Proof: Let $f(x) = \prod_{i=1}^{p-1}(x-i) - x^{p-1} + 1 = a_{p-2}x^{p-2} + a_{p-1}x^{p-1} + \cdots + a_0$, a polynomial of degree $p-2$. If $1 \le s \le p-1$ then $f(s) \equiv 0 \pmod{p}$ since $s^{p-1} \equiv 1 \pmod{p}$ by Fermat's little theorem. Hence $f(x) \equiv 0 \pmod{p}$ has at least $p-1$ solutions modulo p. By Lagrange's theorem, it must be the case that $a_{p-2} \equiv a_{p-3} \equiv \cdots \equiv a_0 \equiv 0 \pmod{p}$. Hence for any $x, f(x) \equiv 0 \pmod{p}$.

Letting $x = p$, we obtain

$$f(p) = \prod_{i=1}^{p-1}(p - i) - p^{p-1} + 1 \equiv 0 \pmod{p}$$

Hence

$$(p - 1)! \equiv -1 \pmod{p} \qquad\blacksquare$$

Note that the converse of Wilson's theorem is also true: namely, if n is composite, then $(n - 1)! \not\equiv -1 \pmod{n}$. In fact, if $n > 4$ is composite, then $(n - 1)! \equiv 0 \pmod{n}$. We leave the verification as Exercise 2.5.7.

When is it the case that $f(x) \equiv 0 \pmod{p}$ where $\deg f = n$ has exactly n incongruent solutions? Perhaps surprisingly, there is a straight forward answer. Without loss of generality, we may assume $n < p$ in what follows.

Theorem 2.21: Let $f(x)$ be a monic polynomial (i.e., $a_n = 1$) of degree n. Then $f(x) \equiv 0 \pmod{p}$ has precisely n incongruent solutions if and only if $f(x) \mid x^p - x \pmod{p}$.

Proof: ($\Leftarrow$)Let $f(x) \mid x^p - x \pmod{p}$. So $x^p - x = f(x) \cdot g(x) + p \cdot h(x)$ where $\deg f = n$, $\deg g = p - n$, and f and g are monic polynomials with integer coefficients. Here $h(x)$ is either identically zero or is a polynomial with integer coefficients of degree less than n. By Fermat's little theorem, $x^p - x \equiv 0 \pmod{p}$ has the p solutions $0, 1, \ldots, p-1$. Hence $f(x) \cdot g(x) \equiv 0 \pmod{p}$ has p solutions. Suppose $x = c$ is one such solution. Then either $p \mid f(c)$ or $p \mid g(c)$ by Euclid's lemma. But Lagrange's theorem guarantees there are at most n solutions to $f(x) \equiv 0 \pmod{p}$ and $p - n$ solutions to $g(x) \equiv 0 \pmod{p}$. Thus, there are exactly n solutions to $f(x) \equiv 0 \pmod{p}$ as required as well as $p - n$ solutions to $g(x) \equiv 0 \pmod{p}$.

($\Rightarrow$) Suppose $f(x) \equiv 0 \pmod{p}$ has precisely n incongruent solutions.

Dividing $x^p - x$ by $f(x)$ we get

$$x^p - x = f(x) \cdot g(x) + r(x)$$

where $r(x)$ is either identically zero or a polynomial of degree less than n. Since any solution $x = c$ to $f(x) \equiv 0 \pmod{p}$ is also a solution to $x^p - x \equiv 0 \pmod{p}$ by Fermat's little theorem, $r(c) \equiv 0 \pmod{p}$ as well. So $r(x) \equiv 0 \pmod{p}$ has at least n solutions and n exceeds the degree of $r(x)$. By Lagrange's theorem, all the coefficients of $r(x)$ are divisible by p and hence there is a polynomial $h(x)$ with integer coefficients such that

$$x^p - x = f(x) \cdot g(x) + p \cdot h(x).$$

$$\text{Hence } f(x) \mid x^p - x \pmod{p} \qquad\blacksquare$$

For example, $(x^3 - 1) \mid (x^7 - x)$ and $f(x) = x^3 + 14x - 1 \equiv x^3 - 1 \pmod 7$.

Hence $f(x) \equiv 0 \pmod 7$ has precisely three incongruent solutions $\pmod 7$. Verify that they are in fact 1, 2, and 4.

The next corollary (due to Lagrange) will prove useful in Chapter 4.

Corollary 2.21.1: Let p be prime and $k \mid p-1$. Then $x^{k-1} \equiv 0 \pmod{p}$ has exactly k incongruent solutions.

Proof: Let $p-1 = kt$. Then $(x^k - 1) x (x^{k(t-1)} + x^{k(t-2)} + \cdots + 1) = x^p - x$. ∎

Exercise 2.5

1. Determine $\phi(n)$ for $n = 10, 12, 18, 24, 35, 97$.
2. (a) Which of the following sets form a reduced set of residues modulo 9?

$$S_1 = \{1, 2, 4, 5, 7, 8\}, S_2 = \{2, 3, 5, 6, 8, 9\},$$

$$S_3 = \{1, 4, 7, 11, 14, 17, 19\}, S_4 = \{10, 11, 13, 14, 16\}$$

   (b) Which of the following sets form a reduced set of residues modulo 12?

$$S_1 = \{1, 2, 3, 4, 5, 6, 7, 8, 9, 10, 11, 12\}, S_2 = \{1, 5, 7, 11\}$$

$$S_3 = \{25, 47, 127, 185\}, S_4 = \{13, 18, 19\}, S_5 = \{1, 17, 19, 35, 37\}$$

3. (a) Use Fermat's little theorem to determine $3^{96} \pmod{97}, 8^{102} \pmod{11}$, and $(-5)^{12002} \pmod{13}$.
   (b) Use the Euler-Fermat theorem to determine $71234 \pmod{10}$, $5^{1111} \pmod{12}$, and $3^{4000} \pmod{20}$.
4. Use the binary exponentiation algorithm to determine
   (a) $2^{23} \pmod{37}$
   (b) $5^{44} \pmod{61}$
   (c) $3^{32} \pmod{101}$
5. Compute the following:
   (a) $2^{3011} \pmod{31}$
   (b) $3^{52009} \pmod{53}$
   (c) $(-59)^{1000} \pmod{20}$
   (d) $348^{348} \pmod{17}$
6. (a) What is the remainder when we divide 3^{2402} by 13?
   (b) What is the last digit of 17^{3241}?
   (c) What are the last two digits of 17^{3241}?
   (d) What are the last two digits of 49^{4802}?
7. Prove the converse of Wilson's theorem. Does this give us a good primality test?
8. (a) Show that n^5 has the same last digit as n for any integer n.
   (b) Show that if $\gcd(n, 100) = 1$ then n^{41} has the same last two digits as n. Is it true that for all n, n and n^{41} have the same last two digits?
9. Show that if p is prime, then $1 + 1^{p-1} + 2^{p-1} + \cdots + (p-1)^{p-1} \equiv 0 \pmod{p}$.

Whether the converse is true is unknown, though it has been verified up to 10^{1700} by E. Bedocchi (1985).

10. Show that $1^p + 2^p + \cdots + (p-1)^p \equiv 0 \pmod{p}$ for any odd prime p.

11. Let p be prime. Show that $p^2 \mid [(p-1)^{p-1} - 1][(p-1)! + 1]$.

12. If p is prime and $a + b = p - 1$, then show $a! \, b! \equiv (-1)^{b+1} \pmod{p}$.

13. Let $A = 97^{9797}$, let B be the sum of the digits of A, let C be the sum of the digits of B, and let D be the sum of the digits of C. What is D? (Hint: Apply Exercise 1.3.3(a) and then obtain upper bounds for A, B, and C, respectively.)

14. (a) Show if $\gcd(n, 10) = 1$, then n divides infinitely many numbers of the set $\{1, 11, 111, 1111, 11111, \ldots\}$. Such numbers are called *repunits*.

    (b) Show that beyond 1, no repunit is a square.

    (c) Show that if $m > 1$ is such that m^3 is a repunit, then $m \equiv 1 \pmod{10}$, 71 $\pmod{100}$, 471 $\pmod{1000}$, and $\equiv 8471 \pmod{10000}$.

      (A. Rotkiewicz (1987) has shown that, in fact, no repunit beyond 1 is a cube.) It is unknown if infinitely many repunits are prime. However, some large ones are prime – e.g., 1031 1's (H.C. Williams and H. Dubner.)

15. Let m and n be positive integers with $\gcd(n, 10) = 1$. Let $\bar{m}\bar{m}$ denote juxtaposition of the digits of m. Show that n divides infinitely many numbers of the set $\{\bar{m}, \bar{m}\bar{m}, \bar{m}\bar{m}\bar{m}, \ldots\}$.

16. (An alternate proof of Wilson's theorem)

    (a) For $2 \le a \le p - 2$, let a^* be the arithmetic inverse of a modulo p. Show that $2, 3, \ldots, p - 2$ can be partitioned into $(p - 3)/2$ pairs (r_i, r_i^*) where r_i and $r_i^{\ *}$ are arithmetic inverses modulo p for all i. (You must show that $r_i \ne r_i^*$.)

    (b) Consider $(p-1)! = 1 \cdot (p-1) \cdot 2 \cdot 3 \cdot \cdots \cdot (p-2) = (p-1) \sum_{i=1}^{(p-3)/2} r_i r_i^*$ to obtain Wilson's theorem.

17. Verify for $p = 5$ and $p = 13$ that $(p - 1)! \equiv -1 \pmod{p^2}$. Primes satisfying this congruence are called Wilson primes. The only other known Wilson prime is $p = 563$ (discovered by K. Goldberg in 1953.) Computations by E. Costa, R. Gerbicz, and D. Harvey confirm that here are no others below $2 \cdot 10^{13}$.

18. How many incongruent solutions $\pmod{101}$ are there to the congruence relation $x^{25} + 1 \equiv 0 \pmod{101}$? Why?

19. (a) Show that the congruence relation $x^5 - 5x^3 + 4x \equiv 0 \pmod{p}$ has exactly the solutions $0, 1, 2, p - 2$, and $p - 1 \pmod{p}$ for $p \ge 5$.

    (b) Find all solutions to $2x^5 - 20x^3 + 18x \equiv 0 \pmod{p}$ for $p \ge 7$.

    (c) Find all solutions to $x^4 - 17x^2 + 16 \equiv 0 \pmod{p}$ for $p \ge 11$.

20. (a) Use Theorem 2.20 with $f(x) = x^{p-1} - 1$ to derive another proof of Wilson's theorem.

    (b) Use Theorem 2.20 with $f(x) = x^{p-1} - 1$ to show that if P_k represents the sum of all products of k integers from the set $1, 2, \ldots, p - 1$, then $P_k \equiv 0 \pmod{p}$.

21. Wolstenholme's theorem states that the numerator of the fraction $1 + \frac{1}{2} + \frac{1}{3} + \ldots + \frac{1}{p-1}$ is divisible by p^2 for any prime $p \ge 5$. Verify the result for $p = 5$, $p = 7$. (The theorem was actually discovered by E. Waring – 1782.)

22. Thue's Lemma (Axel Thue – 1909) Use the pigeonhole principle to establish that if $n > 1$ and $\gcd(a, n) = 1$, then there exist x and y with $1 \le x \le \sqrt{n}$ and $1 \le |y| \le \sqrt{n}$ such that $ax \equiv y \pmod{n}$.

23. Show that if p is prime, then p divides $(p-2)! - 1$. Is the converse true?

24. For $p \ge 5$ prime and $2 \le n \le p - 3$ with $n \ne p - 4$, show that there exist $1 < a_1 < a_2 < \cdots < a_n < p$ such that $\prod_{i=1}^{n} a_i \equiv 1 \pmod{p}$.

2.6 General polynomial congruences: Hensel's lemma

Let $f(x)$ be a nonzero polynomial and n a positive integer. In this section we make some general comments concerning the solution of the congruence

$$f(x) \equiv 0 \pmod{n} \tag{2.11}$$

We have already made some contributions to the study of eq. (2.11). On the one hand, if $f(x) = ax + b$ and n is arbitrary, then the situation is completely described by the linear congruence theorem, Proposition 2.5. On the other hand, if $f(x)$ is arbitrary and n is prime, then Lagrange's theorem gives an upper bound for the number of solutions to eq. (2.11) modulo n.

Suppose that $n = \prod_{i=1}^{t} p_i^{a_i}$ is the canonical prime factorization of n. If s is a solution to eq. (2.11), then s is also a solution to all the congruences

$$f(x) \equiv 0 \pmod{p_i^{a_i}} \text{ for all } i = 1, \ldots, t \tag{2.12}$$

Conversely, if eq. (2.12) is solvable with s_i for each $i = 1, \ldots, t$, then by the CRT there is a simultaneous solution s to all congruences (2.12) unique modulo n. But then s is a solution to eq. (2.11) by Corollary 2.1.1(c) or Exercise 2.1.17(c). Hence the problem of solving eq. (2.11) reduces to solving each of the congruences in eq. (2.12).

Next we show that if $f(x) \equiv 0 \pmod{p^a}$ can be solved, then there is an effective algorithm to determine the solutions to $f(x) \equiv 0 \pmod{p^{a+1}}$. Thus, to solve eq. (2.11), it suffices to solve $f(x) \equiv 0 \pmod{p}$ for each $p \mid n$ as outlined above. The result below is due to the German mathematician Kurt Hensel (1861–1941) who was one of the creators of p-adic analysis.

Theorem 2.22 (Hensel's Lemma): Let $f(x)$ be a polynomial with integral coefficients and $a \ge 1$. Consider the congruence

$$f(x) \equiv 0 \pmod{p^a} \tag{2.13}$$

and let $x = s$ be a solution.

(a) Let $f'(s) \not\equiv 0 \pmod{p}$. Then there is a unique $t \pmod{p}$ such that

$$f(s + tp^a) \equiv 0 \pmod{p^{a+1}}$$

(b) Let $f'(s) \equiv 0 \pmod{p}$.
  (i) If $f(s) \equiv 0 \pmod{p^{a+1}}$ then $f(s + tp^a) \equiv 0 \pmod{p^{a+1}}$ for all t.
  (ii) If $f(s) \not\equiv 0 \pmod{p^{a+1}}$ then there are no solutions to $f(x) \equiv 0 \pmod{p^{a+1}}$.

Case (a) is called the nonsingular case, whereas case (b) is called the singular case. To summarize, in the nonsingular case, a solution to eq. (2.13) lifts uniquely to a solution of $f(x) \equiv 0 \pmod{p^{a+1}}$. In the singular case, a solution to eq. (2.11) either lifts to p incongruent solutions modulo p^{a+1} or to none at all.

Verify that $s + t_1 p^a \equiv s + t_2 p^a \pmod{p^{a+1}}$ for $0 \le t_1 < t_2 \le p - 1$.

Lemma 2.22.1 (Taylor's Expansion): Let $f(x)$ be a polynomial of degree n with integral coefficients. Then

$$f(x + y) = f(x) + f'(x)y + \frac{f''(x)}{2!}y^2 + \cdots + \frac{f^{(n)}(x)}{n!}y^n$$

In addition, the coefficients of y are polynomials in x with integral coefficients.

Proof: Let $P_m(x) = x^m$. Then

$$P_m(x + y) = \sum_{k=0}^{m} \binom{m}{k} x^{m-k} y^k \text{ by Theorem 1.8}$$

But $P_m^{(k)}(x) = m(m-1) \cdots (m - k + 1)x^{m-k}$. Hence,

$$\frac{P_m(k)(x)}{k!} = \binom{m}{k} x^{m-k} \quad \text{where} \quad \binom{m}{k} \text{ are integers}$$

So the result follows for $P_m(x) = x^m$.

Recall that differentiation is a linear operator, i.e. $(f + g)' = f' + g'$ and $(rf)' = rf'$ for all polynomials f and g and integers r. By induction, $(f + g)^{(k)} = f^{(k)} + g^{(k)}$ and $(rf)^{(k)} = rf^{(k)}$ for all k. Any polynomial f of degree n can be expressed as a linear combination of P_m's for $m \le n$. This establishes the lemma for all such polynomials f. ∎

Proof of Theorem 2.22: Since $x = s$ is a solution to eq. (2.13), any solution to

$$f(x) \equiv 0 \pmod{p^{a+1}} \tag{2.14}$$

must be of the form $x = s + tp^a$ for some t. Next, we determine appropriate conditions on t.

Let f be of degree n. We may assume $n \geq 1$ since the result is clearly true if $f(x) \equiv r$ for some integer r (even zero). Now

$$f(s + p^a) = f(s) + f'(s)t^a + \frac{f''(s)}{2!}(t^a)^2 + \ldots + \frac{f^{(n)}(s)}{n!}(t^a)^n$$

by Lemma 2.19.1 with $x = s$ and $y = tp^a$. Since $a \geq 1$, $a + 1 \leq ma$ for $2 \leq m \leq n$ and so $p^{a+1} \mid p^{ma}$ for $2 \leq m \leq n$. Furthermore $\frac{f^{(k)}(s)}{k!} \in \mathbb{Z}$ for all k and hence

$$f(s + tp^a) \equiv f(s) + f'(s)tp^a \pmod{p^{a+1}}$$

If $x = s + tp^a$ is a solution to eq. (2.14), then $f'(s)tp^a \equiv -f(s) \pmod{p^{a+1}}$. But $f(s) \equiv 0 \pmod{p^a}$ and hence we can divide by p^a, obtaining

$$f'(s)t \equiv \frac{f(s)}{p^a} \pmod{p}$$

This may be rewritten as

$$f'(s)t + \frac{f(s)}{p^a} \equiv 0 \pmod{p} \tag{2.15}$$

This is just a linear congruence in t. Apply Proposition 2.5 appropriately:

$$\text{Let } d = \gcd\ (f'(s), p).$$

If $f'(s) \not\equiv 0 \pmod{p}$ then $d = 1$ and by Proposition 2.5(a), eq. (2.15) has a unique solution modulo p. In fact, $t = \frac{-f(s)}{p^a} \cdot f'(s)^*$ is that solution where $f'(s)^*$ is the arithmetic inverse of $f'(s)$ modulo p. Thus part (a) is proved.

If $f'(s) \equiv 0 \pmod{p}$ then $d = p$. By Proposition 2.5(b), if $p \mid \frac{f(s)}{p^a}$ (i.e., $f(s) \equiv 0 \pmod{p^{a+1}}$) then $x = s + tp^a$ is a solution to eq. (2.14) for all $t = 0, \ldots, p-1 \pmod{p}$. If $p \nmid \frac{f(s)}{p^a}$, i.e., $f(s) \not\equiv 0 \pmod{p^{a+1}}$) then there are no solutions to eq. (2.14). This establishes part (b) and completes our proof. ∎

In the nonsingular case, we may lift a solution $x = s_1$ of $f(x) \equiv 0 \pmod{p}$ to a solution $x = s_2$ of $f(x) \equiv 0 \pmod{p^2}$ where $s_2 = s_1 - f(s_1) \cdot f'(s_1)^*$. But if $f'(s_1) \not\equiv 0 \pmod{p}$, then $f'(s_2) \not\equiv 0 \pmod{p}$ since $f(s_1) \equiv 0 \pmod{p}$. Hence we can apply Hensel's lemma again to obtain a solution $x = s_3$ to $f(x) \equiv 0 \pmod{p^3}$. Furthermore, since $s_2 \equiv s_1 \pmod{p}$, it follows that $f'(s_2) \equiv f'(s_1) \pmod{p}$. Hence $s_3 = s_2 - f(s_2) \cdot f'(s_1)^*$. This may be repeated ad infinitum. In general, if $f'(s_1) \not\equiv 0 \pmod{p}$, then $x = s_i$ is a solution to $f(x) \equiv 0 \pmod{p^i}$ where

$$s_{i+1} \equiv s_i - f(s_i) \cdot f'(s_1)^* \pmod{p^{i+1}} \text{ for } i \geq 1 \tag{2.16}$$

It is instructive to compare this procedure with Newton's recursive formula for locating roots of polynomials.

Let us work through an example.

Example 2.16: Solve $2x^2 - x + 14 \equiv 0 \pmod{125}$.

Solution: First consider $f(x) \equiv 0 \pmod 5$ where $f(x) = 2x^2 - x + 14$. This reduces to $2x^2 - x - 1 \equiv 0 \pmod 5$ which has solutions $x = 1$ and $x = 2$ by inspection. Let $s_1 = 1$. $f'(1) = 3 \not\equiv 0 \pmod 5$. So s_1 lifts to a unique solution $x = s_2$ of $f(x) \equiv 0 \pmod{25}$.

By eq. (2.16) $s_2 \equiv s_1 - f(s_1) \cdot f'(s_1)^* \pmod{25}$. But $f(1) = 15$ and $f'(1)^* = 3^* = 2 \pmod 5$. Thus $s_2 \equiv -29 \equiv -4 \pmod{25}$. Similarly, $s_3 \equiv s_2 - f(s_2) \cdot f'(s_1)^* \pmod{125}$. Since $f(-4) = 50$, $s_3 \equiv -104 \equiv 21 \pmod{125}$.

Now let $r_1 = 2$ be the other solution of $f(x) \equiv 0 \pmod 5$. $f'(2) = 7 \not\equiv 0 \pmod 5$ and so r_1 lifts in an analogous manner to a solution $x = r_2$ of $f(x) \equiv 0 \pmod{25}$ and then to a solution $x = r_3$ of $f(x) \equiv 0 \pmod{125}$. We find $r_2 = -8$ and $r_3 = 42$. Hence the only solutions to $2x^2 - x + 14 \equiv 0 \pmod{125}$ are $x = 21$ and $x = 42$.

We complete this section with two additional observations on solving the polynomial congruence (2.11). One is that the coefficients of $f(x)$ can be reduced modulo n by Proposition 1.7. For example, $22x^5 + 103x^3 + 54x \equiv 2x^5 + 3x^3 + 4x \pmod{10}$.

The other observation is that if n is prime, then the degree of $f(x)$ can often be reduced by applying Fermat's little theorem. For example, $x^{100} = x^{19 \cdot 5 + 5} \equiv x^5 \cdot x^5 \equiv x^{10} \pmod{19}$ for all x whether or not $19 \mid x$. Let's look at one final example.

Example 2.17: Solve $x^{94} + 1200x^{38} + 400x^{12} + 100x^2 + 700 \equiv 0 \pmod 7$.

Solution: The congruence $x^{94} + 1200x^{38} + 400x^{12} + 100x^2 + 700 \equiv 0 \pmod 7$ is equivalent to $x^{94} + 3x^{38} + x^{12} + 2x^2 \equiv 0 \pmod 7$ by reducing the coefficients $\pmod 7$. But $x^{94} = x^{7 \cdot 13 + 3} \equiv x^{13} \cdot x^3 \equiv x^{7 \cdot 2 + 2} \equiv x^{2 \cdot 2} = x^4 \pmod 7$ by Fermat's little theorem. Analogously, $x^{38} \equiv x^2 \pmod 7$ and $x^{12} \equiv x^6 \pmod 7$. It follows that

$$x^{94} + 1200x^{38} + 400x^{12} + 100x^2 + 700 \equiv x^4 + 3x^2 + x^6 + 2x^2$$

$$\equiv x^6 + x^4 + 5x^2 \pmod 7$$

If $7 \mid x$, then $x \equiv 0 \pmod 7$ and $x^6 + x^4 + 5x^2 \equiv 0 \pmod 7$. Hence $x = 0$ is a solution to the congruence relation.

If $7 \nmid x$, then $x^6 \equiv 1 \pmod 7$. Thus if $x \not\equiv 0 \pmod 7$, then the problem reduces to solving $x^4 + 5x^2 + 1 \equiv 0 \pmod 7$. We only need to check $x = \pm 1, \pm 2, \pm 3$. In fact, since all terms are of even degree, it suffices to check $x = 1, x = 2$, and $x = 3$ since the negatives behave identically. In fact, of these only $x = 1$ satisfies the congruence. Therefore, the solutions to $x^{94} + 1200x^{38} + 400x^{12} + 100x^2 + 700 \equiv 0 \pmod 7$ are $0, 1, 6 \pmod 7$.

Exercise 2.6

1. Reduce the following polynomial congruences to those of degree at most 10 with nonnegative coefficients at most 10.

    (a) $x^{1248} + 309x^{765} - 144x^{271} \equiv 0 \pmod{11}$.

    (b) $4x^{900} + 3x^{800} + 2x^{700} + x^{600} \equiv 0 \pmod{11}$.

2. Show that $x^6 + 98x^5 + 35x^4 + 84x^3 + 21x^2 + 133x + 1 \equiv 0 \pmod{147}$ is unsolvable by reasoning modulo 7.

3. Explain why $9x^4 + 33x^3 - 21x^2 + 15x + 22 \equiv 0 \pmod{75}$ is unsolvable.

4. Verify Lemma 2.20.1 on the following functions:

    (a) $f(x) = x^3$.

    (b) $f(x) = x^5 + 3x$.

    (c) $f(x) = x^4 + x^3 + x^2 + x + 1$.

5. Let $f(x)$ be a polynomial of degree n and p be prime. Let $m = min\{n, p\}$. Show that there is a monic polynomial $g(x)$ of degree at most m such that $g(x) \equiv 0 \pmod{p}$ has the same solutions as $f(x) \equiv 0 \pmod{p}$.

6. (a) Find all solutions to $x^2 + x \equiv 0 \pmod{27}$.

    (b) Solve $x^2 + x + 1 \equiv 0 \pmod{27}$.

7. Find all solutions to $x^3 + 3x^2 - 5 \equiv 0 \pmod{49}$.

8. Find all solutions to $x^2 - 3x + 1 \equiv 0 \pmod{121}$.

9. Determine all solutions to $x^3 + 2x - 3 \equiv 0 \pmod{100}$.

10. Determine all solutions to $5x^2 + x + 9 \equiv 0 \pmod{153}$.

Chapter 3
Arithmetic functions

3.1 Important arithmetic functions

In this section we introduce several important examples of arithmetic functions. Many of them are ubiquitous throughout number theory. Our goal in this chapter is to develop closed formulas for these functions which can then be applied to further investigations.

An arithmetic function is a mathematical function having the natural numbers as its domain. The range of values of an arithmetic function is often a set of integers, but occasionally is the set of reals or complexes.

Here are some notable examples (in all examples, divisor means positive divisor):

1. $P_r(n) =: n^r$ for a fixed integer r
2. $\tau(n) =:$ the number of distinct divisors of n
3. $\sigma(n) =:$ the sum of distinct divisors of n
4. $\sigma_r(n) =:$ the sum of the rth powers of the divisors of n
5. $\omega(n) =:$ the number of distinct prime factors of n
6. $\Omega(n) =:$ the total number of prime factors of n (counting multiplicity)
7. $\phi(n) =:$ the number of positive integers at most n that are relatively prime to n
8. $\mu(n) =: \begin{cases} 1 \text{ if } n=1 \\ (-1)^r \text{ if } n = p_1 \cdot p_2 \ldots p_r \text{ (a product of distinct primes)} \\ 0 \text{ if } n \text{ is not square-free} \end{cases}$
9. $I(n) =: \begin{cases} 1 \text{ if } n=1 \\ 0 \text{ if } n>1 \end{cases}$

For instance, let $n = 180 = 2^2 \times 3^2 \times 5$. Then $P_2(180) = 180^2 = 32400$, $\omega(180) = 3$, $\Omega(180) = 5$, and $\mu(180) = 0$. The positive divisors of n are $1, 2, 3, 4, 5, 6, 9, 10, 12, 15, 18, 20, 30, 36, 45, 60, 90$, and 180. So $\tau(180) = 18$ and $\sigma(180) = 546$.

We will develop formulas for $\tau(n)$ and $\sigma(n)$ which will make their computation much easier. The function $\phi(n)$ was introduced in Chapter 2. For now the calculation of $\phi(180)$ might seem rather time-consuming. In fact, $\phi(180) = 48$. Again, we will develop a formula for $\phi(n)$.

Check your understanding by noting $\sigma_0(n) = \tau(n)$ and $\sigma_1(n) = \sigma(n)$ for all n. Also $\tau(p) = 2$ and $\sigma(p) = p+1$ if and only if p is prime. Furthermore, both $\tau(n) = 1$ and $\sigma(n) = 1$ if and only if $n = 1$. Also note that $\omega(1) = 0$ and $\Omega(1) = 0$.

$I(n)$ is the *identity function* and $\tau(n)$ is called the *divisor function*. We will call $P_r(n)$ the rth *power function*.

The function $\mu(n)$ is called the *Möbius mu function*. It is named after August Ferdinand Möbius (1790–1868), who was a celebrated student of Gauss' and later director

https://doi.org/10.1515/9783111579283-003

of the Leibzig Astronomical Observatory. Möbius created the function for the purpose of inverting a particular class of summation formulae. Its significance will be apparent when we introduce the MIF (Theorem 3.5).

Exercise 3.1

1. (a) Calculate $\tau(108)$, $\sigma(108)$, $\phi(108)$, $\Omega(108)$, $\omega(108)$.
    (b) Calculate $\tau(210)$, $\sigma(210)$, $\Omega(210)$, $\omega(210)$, $\mu(210)$.
2. (a) Calculate $\tau(4)$, $\tau(27)$, $\tau(6)$, $\tau(18)$ and compare with $\tau(108)$. Care to make a conjecture?
    (b) Calculate $\sigma(4)$, $\sigma(27)$, $\sigma(6)$, $\sigma(18)$ and compare with $\sigma(108)$. Any conjectures?
3. (a) Calculate $\sigma_r(10)$ for $r = 1, 2, 3$, and 4.
    (b) Calculate $\mu(n)$ for $n = 17$, 105, 277, and 1234567890.
4. (a) Calculate $\tau(2^n)$, $\sigma(2^n)$, $\phi(2^n)$.
    (b) Calculate $\tau(2^n \cdot 3)$, $\sigma(2^n \cdot 3)$, $\phi(2^n \cdot 3)$, $\mu(2^n \cdot 3)$.
5. Under what conditions is $\omega(n) = \Omega(n)$? What about $2\omega(n) = \Omega(n)$?
6. Show that if $\Omega(n) = k$, then n is divisible by a prime $p \le \sqrt[k]{n}$.
7. Prove that for any fixed $m \ge 2$, the equation $\tau(n) = m$ has infinitely many solutions for positive integers n.
8. Show that there are infinitely many positive integers m such that $\sigma(n) = m$ has no solutions.
9. (a) Find all n such that $\phi(n) = 2$.
    (b) Find all n such that $\phi(n) = 3$.
10. Find all n such that $\sigma(n) = 56$.
11. (a) Find all n for which $\tau(n)\sigma(n) = 12$.
    (b) Show that $\tau(n)\sigma(n) = 20$ is unsolvable.
    (c) Show that $\tau(n) + \sigma(n) = 12$ is unsolvable.
    (d) Find all n for which $\tau(n) + \sigma(n) = 22$.
12. (a) Show that n is prime $\Leftrightarrow \sigma(n) + \mu(n) = n$.
    (b) Show that n is prime $\Leftrightarrow \tau(n) + \mu(n) = 1$.
13. (a) Show that if n is prime, then $n \mid (\tau(n)\phi(n) + 2)$. It is unknown if any composite number $n > 4$ satisfies this condition.
    (b) Show that if n is prime, then $n \mid (\tau(n)\sigma(n) - 2)$. Find a composite n satisfying this condition.
14. (a) Find all three values of $n < 100$ such that $\omega(n) = \omega(n+1) = \omega(n+2) = 2$ and $\Omega(n) = \Omega(n+1) = \Omega(n+2) = 2$. Put into words what the above conditions mean. (Note: It is unknown if there are infinitely many such consecutive triples of numbers).
    (b) Show that there are no consecutive quadruples of such numbers.
15. (a) Find all sets of $n < 100$ for which $\Omega(n+k) \le 2$ for $k = 1, \ldots, 7$.
    (b) Show that there are never eight consecutive such numbers.

3.2 Multiplicativity

An important class of arithmetic functions are the multiplicative functions. In this section we develop some of the key consequences of multiplicativity.

Definition 3.1: An arithmetic function f is *multiplicative* if f is not identically zero and $f(mn) = f(m)f(n)$ whenever $\gcd(m, n) = 1$. If $f(mn) = f(m)f(n)$ for all m, n then f is said to be *completely multiplicative*.

Multiplicative functions are easier to evaluate than those that are not. If f is multiplicative and $n = p_1^{a_1} \cdots p_t^{a_t}$, then $f(n) = f(p_1^{a_1}) \cdots f(p_t^{a_t})$. Hence the problem of evaluating a multiplicative function f at a positive integer is reduced to that of evaluating f at prime powers. Furthermore, if f is completely multiplicative, then $f(n) = f(p_1)^{a_1} \ldots f(p_t)^{a_t}$. So, the evaluation of a completely multiplicative function f at a positive integer n is reduced to that of evaluating f at the primes dividing n.

For example, $P_r(n) = n^r$ for all n is a trivial example of a completely multiplicative function as follows from the law of exponents from elementary algebra. Note that the unit function, $P_0(n) = 1$ for all n, is completely multiplicative.

The next theorem describes a useful method of building new multiplicative functions from old ones. Notice that all sums in the proof of Theorem 3.1 are finite. Essentially, we are simply multiplying polynomials together. The above observation should help to make many of the proofs in this chapter appear less cumbersome.

Theorem 3.1: Let f be a multiplicative function and suppose

$$g(n) = \sum_{d|n} f(d) \tag{3.1}$$

Then g is multiplicative.

Proof: Here, we will make use of the fact that if $d|m$ and $k|n$ then $dk|mn$. Furthermore, if $\gcd(m, n) = 1$, then dk runs through the divisors of mn exactly once as d runs through the divisors of m and k runs through the divisors of n.

Let $\gcd(m, n) = 1$. Then

$$g(m)g(n) = \left(\sum_{d|m} f(d) \right) \left(\sum_{k|n} f(k) \right)$$

$$= \sum_{d|m|n} \sum_{k|n} f(d)f(k)$$

$$= \sum_{c|mn} f(c)$$

$$= g(mn) \qquad \blacksquare$$

Example 3.1: Notice that $\sigma_r(n) = \sum_{d|n} P_r(d)$. Since P_r is a multiplicative function, so too is σ_r by Theorem 3.1. In particular, both $\tau(n)$ and $\sigma(n)$ are multiplicative.

It is now an easy task to derive a general formula for $\tau(n)$. Note that $\tau(p^a) = a + 1$ since the divisors of p^a are $1, p, \ldots,$ and p^a. From our previous remarks, if $n = \prod_{i=1}^{t} p_i a_i$ then

$$\tau(n) = \prod_{i=1}^{t} (a_i + 1) \tag{3.2}$$

It is straightforward to derive a general formula for $\sigma(n)$ as well. Note that $\sigma(p^a) = 1 + p + \cdots + p^a = \frac{p^{a+1}-1}{p-1}$.
Hence if $n = \prod_{i=1}^{t} p_i a_i$ then

$$\sigma(n) = \prod_{i=1}^{t} \frac{p_i^{a_i+1}-1}{p_i-1} \tag{3.3}$$

For example, if $n = 72000 = 2^6 \times 3^2 \times 5^3$, then $\tau(n) = (6+1)(2+1)(3+1) = 84$ and $\sigma(n) = \frac{(2^7-1)(3^3-1)(5^4-1)}{(2-1)(3-1)(5-1)} = 257556$. Explicit formulas can turn otherwise time-consuming problems into trivial ones! Here is a somewhat more challenging example.

Example 3.2: Find the largest n for which $20 \mid n$, $\tau(n) = 12$, and n is not divisible by any prime $p > 23$.
Solution: Let $n = 2^r \times 5^s \times q$ where $r \geq 2$, $s \geq 1$, and $\gcd(10, q) = 1$. By formula (3.2), $\tau(n) = (r+1)(s+1)\tau(q)$. Since $\tau(n) = 12$, there are four possibilities:
(i) $r+1 = 6$, $s+1 = 2$, $\tau(q) = 1$; (ii) $r+1 = 4$, $s+1 = 3$, $\tau(q) = 1$;
(ii) $r+1 = 3$, $s+1 = 4$, $\tau(q) = 1$; (iv) $r+1 = 3$, $s+1 = 2$, $\tau(q) = 2$.

Recall that $\tau(q) = 1$ if and only if $q = 1$ and that $\tau(q) = 2$ if and only if q is prime. Hence, we have the following results:
Case (i): $n = 2^5 \times 5^1 = 160$, Case (ii): $n = 2^3 \times 5^2 = 200$, Case (iii): $n = 2^2 \times 5^3 = 500$,
Case (iv): $n = 2^2 \times 5 \times q$ where q is a prime other than 2 or 5 less than or equal to 23. In case (iv) the largest possible n is $n = 2^2 \times 5 \times 23 = 460$. Therefore, the largest n satisfying the given conditions is $n = 500$.

Definition 3.2: Let f and g be arithmetic functions. The *(Dirichlet) convolution* of f and g is the function $F = f * g$ defined by $F(n) = \sum_{d|n} f(d)g(n/d)$.

For example, let $f(n) = 3n$ and $g(n) = \tau(n)$. Then

$$(f * g)(12) = \sum_{d|12} 3d \cdot \tau(n/d)) = 3\tau(12) + 6\tau(6) + 9\tau(4) + 12\tau(3) + 18\tau(2) + 36\tau(1)$$

$$= 3(6) + 6(4) + 9(3) + 12(2) + 18(2) + 36(1) = 165$$

If you have difficulty following any of the steps in the proofs that follow, it is often helpful to make up a numerical example as above. Notice in the definition of convolution that as d runs through the divisors of n, so does n/d. The next proposition describes some elementary properties of convolution.

Proposition 3.2: Let f, g, and h be arithmetic functions.
(a) (Commutativity) $f * g = g * f$.
(b) (Associativity) $f * (g * h) = (f * g) * h$.
(c) $f * I = I * f = f$.
(d) $\tau = P_0 * P_0$.
(e) $\sigma = P_1 * P_0$.

Proof: (a) By definition

$$(f * g)(n) = \sum_{d \mid n} f(d) g(n/d)$$

$$= \sum_{ab=n} f(a) g(b)$$

$$= \sum_{d \mid n} g(d) f(n/d)$$

$$= (g * f)(n)$$

(b) Let $F = g * h$. Then

$$(f * (g * h))(n) = (f * F)(n)$$

$$= \sum_{d \mid n} f(d) F(n/d)$$

$$= \sum_{ad=n} f(a) \sum_{bc=d} g(b) h(c)$$

$$= \sum_{abc=n} f(a) g(b) h(c)$$

Now let $H = f * g$. Then

$$((f * g) * h)(n) = (H * h)(n)$$

$$= \sum_{d \mid n} H(d) h(n/d)$$

$$= \sum_{cd=n} h(c) \sum_{b=d} f(a) g(b)$$

$$= \sum_{abc=n} f(a) g(b) h(c)$$

Hence $f * (g * h) = (f * g) * h$.

(c) $(f * I)(n) = f(n) = (I * f)(n)$ for all n.

(d) $(P_0 * P_0)(n) = \sum_{d|n} P_0(d)P_0(n/d) = \sum_{d|n} 1 = \tau(n)$.

(e) $(P_1 * P_0)(n) = \sum_{d|n} P_1(d)P_0(n/d) = \sum_{d|n} d = \sigma(n)$. ∎

Note that Theorem 3.1 can be restated as follows: if f is multiplicative and $g = f * P_0$, then g is multiplicative. Now let us extend this observation.

Theorem 3.3: Let f and g be multiplicative functions. Then

(a) $f \cdot g$ is multiplicative, and

(b) $f * g$ is multiplicative.

Proof: Let $\gcd(m, n) = 1$.

(a) Then $fg(mn) = f(mn) \cdot g(mn) = f(m)f(n)g(m)g(n) = fg(m) \cdot fg(n)$.

(b) $(f * g)(mn) = \sum_{d|mn} f(d)g(mn/d)$. Since $\gcd(m, n) = 1$, as a runs through the divisors of m and b runs through the divisors of n, ab runs exactly once through the divisors of mn. Hence

$$(f * g)(mn) = \sum_{a|m; b|n} f(ab)g(mn/ab)$$

$$= \sum_{a|m; b|n} f(a)f(b)g(m/a)g(n/b)$$

$$= \sum_{a|m} f(a)g\left(\frac{m}{a}\right) \sum_{b|n} f(b)g(n/b)$$

$$= (f * g)(m) \cdot (f * g)(n)$$ ∎

Exercise 3.2

1. (a) Calculate $\tau(n)$ for $n = 66, 175, 6750$, and 10000.
    (b) Calculate $\sigma(n)$ for $n = 18, 45, 112$, and 6600.
2. If $n = \prod_{i=1}^{t} p_i a_i$, derive a formula for $\sigma_r(n)$.
3. Show that if f is a multiplicative function then $f(1) = 1$.
4. (a) Generalize Proposition 3.2(d, e) by showing that $\sigma_r = P_r * P_0$.
    (b) If f is completely multiplicative, must $g = f * P_0$ be completely multiplicative?
5. If f and g are completely multiplicative, must $f \cdot g$ be completely multiplicative?
6. Show that $f * (g + h) = f * g + f * h$. (Together with Proposition 3.2, this shows that the set of arithmetic functions form a commutative ring with identity with respect to addition and convolution – E.T. Bell (1915)).
7. Show that if f and g are multiplicative then $h = f/g$ is multiplicative.

8. Describe all n for which $\tau(n) = 4$.
9. (a) Find n for which $12 \,|\, n$ and $\tau(n) = 10$.
    (b) Show that there are infinitely many n for which $36 \,|\, n$ and $\tau(n) = 18$.
10. (a) Find n for which $\sigma(n) = 32$.
    (b) Find n for which $12 \,|\, n$ and $\sigma(n) = 168$.
11. Show that for any k there are only finitely many solutions to $\sigma(n) = k$.
12. If $n = \prod_{i=1}^{t} p_i a_i$, define the Liouville lambda function by

$$\lambda(n) = (-1)^{a_1 + \cdots + a_t} = (-1)^{\Omega(n)}$$

    (a) Calculate $\lambda(77), \lambda(100), \lambda(105), \lambda(147)$.
    (b) Calculate $\sum_{d|n} \lambda(d)$ for $n = 77, 100, 105, 147$.
    (c) Make a conjecture based on part (b) and prove it.
13. At the prime number penitentiary, the prisoners are kept in individual cells numbered 1 to 10,000. There are 10,000 jailors as well. Turning a key in a cell lock alternates between locking and unlocking the cell. Jailor #1 turns all the keys to lock in all the prisoners. Next jailor #2 turns the key in every other lock beginning with cell #2. In general, jailor #n turns the key in locks of cells divisible by n. After all 10,000 jailors are done "locking up," how many prisoners have unlocked cells?
14. Let $n = p_1{}^{a_1} \cdots p_t{}^{a_t}$. Derive the formula for $\sigma_r(n)$.
15. Is Theorem 3.1 still true if the word "multiplicative" is replaced by "completely multiplicative" throughout? Explain.
16. Define the inverse of f to be a function g for which $f * g = I$. Show that f has an inverse if and only $f(1) \neq 0$.
    (Hint: Let $g(1) = 1/f(1)$ and $g(n) = -g(1)\sum_{d|n} f(d)g(n/d)$ over divisors $d > 1$).
    Furthermore, show that the inverse of f is unique.
17. Let $F(n) = \sum_{k|n} \tau(k)$. Derive a formula for $F(p_1 \cdots p_t)$ where the p_i's are distinct primes.
18. Let $F(n) = \sum_{k|n} \sigma(k)$. Derive a formula for $F(p_1 \cdots p_t)$ where the p_i's are distinct primes.
19. Show that if $\sigma(n)$ is prime, then $\tau(n)$ is prime (Ross Honsberger – *More Mathematical Morsels*).
20. Find $\Omega(n)$ if $\omega(n) = 2$ and $\tau(n) = pq$ where p and q are distinct primes.
21. Find all $n < 100$ for which $\tau(n) = \tau(n+1)$. (D.R. Heath-Brown (1984) has proven that there are infinitely many such n).

3.3 Möbius inversion and some consequences

In this section we prove the Möbius inversion formula (MIF), which is an extremely useful theorem due to Möbius (1832). Next, we derive several consequences of it including a formula for $\phi(n)$. Finally, we make some generalizations of the MIF.

Theorem 3.4: (a) The Möbius mu function $\mu(n)$ is multiplicative.

(b) $\sum_{d|n} \mu(d) = \begin{cases} 1 \text{ if } n = 1 \\ 0 \text{ if } n > 1 \end{cases}$

Proof:

(a) Let $\gcd(m, n) = 1$. If there exists a prime p such that $p^2 \mid m$ or $p^2 \mid n$ then $\mu(mn) = 0 = \mu(m)\mu(n)$.

   If either $m = 1$ or $n = 1$ then clearly $\mu(mn) = \mu(m)\mu(n)$ since $\mu(1) = 1$.

   Suppose that m and n are both square-free and larger than 1. Specifically, let $m = p_1 \cdots p_s$ and $n = q_1 \cdots q_t$ where the p's and q's are distinct primes. Then, $\mu(mn) = (-1)^{s+t} = (-1)^s (-1)^t = \mu(m)\mu(n)$.

(b) If $n = 1$ then $\sum_{d|n} \mu(d) = \mu(1) = 1$.

   Suppose $n > 1$ and let $n = \prod_{i=1}^{t} p_i a_i$. By part (a), $\mu(n)$ is a multiplicative function. Now let $g(n) = \sum_{d|n} \mu(d)$. By Theorem 3.1, $g(n)$ is multiplicative.

   As before let $n = p_1{}^{a_1} \cdots p_t{}^{a_t}$. Then $g(n) = \prod_{i=1}^{t} g(p_i{}^{a_i})$. But

$$g(p^a) = \mu(1) + \mu(p) + \cdots + \mu(p^a) = 1 + (-1) + 0 + \cdots + 0 = 0$$

Thus if $n > 1$, then $g(n) = 0$. ■

We may rewrite Theorem 3.4 (b) as $\mu * P_0 = I$.

Theorem 3.5 (MIF): Let $f(n)$ be any arithmetic function and suppose

$$F(n) = \sum_{d|n} f(d) \tag{3.4}$$

Then

$$f(n) = \sum_{d|n} \mu(d) F(n/d) \tag{3.5}$$

Conversely, if $f(n) = \sum_{d|n} \mu(d) F(n/d)$, then $F(n) = \sum_{d|n} f(d)$.

Proof: Equation (3.4) may be rewritten as $F = f * P_0$. Convolution by μ gives $F * \mu = (f * P_0) * \mu = f * (P_0 * \mu) = f * (\mu * P_0) = f * I = f$, which is eq. (3.5).

   Conversely, let $f = F * \mu$. Then convolution by P_0 gives

$$f * P_0 = (F * \mu) * P_0 = F * (\mu * P_0) = F * I = F$$ ■

Example 3.3: Let $P_r(n) = n^r$. Since $\sigma_r(n) = \sum_{d|n} P_r(n)$, MIF implies that

$$n^r = \sum_{d|n} \mu(d)\sigma_r(n/d) \tag{3.6}$$

In particular if $r = 0$ in formula (3.6), then we get

$$1 = \sum_{m|n} \mu(m)\tau(n/m).$$

If $r = 1$ in formula (3.6), then

$$n = \sum_{d|n} \mu(d)\sigma(n/d).$$

Example 3.4: Let $n = \prod_{i=1}^{t} p_i a_i$. If $d|n$, then $\mu^2(d) = 1$ if and only if all prime divisors of d appear to the first power, that is, $\mu^2(d) = 1$ if and only if d is square-free. Otherwise, $\mu^2(d) = 0$. So $f(d) = \mu^2(d)$ is the square-free indicator function. The square-free divisors d of n are in one-to-one correspondence with all the subsets of $P = \{p_1, \ldots, p_t\}$. But there are 2^t such subsets. Hence $\sum_{d|n} \mu^2(d) = 2^t = 2^{\omega(n)}$.

Applying MIF we obtain

$$\mu^2(n) = \sum_{d|n} \mu(d)2^{\omega(n/d)} \tag{3.7}$$

Try to derive formula (3.7) some other way!

We now turn to a more substantive application of MIF, namely to a derivation of a formula for $\phi(n)$.

Proposition 3.6: For all $n \geq 1$, $n = \sum_{d|n} \phi(d)$.

Proof: Consider the n fractions $\frac{1}{n}, \frac{2}{n}, \ldots, \frac{n}{n}$. Now reduce them to lowest terms obtaining $f_1, f_2, \ldots, f_n$, that is if $f = a/b$ then $\gcd(a, b) = 1$. The set of all denominators among the f_i's is exactly the set of positive divisors of n. Furthermore, if $d|n$, then there are $\phi(d)$ fractions among the reduced fractions f_i having denominator d. Since there are n fractions in all, $n = \sum_{d|n} \phi(d)$. ∎

Since $P_1(n) = n$ is a multiplicative function, it will follow that so too is $F(n) = \phi(n)$ by Proposition 3.6 and Exercise 3.3.2 at the end of this section. However, below we show directly that $F(n) = \phi(n)$ is multiplicative.

Applying MIF to Proposition 3.6 yields

$$\phi(n) = \sum_{d|n} \mu(d)\frac{n}{d} = n \sum_{d|n} \frac{\mu(d)}{d} \tag{3.8}$$

Since $f(n) = \mu(n)$ and $P_{-1}(n) = 1/n$ are multiplicative, so too is $g(n) = \mu(n)/n$ by

Theorem 3.3(a). But then $G(n) = \sum_{d|n} \frac{\mu(d)}{d}$ is multiplicative by Theorem 3.1.

Finally, $\phi(n) = nG(n)$ is multiplicative.

By formula (3.8), $\phi(p^a) = p^a \sum_{d|p^a} \frac{\mu(d)}{d}$. But $\mu(1) = 1, \mu(p) = -1$, and $\mu(p^t) = 0$

for $t > 1$. Hence $\phi(p^a) = p^a\left(1 - \frac{1}{p}\right)$.

If we let $n = \prod_{i=1}^{t} p_i^{a_i}$ then $\phi(n) = \prod_{i=1}^{t} p_i^{a_i}\left(1 - \frac{1}{p_i}\right)$. Alternately,

$$\phi(n) = \prod_{i=1}^{t} p_i^{a_i-1}(p_i - 1) \tag{3.9}$$

For example, if $n = 18000 = 2^4 3^2 5^3$, then $\phi(n) = 2^3(2-1)3^1(3-1)5^2(5-1) = 4800$.

Imagine actually counting all appropriate integers relatively prime to 18000. One final example may be helpful.

Example 3.5: Find all n for which $\phi(n) = 8$.
Solution: By formula (3.9) no prime larger than 9 divides n since otherwise $\phi(n) > 8$. However, $7 \nmid n$ since if $7 \mid n$, then $6 \mid \phi(n)$. This would be contrary to the fact $\phi(n) = 8$. Hence, we may write $n = 2^a 3^b 5^c$ where $a, b, c \geq 0$. There are two cases: $c = 0$ and $c = 1$ (why?). Furthermore, $b \leq 1$ since $3 \nmid 8$.

If $c = 0$, then $n = 2^a 3^b$ and either $a = 4$, $b = 0$ or $a = 3$, $b = 1$. If $c = 1$, then $n = 2^a 3^b 5$. Here there are three possibilities: $a = 2$, $b = 0$, or $a = 1$, $b = 1$, or $a = 0$, $b = 1$. Combining our results there are five values of n for which $\phi(n) = 8$, namely $n = 15, 16, 20, 24$, and 30.

Exercise 3.3

1. Calculate $\phi(n)$ for $n = 91, 1000, 3125, 49000$, and $1,000,000$.
2. Prove the converse of Theorem 3.1, that is, if g is multiplicative and $g(n) = \sum_{d \mid n} f(d)$, then f is multiplicative.
3. (a) Let $a, b \in \mathbf{N}$ such that $p \mid a \Rightarrow p \mid b$. Show that $\phi(ab) = a\phi(b)$.
   (b) Let $a, b \in \mathbf{N}$ such that $p \mid a \Leftrightarrow p \mid b$. Show that $\frac{a}{b} = \frac{\phi(a)}{\phi(b)}$.
4. Show that there are infinitely many positive integers m such that $\phi(n) = m$ has no solutions.
5. (a) Find all n for which $\phi(n) = 6$.
   (b) Find all n for which $\phi(n) = 12$.
6. Find n for which $\phi(n) = 96$ and $\tau(n) = 24$.
7. Let $n^2 = \sum_{d \mid n} g(d)$. Use the MIF to find a formula for $g(n)$ in terms of the canonical prime factorization of n.
8. Let p, q, and r be distinct primes dividing N. Derive a formula for the number of positive integers less than N relatively prime to pqr.
9. Show that formula (3.8) can be rewritten as $\phi(n) = \sum_{dk=n} k\mu(d)$, where the sum is over all ordered pairs of naturals (d, k) where $dk = n$.
10. Show that formula (3.9) can be rewritten as $\phi(n) = n \prod_{p \mid n} (1 - 1/p)$.

11. Show that $\sum_{d|n} |\mu(d)| = 2^{\omega(d)}$ for all $n \geq 1$ (note that $\omega(1) = 0$).

12. Define the von Mangoldt lambda function by

$$\Lambda(n) = \begin{cases} \log p \text{ if } n \text{ is a prime power } p^t \text{ for } t \geq 1 \\ 0 \text{ otherwise} \end{cases}$$

   (a) Prove that $\sum_{d|n} \Lambda(d) = \log n$.
   (b) Show that $\Lambda(n) = -\sum_{d|n} \mu(d)\log d$.

13. (a) What are the last three digits of 73603?
   (b) What are the last four digits of 63108001?

14. Let $n = 2^{46} \cdot 3^{20} \cdot 5^9 \cdot 7^6 \cdot 11^2 \cdot 13^3 \cdot 17 \cdot 19 \cdot 23 \cdot 31 \cdot 41 \cdot 47 \cdot 59 \cdot 71$ (the order of the Monster group).
   (a) What is the largest prime p which divides $\phi(n)$?
   (b) What is the largest odd cube which divides $\phi(n)$?

15. A. Makowski and A. Schinzel have conjectured that $n \leq 2\sigma(\phi(n))$ for all $n > 1$.
   (a) Prove the conjecture for the case where n is the product of at most three primes (not necessarily distinct).
   (b) Verify the conjecture for all $n < 200$.

16. Let $f(n)$ be any arithmetic function. Argue as in the proof of Proposition 3.6 to show that

$$\sum_{1 \leq k \leq n} f\left(\frac{k}{n}\right) = \sum_{d|n} \sideset{}{'}\sum_{1 \leq a \leq d} f\left(\frac{a}{d}\right)$$

where Σ' denotes a sum over all a relatively prime to d.

17. (a) Let $F(n) = \sum_{d|n} f(d)$ for all n. Show that $\sum_{r=1}^n F(r) = \sum_{r=1}^n \left[\frac{n}{r}\right] \cdot f(r)$.
   (b) Show that $\sum_{r=1}^n \left[\frac{n}{r}\right] \cdot \mu(r) = 1$ for all positive integers n.
   (c) Show that $\sum_{r=1}^n \left[\frac{n}{r}\right] \cdot \phi(r) = n(n+1)/2$, the nth triangular number.

18. (a) Show that if $2|n$ but $4 \nmid n$, then $\phi(n) = \phi(n/2)$.
   (b) Suppose $n = \prod_{i=1}^t p_i^{a_i}$ where $p_1 = 2$ and $a_1 \geq 2$. Let $0 \leq b_i < a_i$ for all $i = 1, \ldots, t$. Show that if $q = \prod_{i=1}^t p_i^{b_i} + 1$ is prime and $q \nmid n$, then $\phi(n) = \phi\left(\frac{nq}{q-1}\right)$.

19. (a) Carmichael's conjecture states that for every $k \geq 1$, the equation $\phi(n) = k$ has either no solution or at least two solutions. R. D. Carmichael published an erroneous proof of this proposition in 1907 but acknowledged its inadequacy in 1922. Use Exercise 3.3.18 to show that if there is a counterexample n to Carmichael's conjecture, then $2^2 3^2 7^2 43^2 \,|\, n$.
   (b) Show that if n is the smallest counterexample, then $8 \nmid n$. Aaron Schafly and Stan Wagon (1994) have proven that if there is a counterexample n to Carmichael's conjecture, then $n > 10^{10,900,000}$.

20. (a) Show that for every m, there are infinitely many n for which $m! \,|\, \phi(n)$.
   (b) Show that for every m, there is an n for which $\tau(n) = m!$ and $! \,|\, \phi(n)$.

21. Find all values of $n \le 100$ for which $\phi(n) = \phi(n+1)$. It is unproven that there are infinitely many such n (cf. Exercise 3.2.21).
22. In 1897, F. Mertens conjectured that if $M(N) = \sum_{i=1}^{N} \mu(n)$, then $|M(N)| < \sqrt{N}$ for all $N > 1$. Confirm Mertens' conjecture for all $N < 100$. (In fact, Mertens' conjecture was disproven by A. Odlyzko and H. te Riele in 1985). However, it is still unknown if there is a constant C such that $|M(N)| < C\sqrt{N}$ for all $N > 1$. If so, the Riemann hypothesis follows (see Section 9.1).

3.4 Perfect numbers and amicable pairs

Two number-theoretic concepts that date back to the Pythagoreans are that of perfect number and that of an amicable pair of numbers. To the ancient Greek and Jewish scholars, such numbers had mystical and divine significance. In this section we derive some of their mathematically interesting properties and then present some open questions.

If n is a natural number, then the *proper divisors* of n are the set of all positive divisors less than n including 1.

Definition 3.3: The natural number n is said to be *perfect* if the sum of the proper divisors of n is equal to n itself.

Equivalently, n is perfect if and only if $\sigma(n) = 2n$. If $\sigma(n) < 2n$ then n is said to be *deficient* while if $\sigma(n) > 2n$ then n is said to be *abundant*. The first four perfect numbers were known to the ancient Greeks. They are 6, 28, 496, and 8128. Their prime factorizations are significant: $6 = 2 \times 3$, $28 = 2^2 \times 7$, $496 = 2^4 \times 31$, and $8128 = 2^6 \times 127$. In each case, the perfect number is expressible as $2^{p-1}(2^p - 1)$ where p and $2^p - 1$ are primes. Euclid established that this is no coincidence.

Theorem 3.7 (*Elements* – Book IX, Prop. 36): If $2^p - 1$ is prime, then $n = 2^{p-1} \cdot (2^p - 1)$ is perfect.

Proof: Let $2^p - 1$ be prime and let $n = 2^{p-1} \cdot (2^p - 1)$. Notice that $\gcd(2^{p-1}, 2^p - 1) = 1$. Hence $\sigma(n) = \sigma(2^{p-1}) \cdot \sigma(2^p - 1) = (2^p - 1) \cdot 2^p = 2n$ by formula (3.3). Hence n is perfect. ∎

It is of related interest to realize that if $a^n - 1$ is prime for $n > 1$, then $a = 2$ and n is prime. The first observation follows from the factorization

$$a^n - 1 = (a - 1)(a^{n-1} + a^{n-2} + \cdots + 1)$$

In order for $a^n - 1$ to be prime, $a - 1 = 1$. The second observation is apparent by reasoning that if n were composite, then $n = st$ where $s, t > 1$. But then

$$a^n - 1 = a^{st} - 1 = (a^s - 1)\left(a^{s(t-1)} + a^{s(t-2)} + \cdots + 1 \right)$$

and both factors are greater than 1.

In Chapter 7 we will have more to say about primes of the form $2^p - 1$ and we will derive a useful algorithm for determining whether $2^p - 1$ is prime for given p. For now, note that if p is prime then $2^p - 1$ is not necessarily prime. For example, $2^{11} - 1 = 2047 = 23 \times 89$. The next result is the converse of Theorem 3.7 for even perfect numbers. It is due to Euler but was published posthumously in 1849.

Theorem 3.8: Let n be an even perfect number. Then n is of the form $2^{p-1}(2^p - 1)$ where $2^p - 1$ is prime.

Proof: Let $n = 2^a \cdot b$ where $a \geq 1$ and b is odd. Since n is perfect, $\sigma(n) = 2n = 2^{a+1 \cdot b}$. But σ is multiplicative and hence $\sigma(n) = \sigma(2^a) \cdot \sigma(b)$. Thus

$$2^{a+1 \cdot b} = \left(2^{a+1} - 1 \right) \cdot \sigma(b) \tag{3.10}$$

But $\gcd\left(2^{a+1}, 2^{a+1} - 1 \right) = 1$ and so $2^{a+1} \mid \sigma(b)$ by Corollary 2.1.1(b). Let

$$\sigma(b) = 2^{a+1} \cdot d \tag{3.11}$$

Combining eqs. (3.9) and (3.10) yields

$$b = \left(2^{a+1} - 1 \right) \cdot d \tag{3.12}$$

We now show that $d = 1$. If $d > 1$ then b has at least $1, b,$ and d as distinct positive divisors. Hence $\sigma(b) \geq b + d + 1 = \left(2^{a+1} - 1 \right) \cdot d + d + 1 = 2^{a+1} \cdot d + 1$ which contradicts eq. (3.10). Thus $d = 1$ and $b = 2^{a+1} - 1$ from eq. (3.11).

From eq. (3.10), $\sigma(b) = 2^{a+1}$. So $\sigma(b) = b + 1$ and b is prime. Let $a + 1 = p$ and the proof is complete. ∎

Currently 52 perfect numbers are known. The largest one is of the form $2^{p-1}(2^p - 1)$ with $p = 136, 279, 841$. It is unknown if there are infinitely many perfect numbers. Whether or not there exist odd perfect numbers remains an open question. P. Hagis (1975) and E.Z. Chein (1979) have independently proven that an odd perfect number n would necessarily have at least eight distinct prime factors. This has been further extended most recently by P. Nielsen (2015) who has shown that at least 10 distinct prime factors are necessary. Furthermore, if $3 \nmid n$, then Hagis (1983) showed it has at least 11 prime factors. P. Ochem and M. Rao (2012) have shown that $n > 10^{1500}$ and recent extensions push that bound above 10^{2000}. T. Goto and Y. Ohno have shown that n has a prime factor larger than 10^8. Additionally, D. Iannucci (2000) has proven that the second largest prime factor of n must be larger than 10,000. Here we prove two modest results that are far from optimal but give some flavor of the techniques employed in this area.

In general, we write $p^a \,\|\, n$ to denote the fact that p^a *exactly divides n*; that is, $p^a \,|\, n$ but p^{a+1} does not divide n. If n is an odd perfect number, then $\sigma(n) = 2n$ is exactly divisible by 2. This simple observation led Euler to the following:

Proposition 3.9: If n is an odd perfect number, then $n = p^a \cdot s^2$ where p is a prime with $p \equiv 1 \pmod 4, a \equiv 1 \pmod 4$, and $\gcd(p, s) = 1$.

Proof: Let $n = \prod_{i=1}^{t} p_i a_i$. Then $\sigma(n) = 2n \equiv 2 \pmod 4$. By the multiplicativity of the function σ, there exists a unique $p \,|\, n$ with $p^{a/n}$ such that $\sigma(p^a) \equiv 2 \pmod 4$. If $p \equiv 3 \pmod 4$ then $\sigma(p^a) \equiv 0$ or $1 \pmod 4$ since $\sigma(p^a) = 1 + p + \cdots + p^a$. It follows that $p \equiv 1 \pmod 4$ and $a \equiv 1 \pmod 4$.

Let $m = n/p^a$. Then $\sigma(m) \equiv 1 \pmod 4$. If q is a prime with $q^b \,\|\, m$ then $\sigma(q^b)$ must be odd by the multiplicativity of σ. It follows that b is even. But the product of primes to even powers is a perfect square. Thus $m = s^2$ for some integer s. The result follows. ∎

The next result published in 1888 was the first of a series of significant investigations due to the British mathematician James Joseph Sylvester (1814–1897). Sylvester is best known for his work in algebra, number theory, and combinatorics. Together with Arthur Cayley, he developed invariant theory and the theory of matrices. Being Jewish he was barred from being granted a degree although he was Second Wrangler at Cambridge University in 1837 (that is, finished second in the prestigious mathematical Tripos examination). Sylvester spent some years in America, first at the University of Virginia and later at Johns Hopkins University.

Proposition 3.10: If n is an odd perfect number, then n is divisible by at least four distinct primes.

We precede the proof with two helpful formulas concerning the function σ. The first is an upper bound for $\sigma(n)/n$ while the second is a lower bound. Let $n = r \cdot s$ where $\gcd(r, s) = 1$ and let p and q represent primes:

$$\frac{\sigma(n)}{n} \leq \prod_{p|r} (p/p - 1) \cdot \prod_{q^a\|1} (\sigma(q^a)/q^a) \tag{3.13}$$

If either $r = 1$ or $s = 1$, then define the corresponding empty product to be 1. Formula (3.12) follows from the fact that $\sigma(p^a)/p^a = \frac{p - p^{-a}}{p - 1} < \frac{p}{p - 1}$.

Let $n = \prod_{i=1}^{t} p_i a_i$. Suppose that $0 \le b_i \le a_i$ for $1 \le i \le t$. Then

$$\frac{\sigma(n)}{n} \ge \prod_{i=1}^{t} \left(1 + 1/p_i + \cdots + 1/p_i^{b_i}\right) \tag{3.14}$$

The method of proof will be to assume that n is an odd perfect number consisting of at most three distinct primes. Then we will derive various contradictions from either eq. (3.12) or (3.13).

Proof of Proposition 3.10: Let n be an odd perfect number with prime factorization as above. N perfect implies $\frac{\sigma(n)}{n} = 2$. If $t = 1$ then $\frac{p_1}{p_1-1} \le 3/2$ which contradicts eq. (3.12) with $r = n$. Similarly, if $t = 2$ then

$$\frac{p_1}{p_1-1} \cdot \frac{p_2}{p_2-1} \le (3/2)(5/4) = 15/8$$

again contradicting eq. (3.12).

Hence $t \ge 3$.

Suppose $t = 3$. If $p_2 > 5$ then $\prod_{i=1}^{3} (p_i/p_i - 1) \le (3/2)(7/6)(11/10) = 231/120$ which contradicts eq. (3.12). So $p_1 = 3$ and $p_2 = 5$. If $p_3 \ge 17$ then

$$\prod_{i=1}^{3} (p_i/p_i - 1) \le (3/2)(5/4)(17/16) = 255/128 < 2$$

Thus $n = 3^a 5^b p^c$ where p is either 7, 11, or 13.

We will now show that each of the three possible values for p is untenable.

(i) $p = 7$: By Proposition 3.9, a and c are even since $3 \equiv 7 \equiv 3 \pmod 4$ and so it must be the case that $a \ge 2$ and $c \ge 2$. Let $b_1 = 2, b_2 = 1$, and $b_3 = 2$ in formula (3.13). We get $\frac{\sigma(n)}{n} \ge (1 + 1/3 + 1/9)(1 + 1/5)(1 + 1/7 + 1/49) = 4446/2205 > 2$, a contradiction.

(ii) $p = 11$: As above, a and c are even. But $\sigma(3^2) = 13$ and $13 \nmid \sigma(n)$ since $\sigma(n) = 2 \times 3^a 5^b 11^c$. Hence $a \ge 4$ and $c \ge 2$.

If $b \ge 2$ then we can let $b_1 = 4, b_2 = 2$, and $b_3 = 2$ in (3.13). We obtain $(1 + 1/3 + 1/9 + 1/27 + 1/81)(1 + 1/5 + 1/25)(1 + 1/11 + 1/121) = 4123/2025 > 2$, a contradiction.

If $b = 1$ however, then let $s = 5$ in eq. (3.12) obtaining $(3/2)(11/10)(6/5) = 198/100 < 2$. This contradicts eq. (3.12).

(iii) $p = 13$: Notice that $7 | \sigma(13)$ and yet $\sigma(n) = 2 \times 3^a 5^b 13^c$ which is not divisible by 7. Hence $c \ge 2$.

If $a \ge 4$ and $b \ge 2$ then let $b_1 = 4$, $b_2 = 2$, and $b_3 = 2$ in eq. (3.13). But then

$$\left(1 + \frac{1}{3} + \frac{1}{9} + \frac{1}{27} + \frac{1}{81}\right)\left(1 + 5 + \frac{1}{25}\right)\left(1 + \frac{1}{13} + \frac{1}{169}\right)$$

$$= 228811/114075 > 2, \text{ a contradiction.}$$

If $a = 2$ then let $s = 9$ in eq. (3.12). We obtain

$$(5/4)(13/12)(13/9) = 845/432 < 2, \text{ a contradiction.}$$

If $b = 1$ let $s = 5$ in eq. (3.12). We obtain

$$(3/2)(13/12)(6/5) = 234/120 < 2, \text{ a contradiction}$$

Thus $t \geq 4$ and the proposition is established. ∎

We complete this section with a brief discussion of a related topic.

Definition 3.4: The numbers a and b form an *amicable pair* if the sum of the proper divisors of a is b and the sum of the proper divisors of b is a.

The ancient Greeks knew the amicable pair 220 and 284 but no others. In the ninth century C.E., the Arab scholar and prolific scientific translator Thabit ibn-Qurra (826–901) discovered a remarkable sufficient condition for amicable pairs. We discuss it below.

Proposition 3.11: If p, q, and r are prime numbers of the form $p = 3 \cdot 2^{n-1} - 1$, $q = 3 \cdot 2^n - 1$, and $r = 9 \cdot 2^{2n-1} - 1$ for some integer $n > 1$, then $a = 2^n pq$ and $b = 2^n r$ form an amicable pair.

Proof: It suffices to show that both $\sigma(a) = \sigma(b)$ and $\sigma(b) = a + b$. $\sigma(a) = \sigma(2^n)\sigma(p)\sigma(q)$ and $\sigma(b) = \sigma(2^n)\sigma(r)$. Hence it suffices to demonstrate

$$\sigma(p)\sigma(q) = \sigma(r) \tag{3.15}$$

and

$$\sigma(2^n)\sigma(r) = 2^n(pq + r) \tag{3.16}$$

$$\sigma(p)\sigma(q) = (p+1)(q+1) = 9 \cdot 2^{2n-1} = r + 1 = \sigma(r)$$

which establishes eq. (3.14).

$$\sigma(2^n)\sigma(r) = (2^{n+1} - 1)(r+1) = 2^n\left(9 \cdot 2^{2n} - 9 \cdot 2^{n-1}\right)$$
$$= 2^n\left(9 \cdot 2^{2n} - 6 \cdot 2^{n-1} - 3 \cdot 2^{n-1}\right)$$
$$= 2^n\left(9 \cdot 2^{2n-1} + 9 \cdot 2^{2n-1} - 3 \cdot 2^n - 3 \cdot 2^{n-1}\right) = 2^n(pq + r)$$

Thus eq. (3.15) holds and hence the proposition is established. ∎

The amicable pair 220 and 284 results from $n = 2$ in Proposition 3.11. In 1636, Fermat noted the pair 17296 and 18416 which follows from $n = 4$. In 1638, Descartes and Fermat independently verified another pair when $n = 7$. Recently, historians have verified that the cases $n = 4$ and $n = 7$ were known to Ibn al-Banna in the fourteenth century and Muhammad Baqir Yazdi in the seventeenth century, respectively. No other pairs were discovered until Euler generalized Proposition 3.11 for p, q, and r of other forms around 1750. He then miraculously added 62 more examples (plus 2 erroneous pairs). In 1830, Legendre found one more. It was thus quite a surprise when in 1866, a 16-year-old Italian student B.N.I. Paganini discovered the second smallest amicable pair: 1184 and 1210. Today there are over 1,000,000,000 pairs known.

The question whether all odd amicable numbers are divisible by 3 was answered in the negative way by B. Battiato and W. Borho in 1988. Still there are many unanswered questions concerning amicable pairs. For example, are there infinitely many amicable pairs? Are there any amicable pairs of opposite parity? Can amicable pairs be relatively prime to each other? It's been shown that the product of any relatively prime amicable pairs must exceed 10^{65}.

Another concept, not to be confused with amicable pairs, are sets of friendly numbers.

Definition 3.5: The *abundancy index* of a natural number n is the ratio $\frac{\sigma(n)}{n}$. Two or more numbers having the same abundancy index are called *friends* and are said to be members of the same *club*. The full set of clubs forms a partition of the natural numbers. Numbers that are members of a club with at least one other member are *friendly numbers*, while those that are not are called *solitary* numbers.

Example 3.6: All perfect numbers are friendly numbers belonging to the same club having abundancy index 2. Some examples of other friendly pairs are 12 and 234 with abundancy index 7/3 and 80 and 200 with abundancy index 93/40 (verify in Exercise 3.4.27).

If m and n form a friendly pair, then so do am and an for any a with gcd$(a, mn) = 1$ by the multiplicativity of the σ function. For example, 6 and 28 form a friendly pair since both are perfect. Since $6 = 2 \times 3$ and $28 = 2^2 \times 7$, the numbers $6a$ and $28a$ form a friendly pair for any a relatively prime to $42 = 2 \times 3 \times 7$. Thus, $6a$ and $28a$ form a friendly pair for all $a \equiv 1, 5, 11, 13, 17, 19, 23, 25, 29, 31, 37$, or 41 (mod 42). This example alone shows that a positive proportion of natural numbers are friendly, in fact at least $12/42 = 2/7$ of them.

In the opposite direction, if n is relatively prime to $\sigma(n)$, then its abundancy index $\sigma(n)/n$ will be a reduced fraction with denominator n. If $m < n$, then m must have a different abundancy index. If $m > n$ and m and n have the same abundancy index, then $\frac{\sigma(n)}{n} = \frac{\sigma(m)}{m}$. But the fraction on the left is reduced and hence $n | m$. Let $m = nk$. Then $\frac{\sigma(n)}{n} = \frac{\sigma(nk)}{nk}$ which implies that $\sigma(nk) = k\,\sigma(n)$. But in fact, $\sigma(nk) > k\,\sigma(n)$ since for every divisor d of n, kd is a divisor of kn with the number 1 being an additional divisor of kn. We state this observation as a proposition.

Proposition 3.12: If n and $\sigma(n)$ are relatively prime, then n is solitary.

Example 3.7: Every prime number p is solitary since $\sigma(p) = p + 1$ and so p and $\sigma(p)$ are relatively prime. Similarly, every prime power is a solitary number as well. Proposition 3.12 guarantees many additional solitary numbers, for example, $n = 21$ for which $\sigma(21) = 32$.

Despite our direct observations, much is left unexplored. It is still unknown whether some small numbers such as 10, 14, and 15 are solitary or not. It is also unknown whether any club has infinitely many members. For example, although we expect there are infinitely many perfect numbers, we know only 52 of them. Currently, the largest identifiable club with 280 known members is that of n with $\sigma(n) = 9n$, hence called 9-tuply perfect numbers.

Exercise 3.4

1. Show that all even perfect numbers are triangular.
2. (a) Show that if n is an even perfect number, then $\omega(n) = 2$ and $\Omega(n)$ is prime.
   (b) Show that the converse is false.
3. Show that n is perfect if and only if $2 = \sum_{d|n} \frac{1}{d}$.
4. Prove that every even perfect number >6 is the sum of consecutive odd cubes.
5. Show that all even perfect numbers have last digit 6 or 8.
6. Comment on the case $n = 1$ in Proposition 3.11.
7. An integer is *k-tuply perfect* if $\sigma(n) = kn$. Verify that 120 and 672 are 3-tuply perfect and that 2178540 is 4-tuply perfect.
8. Show that if n is $(2k + 1)$-tuply perfect and n is odd, then n is a perfect square.
9. Use formula (3.12) to prove that an odd 5-tuply perfect number must have at least six distinct prime factors.
10. Show that if n is k-tuply perfect and $\gcd(k, n) = 1$, then kn is $\sigma(k)$-tuply perfect.
11. (a) Show that a multiple of a perfect or abundant number is abundant. (b) Show that a proper divisor of a deficient or perfect number is deficient.
12. Let $s(n)$ denote the sum of all the proper divisors of n. Note that n is perfect $\Leftrightarrow s(n) = n$. Further, n and $s(n)$ are an amicable pair $\Leftrightarrow s(s(n)) = n$. Define $s_k(n)$ recursively by $s_1(n) = s(n)$ and $s_k(n) = s(s_{k-1}(n))$ for $k > 1$. Call an integer n *sociable of order k* if $n = s_k(n)$ for some k.
   (a) Verify that 12496 is a sociable number of order 5 (P. Poulet).
   (b) Verify that 1264460 is a sociable number of order 4 (H. Cohen).
      No sociable numbers of order 3 are known.
13. Find the first four abundant numbers.
14. (a) Show that 945 is the least odd abundant number.
   (b) What is the least odd square-free abundant number?
15. Determine necessary and sufficient conditions for $\sigma(n)$ to be odd. Deduce that no perfect number is a square.

16. The integer n is *superperfect* if $\sigma(\sigma(n)) = 2n$. Show that if $2p^{p-1}$ is prime, then 2^{p-1} is superperfect. (It is unknown if there are any other superperfect numbers).

17. With the notation of Exercise 3.4.10, consider the sequence $\{n, s(n), s_2(n), s_3(n), \ldots\}$, called the aliquot sequence for n.
    (a) Show that for $n \le 40$, the aliquot sequence eventually ends, that is, either terminates or becomes periodic.
    (b) Which $n \le 40$ has the longest aliquot subsequence?
    (c) It is unknown whether all n have aliquot sequences that end. Calculate the aliquot sequence for $n = 276$ as long as your or your computer's stamina will allow! (The first unknown case is $n = 276$).

18. (a) Verify that 46 cannot be expressed as the sum of two abundant numbers.
    (b) Show that all even integers greater than 46 can be expressed as the sum of two abundant numbers. (Hint: Reason modulo 6 and use Exercise 3.4.9(a) and the fact 20 and 40 are abundant numbers).

19. Verify that the following are amicable pairs:
    (a) 2620 and 2924
    (b) 5020 and 5564
    (c) 6232 and 6368
    (d) 10744 and 10856

20. The divisors of n excluding 1 and n are called the non-trivial divisors of n. The integers a and b are a *betrothed pair* if the sum of the non-trivial divisors of a equals b and vice versa.
    (a) Show that a and b are a betrothed pair if and only if $\sigma(a) = \sigma(b) = a + b + 1$.
    (b) Verify that the following are betrothed pairs: (48,75) and (140,195).
        All known betrothed pairs are of opposite parity.

21. The following are amicable numbers in search of a mate. Can you help? (a) 12285, (b) 63020, and (c) 66928.

22. Find n such that the following form an amicable pair: $3^n \times 5 \times 7 \times 71$ and $3^n \times 5 \times 17 \times 31$.

23. Find p for which $m = 2^7 \times 263 \times 4271 \times 280883$ and $n = 2^7 \times 263 \times p$ form an amicable pair (P. Poulet).

24. Determine a lower bound for an odd perfect number utilizing only Propositions 3.9 and 3.10.

25. Is $n = 3^2 \times 7^2 \times 11^2 \times 13^2 \times 22021$ an odd perfect number?! (R. Descartes − 1638)

26. A number n is *untouchable* if there does not exist an m for which $s(m) = n$ where $s(m)$ is the sum of the proper divisors of m.
    (a) Find all untouchable numbers less than 10.
    (b) Show that if Goldbach's conjecture is true, then 5 is the only odd untouchable number.

27. (a) Verify that 12 and 234 form a friendly pair.
    (b) Verify that 80 and 200 form a friendly pair.

(c) Verify that 600 and 3360 form a friendly pair.

(d) Show that 65, 81, and 2401 are solitary numbers.

28. Find a matching friendly pair for each of the numbers 30, 66, 140, and 150.

29. Amicable pairs a, b satisfy $\sigma(a) = \sigma(b) = a + b$. *Amicable triples* a, b, c satisfy $\sigma(a) = \sigma(b) = \sigma(c) = a + b + c$. Verify that (1980, 2016, 2556) form an amicable triple.

30. *Superabundant* numbers n are those for which $\frac{\sigma(n)}{n} > \frac{\sigma(k)}{k}$ for all $k < n$. They were initially studied but unpublished by S. Ramanujan (1915) and later by L. Alaoglu and P. Erdös (1944). Find all superabundant numbers less than 200.

Chapter 4
Primitive roots and quadratic reciprocity

4.1 Primitive roots

In this section the notion of primitive root is explored. Several examples are given, which show how to find primitive roots and we demonstrate their utility in solving congruence relations. Our main goal here is to classify which integers n have primitive roots.

Let n be a fixed positive integer. Let a be such that $\gcd(a, n) = 1$. By the Euler-Fermat theorem, $a^{\phi(n)} \equiv 1 \pmod{n}$. It is of interest to determine whether there is an $m < \phi(n)$ for which $a^m \equiv 1 \pmod{n}$. The following definition allows us to speak more precisely.

Definition 4.1: Let $\gcd(a, n) = 1$. The smallest positive integer m for which $a^m \equiv 1 \pmod{n}$ is the *order of a modulo n*, denoted $\operatorname{ord}_n a$.

The next example is instructive and serves to demonstrate much of what will be proven in this section.

Example 4.1: (a) Let $n = 13$. One can readily verify that $\operatorname{ord}_{13} 1 = 1$, $\operatorname{ord}_{13} 12 = 2$, $\operatorname{ord}_{13} 3 = \operatorname{ord}_{13} 9 = 3$, $\operatorname{ord}_{13} 5 = \operatorname{ord}_{13} 8 = 4$, $\operatorname{ord}_{13} 4 = \operatorname{ord}_{13} 10 = 6$, $\operatorname{ord}_{13} 2 = \operatorname{ord}_{13} 6 = \operatorname{ord}_{13} 7 = \operatorname{ord}_{13} 11 = 12$. Notice that $\phi(13) = 12$ and that $\operatorname{ord}_{13} a \,|\, 12$ for all a. Furthermore, in this case, for every $d \,|\, 12$ there is an a such that $\operatorname{ord}_{13} a = d$.
(b) Let $n = 8$. Then $\operatorname{ord}_8 1 = 1$ and $\operatorname{ord}_8 3 = \operatorname{ord}_8 5 = \operatorname{ord}_8 7 = 2$. In this case, there is no a for which $\operatorname{ord}_8 a = 4 = \phi(8)$.

Definition 4.2: Let $\gcd(g, n) = 1$. If $m = \phi(n)$ is the smallest integer for which $g^m \equiv 1 \pmod{n}$, then g is called a *primitive root modulo n*. Equivalently, g is a primitive root modulo n if $\operatorname{ord}_n g = \phi(n)$.

Consequently, in Example 4.1, 2, 6, 7, and 11 are primitive roots modulo 13. On the other hand, there are no primitive roots modulo 8. Our next observation shows why primitive roots are considered so important.

Proposition 4.1: Let g be a primitive root modulo n. Then $g, g^2, \ldots, g^{\phi(n)}$ form a reduced residue system modulo n.

Proof: Since g is a primitive root modulo n, $\gcd(g, n) = 1$ and hence $\gcd(g^m, n) = 1$ for all $m \geq 1$. Suppose that $g^a \equiv g^b \pmod{n}$ with $a < b$. Then $n \,|\, (g^a - g^b)$. But $g^a - g^b = g^a(1 - g^{b-a})$. Hence $n \,|\, (1 - g^{b-a})$ by Corollary 2.1.1(b) and so $g^{b-a} \equiv 1 \pmod{n}$. But since g is a primitive root, $b - a \geq \phi(n)$ and the result follows. ■

https://doi.org/10.1515/9783111579283-004

The next proposition presents some of the basic results concerning the order of a modulo n. We will refer repeatedly to these results in our discussion of primality tests in Chapter 7.

Proposition 4.2: Let $\gcd(a, n) = 1$. Then
(a) $\text{ord}_n a$ divides $\phi(n)$.
(b)   If $a^m \equiv 1 \pmod n$ then $\text{ord}_n a$ divides m.
(c) $\text{ord}_n(a^s) = \text{ord}_n a / \gcd(s, \text{ord}_n a)$.

Proof: Let $k = \text{ord}_n a$. Note that (b) implies (a) by the Euler-Fermat theorem.
(b) By the division algorithm, $m = kq + r$ for some r satisfying $0 \le r < k$. Thus

$$1 \equiv a^m = a^{kq+r} = (a^k)^q a^r \equiv 1^q a^r \equiv a^r \pmod n.$$

The minimality of k implies that $r = 0$. Hence $m = kq$ and $k \mid m$.
(c) Let $t = \gcd(s, k)$. By (b),
$(a^s)^u = a^{su} \equiv 1 \pmod n$ if and only if $k \mid su$. Thus
$(a^s)^u \equiv 1 \pmod n$ if and only if $\frac{k}{t} \mid \frac{su}{t}$. But $\gcd(k/t, s/t) = 1$ and so
$(a^s)^u \equiv 1 \pmod n$ if and only if $\frac{k}{t} \mid u$.
   Hence k/t is the least integer u for which $(a^s)^u \equiv 1 \pmod n$ by Proposition 1.3(e). That is, $\text{ord}_n(a^s) = \text{ord}_n a / \gcd(s, \text{ord}_n a)$. ■

For example, refer back to Example 4.1. If $n = 13$ and $a = 2$, then $\text{ord}_{13} 2 = 12$. If $s = 3$, then Proposition 4.2(c) says that $\text{ord}_{13} 8 = \frac{\text{ord}_{13} 2}{(3, \text{ord}_{13} 2)} = 12/3 = 4$. Similarly, if $s = 4$, then $\text{ord}_{13} 3 = \text{ord}_{13} 16 = \frac{\text{ord}_{13} 2}{(4, \text{ord}_{13} 2)} = 12/4 = 3$.

Corollary 4.2.1: Let p be a prime and g be a primitive root modulo p. Then $g^i \equiv g^j \pmod p$ if and only if $i \equiv j \pmod{p-1}$.

Proof: Without loss of generality suppose $i \ge j$.
If $g^i \equiv g^j \pmod p$, then $g^{i-j} \equiv 1 \pmod p$.
Proposition 4.2(b) implies that $(p-1) \mid (i-j)$ and so $i \equiv j \pmod{p-1}$.
On the other hand, if $i \equiv j \pmod{p-1}$ then $i = j + k(p-1)$ for some k. Then $g^i = g^j$.
$(g^p - 1)k \equiv g^j \cdot 1^k \equiv g^j \pmod p$. ■

Example 4.2: We apply our results to a natural question involving card shuffling. How many riffle shuffles are necessary to return a deck of cards to its original order? In particular, we will assume that we have a deck consisting of $2n$ cards and that we repeatedly execute perfect out-shuffles. If we number the cards from 1 on top to $2n$ on the bottom, a perfect out-shuffle means that we cut the cards into two equal halves 1 through n and $n+1$ through $2n$, then carefully interlace them resulting in the order 1, $n+1$, 2, $n+2$, ..., n, $2n$ from top to bottom. Note that in an out-shuffle, the top and bottom cards keep their original positions.

Let $s(k)$ be the position of card k after such a shuffle. So $s(1) = 1$, $s(n+1) = 2$, $s(2) = 3$, $s(n+2) = 4, \ldots, s(2n) = 2n$. For $1 \leq k \leq n$, $s(k) = 2k-1$ and for $n+1 \leq k \leq 2n$, $s(k) = 2k - 2n$. We can combine these by noting

$$s(k) \equiv 2k - 1 \ (\text{mod } 2n - 1) \text{ for all } k$$

It is convenient to let $k = 1 + t$, so that $s(1+t) \equiv 1 + 2t \ (\text{mod } 2n - 1)$. We are interested in studying $s(1+t)$ for $1 \leq t \leq 2n - 2$.

Let $s^2(k) = s(s(k))$ and let $s^r(k)$ denote the r-term composition of $s(k)$ for $r \geq 2$ which gives the position of card k after r shuffles. Then $s^2(1+t) = s(s(1+t)) \equiv s(1+2t) = 2(1+2t) - 1 \equiv 1 + 4t \ (\text{mod } 2n - 1)$. In general,

$$s^r(1+t) \equiv 1 + 2^r t \ (\text{mod } 2n - 1)$$

If $2^r \equiv 1 \ (\text{mod } 2n - 1)$, then $s^r(1+t) = 1 + t$. By Fermat's little theorem, $r = \phi(2n-1)$ solves the congruence. To find the least number of shuffles that returns all cards to their original position, it suffices to find $\text{ord}_{2n-1}(2)$.

For the sake of concreteness, let $2n = 52$, the standard number of cards in a deck. By Proposition 4.2(a), $\text{ord}_{51}(2)$ divides $\phi(51)$. But $\phi(51) = 32$. We readily discover that $\text{ord}_{51}(2) = 8$ since $2^8 = 256 \equiv 1 \ (\text{mod } 51)$. Hence, eight shuffles will return the deck to its original order.

Example 4.3: Solve the congruence $x^5 \equiv 7 \ (\text{mod } 13)$.
Solution: We know that 2 is a primitive root modulo 13. Table 4.1 is useful.

Table 4.1: Powers of 2 modulo 13.

a	0	1	2	3	4	5	6	7	8	9	10	11	12
2^a	1	2	4	8	3	6	12	11	9	5	10	7	1

We see from Table 4.1 that $7 \equiv 2^{11} \ (\text{mod } 13)$. We write $x^5 \equiv 2^{11} \ (\text{mod } 13)$. If x is a solution then $x \equiv 2^y$ for some y by Proposition 4.1. Hence $2^{5y} \equiv 2^{11} \ (\text{mod } 13)$. By Corollary 4.2.1 it must be the case that $5y \equiv 11 \ (\text{mod } 12)$. But $5^* \equiv 5 \ (\text{mod } 12)$ and so $y \equiv 5^* \cdot 11 \equiv 55 \equiv 7 \ (\text{mod } 12)$.

Using Table 4.1 again, $x = 2^7 \equiv 11 \ (\text{mod } 13)$ is a solution.

The method used here is especially useful if the exponent or modulus is large.

Proposition 4.3: Let $a_1, a_2, \ldots, a_m$ be relatively prime to n. Let $\text{ord}_n a_i = k_i$ and suppose the k_i are pairwise relatively prime. Then $\text{ord}_n(a_1 a_2 \cdots a_m) = k_1 k_2 \cdots k_m$.

Before proving Proposition 4.3, we first establish the following lemma:

Lemma 4.3.1: If $\gcd(a, n) = 1$, then $\text{ord}_n a = \text{ord}_n a^*$.

Proof: Let $k = \text{ord}_n a$ and $t = \text{ord}_n a^*$.
Since $a^k \equiv 1 (\text{mod } n)$, it follows that

$$(a^*)^k = 1(a^*)^k \equiv a^k a^{*k} = (aa^*)^k \equiv 1^k \equiv 1 \ (\text{mod } n).$$

By Proposition 4.2(b), $t \mid k$. Similarly, $(a^*)^t \equiv 1 \ (\text{mod } n)$ implies

$$a^t = 1(a^t) \equiv (a^*)^t a^t = (a^*a)^t \equiv 1^t \equiv 1 \ (\text{mod } n).$$

Hence $k \mid t$. By Proposition 1.3(f), $k = t$. ∎

Proof of Proposition 4.3: Certainly, the proposition is trivially true when $m = 1$. Now assume that it is true for some m. Let $P = a_1 a_2 \cdots a_m$ and $Q = k_1 k_2 \cdots k_m$. We will demonstrate that the proposition holds for $m + 1$.

Let $\text{ord}_n a = k$ with $\gcd(k, Q) = 1$. We have $(Pa)^{Qk} = (P^Q)^k (a^k)^Q \equiv 1^k 1^Q \equiv 1(\text{mod } n)$. By Proposition 4.2(b), $\text{ord}_n (Pa) \mid Qk$.

Let $\text{ord}_n (Pa) = t$. Then $1 \equiv (Pa)^t \equiv P^t a^t (\text{mod } n)$ and hence $a^t \equiv (P^t)^* (\text{mod } n)$. Lemma 4.3.1 implies that $\text{ord}_n a^t = \text{ord}_n P^t$.

Now Proposition 4.2(c) implies that $Q/\gcd(t, Q) = k/\gcd(t, k)$. Thus, $Q \cdot \gcd(t, k) = k \cdot \gcd(t, Q)$. But $\gcd(k, Q) = 1$ and hence $k \mid \gcd(t, k)$ and $Q \mid \gcd(t, Q)$. Thus $k \mid t$ and $Q \mid t$. But then $Qk \mid t$ by Corollary 2.1.1(c). Combining this with the above, $\text{ord}_n (Pa) = Qk$ and the result follows by induction on m. ∎

For a given prime p, finding a primitive root $(\text{mod } p)$ requires some effort. As Gauss has said, "Skillful mathematicians know how to reduce tedious calculations by a variety of devices, [but] experience is a better teacher than precept."

Example 4.4: Let us find a primitive root modulo 41. Here $\phi(41) = 40 = 2^3 \cdot 5$. We begin with the smallest candidate 2. It must be the case that $\text{ord}_{41} 2 \mid 40$ by Proposition 4.2(a). Successive doubling and reducing $(\text{mod } 41)$ shows that the smallest exponent a for which $2^a \equiv -1 \ (\text{mod } 41)$ is $a = 10$. It necessarily follows that $\text{ord}_{41} 2 = 20$ since $(-1)^2 = 1$.

The next smallest candidate is 3. Since $a = 4$ is the smallest exponent for which $3^a \equiv -1 \ (\text{mod } 41)$, $\text{ord}_{41} 3 = 8$. But $\text{ord}_{41} 2 = 20$ implies $\text{ord}_{41} 16 = 5$ since $16 = 2^4$, so $a = 5$ is the smallest exponent for which $16^a \equiv 1 \ (\text{mod } 41)$. Since $\gcd(5, 8) = 1$, we can apply Proposition 4.3 obtaining $\text{ord}_{41} 48 = 40$. But $48 \equiv 7 \ (\text{mod } 41)$ and so 7 is a primitive root modulo 41.

We have been discussing $\text{ord}_n a$ for arbitrary n. Now we must specialize to the case where n is prime. Notice how the next theorem (proved by L. Poinsot – 1845) jibes with Example 4.1.

Theorem 4.4: Let p be prime and let $d \mid p - 1$. Then there exists $\phi(d)$ integers a modulo p for which $\text{ord}_p a = d$.

Proof: If $d=1$ then $\phi(1)=1$ and $a=1$ is the only integer of order 1. So assume $d>1$. Let $d=p_1^{a_1}p_2^{a_2}\cdots p_t^{a_t}$ be its canonical prime factorization. The crux of the matter is to demonstrate the existence of any integer a for which $\mathrm{ord}_p a = d$. Counting the number of such a will then be an easy matter. To construct an integer a with $\mathrm{ord}_p a = n$ it suffices to find integers $b_1, \ldots, b_t$ such that $\mathrm{ord}_p b_i = p_i^{a_i}$ for all $i=1, \ldots, t$; for then if $a = \prod_{i=1}^t b_i$ then $\mathrm{ord}_p a = d$ by Proposition 4.3.

In order for b_i to be such that $\mathrm{ord}_p b_i = p_i^{a_i}$ it must be the case that b_i is a solution to

$$x^{p_i^{a_i}} - 1 \equiv 0 \pmod{p} \tag{4.1}$$

Furthermore, b_i must not be a solution to

$$x^{p_i^{a_i-1}} - 1 \equiv 0 \pmod{p} \tag{4.2}$$

In fact, these two conditions characterize those b_i for which $\mathrm{ord}_p b_i = p_i^{a_i}$.

To see this notice that since b_i satisfies eq. (4.1), it follows that $\mathrm{ord}_p b_i \mid p_i^{a_i}$ by Proposition 4.2(b). But then $\mathrm{ord}_p b_i = p_i^r$ for some $r \le a_i$. If $r \le a_i - 1$ then b_i would satisfy eq. (4.2) by Proposition 4.2(b). Hence those b_i which satisfy eq. (4.1) but do not satisfy eq. (4.2) are precisely the b_i with $\mathrm{ord}_p b_i = p_i^{a_i}$.

But by Corollary 2.19.1, there are $p_i^{a_i}$ solutions mod p to eq. (4.1) and $p_i^{a_i-1}$ solutions mod p to eq. (4.2). Further, all solutions of eq. (4.2) are solutions of eq. (4.1) since $p_i^{a_i-1} \mid p_i^{a_i}$. Thus there are exactly $p_i^{a_i} - p_i^{a_i-1} = \phi\left(p_i^{a_i}\right)$ numbers $b_i \pmod p$ for which $\mathrm{ord}_p b_i = p_i^{a_i}$.

Now apply Proposition 4.3 to obtain $\phi(d) = \prod_{i=1}^t \phi\left(p_i^{a_i}\right)$ integers a modulo p with $\mathrm{ord}_p a = d$. ∎

Corollary 4.4.1: If p is prime, then there are $\phi(\phi(p)) = \phi(p-1)$ primitive roots modulo p.

Proof: Let $d = p-1$ in Proposition 4.4. ∎

For example, there are $\phi(18) = 6$ primitive roots $\pmod{19}$.

Our next three results answer the question, "For which values of n do there exist primitive roots modulo n?"

Proposition 4.5: If p is prime then there are $(p-1)\phi(p-1)$ primitive roots modulo p^2.

Proof: If h is a primitive root $\pmod{p^2}$ then h^k runs through a reduced residue system $\pmod{p^2}$ by Proposition 4.1 and h^k is a primitive root if and only if $\gcd(k, p(p-1)) = 1$ by Proposition 4.2(c). But there are $\phi(p(p-1)) = (p-1)\phi(p-1)$ such k and so the result will follow once we establish the existence of any primitive root $\pmod{p^2}$. In fact, we will construct the primitive roots $\pmod{p^2}$ fairly explicitly (given we have found a primitive root modulo p).

By Corollary 4.4.1, it suffices to establish that if g is a primitive root $(\bmod\, p)$ then $g + mp$ is a primitive root$(\bmod\, p^2)$ for precisely $p - 1$ values of m $(\bmod\, p)$.

Let $r = \mathrm{ord}_{p^2}(g + mp)$. (Here r may be a function of m.)

Since $(g + mp)^r \equiv 1 \pmod{p^2}$, it follows that $(g + mp)^r \equiv 1 \pmod{p}$.

By the binomial theorem, $(g + mp)^r = g^r + p \cdot k$ for some integer k and hence $g^r \equiv 1 \pmod{p}$. But then $p - 1 \mid r$ by Proposition 4.2(b).

On the other hand, $r \mid \phi(p^2) = p(p - 1)$ by Proposition 4.2(a).

So either (i) $r = p - 1$ or (ii) $r = p(p - 1)$.

Now let $f(x) = x^{p-1} - 1$ and consider the congruence

$$f(x) \equiv 0 \pmod{p^2} \tag{4.3}$$

In case (i), $g + mp$ is a solution of eq. (4.3) lying above g $(\bmod\, p)$. But $f'(g) = (p - 1)g^{p-2}$ and so $f'(g) \equiv 0 \pmod{p}$. By Hensel's lemma (nonsingular case), there is a unique m $(\bmod\, p)$ for which $x = g + mp$ satisfies eq. (4.3).

So for all the other $p - 1$ values of m $(\bmod\, p)$, $h = g + mp$ does not satisfy eq. (4.3) and hence case (ii) applies. But then $r = \mathrm{ord}_{p^2}(g + mp) = \phi(p^2)$. Hence there are $(p - 1)\phi(p - 1)$ primitive roots modulo p^2. ∎

Proposition 4.6: If p is an odd prime and g is a primitive root modulo p^2, then g is a primitive root modulo p^a for all $a \geq 2$.

Proof: Let g be a primitive root modulo p^2 and let $r = \mathrm{ord}_{p^a} g$ for some $a \geq 2$. Since $g^r \equiv 1 \pmod{p^a}$ it follows that $g^r \equiv 1 \pmod{p^2}$.

Hence $\phi(p^2) = p(p - 1) \mid r$ by Proposition 4.2(b).

On the other hand, $r \mid \phi(p^a) = p^{a-1}(p - 1)$ by Proposition 4.2(a). Thus $r = p^b(p - 1)$ for some $b = 1, 2, \ldots, a - 1$. It remains to show that in fact $b = a - 1$. It suffices to demonstrate that

$$g^s \not\equiv 1 \pmod{p^a} \quad \text{where } s = p^{a-2}(p - 1) \tag{4.4}$$

By Fermat's little theorem, $g^{p-1} \equiv 1 \pmod{p}$ and thus $g^{p-1} = 1 + cp$ where $p \nmid c$ (since g is a primitive root modulo p^2). The binomial theorem says $g^s = (1 + cp)^{p^{a-2}(p-1)} = 1 + p^{a-1}c(p - 1) + p^a k$ for some integer k. Hence $g^s \equiv 1 + p^{a-1}c(p - 1) \pmod{p^a}$ and so $g^s \not\equiv 1 \pmod{p^a}$ because $p \nmid c(p - 1)$.

Since a was arbitrary, this establishes eq. (4.4) for all $a \geq 2$. ∎

Theorem 4.7 (Primitive Root Theorem): There exists a primitive root modulo n if and only if $n = 1, 2, 4, p^a$, or $2p^a$ for odd primes p. In addition, if n has a primitive root, then there are $\phi(\phi(n))$ primitive roots modulo n.

We begin by establishing a technical lemma.

Lemma 4.7.1: If b is odd and $m \geq 3$, then $2^m \mid b^r - 1$ where $r = 2^{m-2}$.

Proof (Induction on m): Let $m = 3$. Then $r = 2$ and $b^2 - 1 = (b+1)(b-1)$. But b odd implies that either $b+1$ or $b-1$ is exactly divisible by 2 (being congruent to $2 \bmod 4$) and the other must be divisible by 4. Hence $8 \mid b^2 - 1$.

Assume now that $2^m \mid b^r - 1$ where $r = 2^{m-2}$. But
$b^{2r} - 1 = (b^r + 1)(b^r - 1)$. Since $b^r + 1$ is even, it follows that $2^{m+1} \mid b^{2r} - 1$. ∎

Proof of Theorem 4.7: Integers 1 and 2 have 1 as a primitive root and 3 is a primitive root $(\bmod 4)$. We have already shown in Example 4.1 that there is no primitive root $(\bmod 8)$. More generally, if $m \geq 3$ and b is odd then $2^m \mid b^r - 1$ where $r = 2^{m-2}$ by Lemma 4.7.1. But $\phi(2^m) = 2^{m-1}$ and hence $b^{\phi(2^m)/2} \equiv 1 \pmod{2^m}$ for all odd b. So there are no primitive roots $(\bmod 2^m)$ for $m \geq 3$.

Suppose p is an odd prime and g is a primitive root modulo p^a. We may assume that g is odd (if not, replace g by $g + p^a$). By Proposition 4.1, the integers $g, g^2, \ldots, g^{\phi(p^a)}$ form a reduced residue system $(\bmod p^a)$. But they are all odd and $\phi(2p^a) = \phi(p^a)$. So $r = \phi(2p^a)$ is the smallest r for which $g^r \equiv 1 \pmod{2p^a}$. Hence all integers of the form $2p^a$ have primitive roots.

Now suppose that $n \neq p^a$ or $2p^a$ for any prime p. Then we can express n as $n = s \cdot t$ where $\gcd(s, t) = 1$ and $s > 2$ and $t > 2$.

Let $c = \mathrm{lcm}[\phi(s), \phi(t)]$. If $\gcd(b, n) = 1$ then $\gcd(b, s) = \gcd(b, t) = 1$.

So $b^{\phi(s)} \equiv 1 \pmod s$ and $b^{\phi(t)} \equiv 1 \pmod t$. But $\phi(s) \mid c$ and $\phi(t) \mid c$. Hence $b^c \equiv 1 \pmod s$ and $b^c \equiv 1 \pmod t$. Since $\gcd(s, t) = 1$, $b^c \equiv 1 \pmod n$.

Since $s > 2$ and $t > 2$, it must be the case that $\phi(s)$ and $\phi(t)$ are even. Hence $2 \mid \gcd(\phi(s), \phi(t))$. By Exercise 2.1.10(a), the product of two integers equals the product of their greatest common divisor and least common multiple.

Hence $c = \frac{\phi(s) \cdot \phi(t)}{\gcd(\phi(s), \phi(t))}$ and so $c < \phi(s) \cdot \phi(t) = \phi(n)$. Therefore, there is no primitive root $(\bmod n)$ in this case.

Hence the integers possessing primitive roots are precisely 1, 2, 4, p^a, and $2p^a$ for odd primes p.

To establish the latter assertion of the theorem, let g be a primitive root $(\bmod n)$. By Proposition 4.1, $g, g^2, \ldots, g^{\phi(n)}$ form a reduced residue system $(\bmod n)$. By Proposition 4.2(c), g^k is a primitive root if and only if $\gcd(k, \phi(n)) = 1$. But there are exactly $\phi(\phi(n))$ such k. ∎

The study of primitive roots has a long history. The term "primitive root" was coined by Euler in 1773 when he gave slightly defective proof that all primes have primitive roots. Such a claim was essentially given by Johann Lambert in 1769. The first correct proof was given by A.M. Legendre in 1785 based on Lagrange's theorem (Corollary 2.19.1) much as in our proof of Theorem 4.4.

Gauss launched an extensive study of primitive roots in the *Disquisitiones Arithmeticae* (1801), including two new proofs of the existence of primitive roots $(\bmod p)$ as well as introducing the notion of $\text{ord}_n a$. Most of the theorems in this section are due to Gauss including Proposition 4.2 and Theorem 4.7.

If $a = -1$ then a is not a primitive root $(\bmod\, p)$ for any prime $p > 3$. Similarly, if $a = n^2$ for some integer n, then $a^{(p-1)/2} \equiv 1(\bmod\, p)$ and so a is not a primitive root for any prime $p > 2$. The *Artin conjecture* states that all other integers a are primitive roots for infinitely many primes. In 1967, Christopher Hooley proved the Artin conjecture subject to the truth of another yet unproven conjecture, the generalized Riemann hypothesis. Since then, others have shown that substantially weaker hypotheses besides the generalized Riemann hypothesis would suffice. However, an unconditional proof of the Artin conjecture has not yet been effected.

Building on the work of Rajiv Gupta and Ram Murty, the best result to date is an amazing theorem of Roger Heath-Brown (1986) which utilizes difficult sieve techniques from analytic number theory. A consequence of his work is that there are at most two primes and at most three positive square-free integers, which are exceptions to the Artin conjecture. Despite this, interestingly, no particular value of a has been proven to be a primitive root for infinitely many primes. However, Heath-Brown's result has tantalizing corollaries such as the following: At least two of the integers 2, 3, 5 are primitive roots for infinitely many primes.

A great deal of research has been done concerning algorithms for finding a primitive root modulo p for a given prime p, including an excellent algorithm of Gauss (article 73 of the *Disquisitiones*). An open question of Erdös asks whether every sufficiently large prime p has a primitive root $q < p$ which is also prime. Much work remains to be done.

Exercise 4.1

1. Prove the converse of Proposition 4.1.
2. (a) Find all primitive roots modulo 11.
    (b) Find all primitive roots modulo 17.
3. (a) Find the smallest primitive root modulo 41.
    (b) Find the smallest primitive root modulo 47.
4. Let $p = 43$. For each $d\,|\,p - 1$, determine how many integers a $(\bmod\, p)$ have $\text{ord}_p a = d$.
5. (a) Verify Proposition 4.2 for $a = 7, n = 40$, $m = 20$, $s = 2$.
    (b) Verify Proposition 4.2 for $a = 3, n = 121$, $m = 20$, $s = 10$.
6. (a) Let g be a primitive root$(\bmod\, p)$. Let $\gcd(a, p) = 1$ and $a \equiv g^s(\bmod\, p)$.
        Show that $\text{ord}_p a = \frac{p-1}{\gcd(s,\, p-1)}$.
    (b) Use the formula in part (a) to compute $\text{ord}_{13}\, a$ for $a = 3,\ 6$, and 8.
7. (a) Verify Theorem 4.4 for all $d\,|\,p - 1$ with $p = 17$.
    (b) Verify Theorem 4.4 for all $d\,|\,p - 1$ with $p = 23$.

8. Determine how many primitive roots the following primes have:
   (a) $p = 29$
   (b) $p = 73$
   (c) $p = 107$
   (d) $p = 337$
   (e) $p = 9973$

9. Determine which of the following integers n have primitive roots. If they do, determine how many distinct primitive roots $(\bmod\ n)$ there are.
   (a) $n = 143$
   (b) $n = 250$
   (c) $n = 729$
   (d) $n = 1372$
   (e) $n = 2662$
   (f) $n = 11979$
   (g) $n = 117649$
   (h) $n = 156250$

10. (a) Solve $x^5 \equiv 6 \pmod{13}$.
    (b) Solve $x^4 \equiv 9 \pmod{13}$.

11. (a) Solve $x^6 \equiv -2 \pmod{17}$.
    (b) Solve $x^4 \equiv 4 \pmod{17}$.

12. (a) Solve $x^4 \equiv 5 \pmod{19}$.
    (b) Solve $x^3 \equiv 11 \pmod{19}$.

13. Solve $x^{17} \equiv 7 \pmod{29}$. (Note: 2 is a primitive root modulo 29.)

14. If $\mathrm{ord}_n a = 12$, $\mathrm{ord}_n b = 5$, and $\mathrm{ord}_n c = 91$ where a, b, and c are pairwise relatively prime, then what is $\mathrm{ord}_n abc$?

15. (a) Determine all primitive roots modulo p^2 where $p = 5$.
    (b) Determine all primitive roots modulo p^2 where $p = 7$. (Compare with Exercise 4.1.27.)

16. Use Proposition 4.5 to determine how many primitive roots there are $(\bmod\ p^2)$ for
    (a) $p = 5$
    (b) $p = 11$
    (c) $p = 73$
    (d) $p = 1201$

17. What is the maximal order of $a \pmod{32}$? How many $a \pmod{32}$ obtain this maximal order?

18. Where did we utilize the fact that p was odd in the proof of Proposition 4.6?

19. Use Proposition 4.1 to prove Wilson's theorem.

20. Make use of the fact $x \equiv x^{25} \pmod{13}$ to solve the congruence $x^5 \equiv 7 \pmod{13}$ in Example 4.2 more quickly.

21. If p is prime and $p \nmid a$, let g be a primitive root$(\bmod\ p)$. Then there exists an i such that $a \equiv g^i \pmod{p}$ with $1 \le i \le p - 1$. Following Gauss, call i the *index of a with respect to g modulo p* and write $i = \mathrm{ind}\ a$. Show the following:

(a) ind $1 = 0$ and ind $g = 1$.

(b) ind $ab \equiv$ ind $a +$ ind $b \pmod{p-1}$ for $p \nmid ab$.

(c) ind $a^r \equiv r \cdot$ ind $a \pmod{p-1}$ for $p \nmid a$.

Notice similar properties between indices and logarithms.

22. Use Proposition 4.6 to show that 2 is a primitive root $\pmod{5^n}$ for all $n \geq 1$.

23. Let p be prime with $p \nmid a$ and A $= \prod_{i=1}^{\mathrm{ord}_p a} a^i$. Show that $A \equiv 1 \pmod{p}$ when $\mathrm{ord}_p a$ is odd and $A \equiv -1 \pmod{p}$ when $\mathrm{ord}_p a$ is even.

24. Prove the following generalization of Wilson's theorem (Gauss *Disquisitiones Art.* 78): For any natural number n, the product of all the integers in a reduced residue system $\pmod n$ is congruent to either $+1$ or $-1 \pmod n$.

25. Let p be prime and $d \mid p - 1$. Let g be a primitive root$\pmod p$.

(a) Show that if $a = g^{r(p-1)/d}$ where r is relatively prime to d, then $\mathrm{ord}_p a = d$.

(b) Explain how this leads to a constructive proof of Theorem 4.4.

26. (a) Show that if g is a primitive root$\pmod n$, then g^m is a primitive root $\pmod n$ for all m with $1 \leq m \leq \phi(n)$ and $\gcd(m, \phi(n)) = 1$.

(b) Show that all primitive roots $\pmod n$ are given by those in part (a).

27. Show that if g is a primitive root$\pmod p$, then either g or $g + p$ is a primitive root $\pmod{p^2}$. Conclude that either g or $g + p$ is a primitive root mod (p^a) for all $a \geq 2$.

28. Investigate the period length of the decimal fraction $1/p$ and how it relates to $\mathrm{ord}_{10} p$. Conclude that there are infinitely many fractions $1/p$ of maximal period if 10 is a primitive root for infinitely many primes.

29. (a) How many perfect out-shuffles are required to return a deck of 16 cards to its original position?

(b) How many perfect out-shuffles are required to return a deck of 54 cards (full deck with two distinguishable jokers) to its original position?

(c) How many perfect in-shuffles (where the card 1 goes to position 2 and so on) are required to return a deck of 52 cards to its original position?

(d) Explain why Artin's primitive root conjecture implies that there are infinitely many values of n for which a deck of $2n$ cards requires $2n - 2$ perfect out-shuffles to return it to its original position.

4.2 Quadratic and *n*th power residues

The concept of nth power residues is defined and developed in this section. We then apply our results to the historically important case $n = 2$ to derive Euler's criterion for quadratic residues.

Definition 4.3: Let p be prime and n a positive integer. Integers a for which

$$x^n \equiv a \pmod{p} \tag{4.5}$$

is solvable are called *n*th *power residues modulo p*. If eq. (4.5) is not solvable, then *a* is called an *n*th *power nonresidue modulo p*.

Notice that if $p|a$ then $x \equiv 0 \pmod p$ is the only solution to eq. (4.5) and if $p \nmid a$ then $x \equiv 0 \pmod p$ is not a solution. In fact, if $p \nmid a$ then eq. (4.5) might be insoluble. For example, $x^3 \equiv a \pmod 7$ has no solution if $a = 2$, 3, 4, or 5 (mod 7). Check it! If $n = 2$ and congruence eq. (4.5) is solvable, then *a* is called a *quadratic residue* (mod *p*). If $n = 3$, then *a* is a *cubic residue* (mod *p*). So 1 and 6 are cubic residues and 2, 3, 4, and 5 are cubic nonresidues (mod 7).

Theorem 4.8: Let *p* be prime and $p \nmid a$. Let $c = \gcd(n, p-1)$. Let *g* be a primitive root (mod *p*) and set $a \equiv g^b \pmod p$. Then eq. (4.5) is solvable if and only if $c|b$. Furthermore, if eq. (4.5) is solvable, then there are exactly *c* incongruent solutions modulo *p*.

It is of interest to note that Theorem 4.8 does not depend on the particular primitive root chosen. An immediate corollary is the following:

Corollary 4.8.1: If $\gcd(n, p-1) = 1$, then congruence equation (4.5) is solvable for all *a*.

Proof of Theorem 4.8: Any solution of eq. (4.5) is not divisible by *p* since $p \nmid a$. By Proposition 4.1, $x \equiv gy \pmod p$ for some *y*. Thus eq. (4.5) is solvable for *x* if and only if $g^{ny} \equiv g^b \pmod p$ is solvable for *y*. By Corollary 4.2.1, the above is equivalent to solving $ny \equiv b \pmod{p-1}$. The last congruence is a linear congruence. By Proposition 2.5(b), it is solvable if and only if $c|b$. In fact, when $c|b$ then there are *c* incongruent solutions (mod *p*). Each of these gives rise to distinct solutions of eq. (4.5). ■

For example, consider $x^3 \equiv 6 \pmod 7$. Since 3 is a primitive root (mod 7) and $6 \equiv 3^3 \pmod 7$, $b|c$ since $b = 3$ and $c = \gcd(3, 6) = 3$. So $x^3 \equiv 6 \pmod 7$ has precisely three solutions (mod 7). Please confirm that they are 3, 5, and 6. Similarly, $x^3 \equiv 1 \pmod 7$ has three solutions (mod 7). Here $b = 6$ and $c|b$. (Note that $x^3 \equiv 1 \pmod 7$ is guaranteed to have precisely three solutions (mod 7) by Corollary 2.19.1 too.)

Theorem 4.8 is useful, but it does necessitate finding a primitive root modulo *p*. The next theorem avoids that requirement.

Theorem 4.9: Let *p* be prime and $p \nmid a$. Let $c = \gcd(n, p-1)$. Then eq. (4.5) is solvable if and only if $a^{(p-1)/c} \equiv 1 \pmod p$.

Proof: Let *g* be a primitive root (mod *p*) and let *b* be such that $a \equiv g^b \pmod p$.
$\Rightarrow$ Suppose eq. (4.5) is solvable with solution *x*. Then

$$a^{(p-1)/c} \equiv (x^n)^{\frac{p-1}{c}} \equiv (x^{p-1})^{\frac{n}{c}} \pmod{p}$$

Fermat's little theorem implies that $x^{p-1} \equiv 1 \pmod{p}$. Thus

$$a^{(p-1)/c} \equiv 1^{n/c} = 1 \pmod{p}$$

($\Leftarrow$) Suppose $a^{(p-1)/c} \equiv 1 \pmod{p}$. If $a \equiv g^b \pmod{p}$ then $1 \equiv (g^b)^{\frac{p-1}{c}} \pmod{p}$.

But $\mathrm{ord}_p g = p-1$ and so $(p-1) \mid \frac{b(p-1)}{c}$ by Proposition 4.2(b). Thus b/c is an integer and $c \mid b$. By Theorem 4.8, the relation (4.5) is solvable. ∎

If $n = 2$ then we obtain Euler's criterion for quadratic residues (first proved by Euler in 1755).

Corollary 4.9.1 (Euler's Criterion): Let p be an odd prime and $p \nmid a$. $x^2 \equiv a \pmod{p}$ is solvable if and only if $a^{(p-1)/2} \equiv 1 \pmod{p}$.

Example 4.5: Determine whether $x^2 \equiv 21 \pmod{37}$ is solvable.
Solution: In this case $p = 37$ and $a = 21$. Since $p > a, p \nmid a$. In order to calculate $21^{18} \pmod{37}$ we invoke the binary exponentiation algorithm. Write 18 in binary: $18 = (10010)_2$. The algorithm yields $1 \xrightarrow{1} 21 \xrightarrow{0} 34 \xrightarrow{0} 9 \xrightarrow{1} 36 \xrightarrow{1} 1$. Hence $21^{18} \equiv 1 \pmod{37}$ and Euler's criterion guarantees that $x^2 \equiv 21 \pmod{37}$ is solvable.

For a fixed p, Euler's criterion is an effective means to determine whether $x^2 \equiv a \pmod{p}$ is solvable. A much more difficult question is the following: given a, for which primes p is $x^2 \equiv a \pmod{p}$ solvable? Much of the rest of this chapter is devoted to obtaining a satisfactory answer to that question. The culmination of our efforts will be Gauss' celebrated *law of quadratic reciprocity*. In addition, we will see other applications of the law of quadratic reciprocity when we discuss primality testing and factoring in Chapter 7. Below we take a modest first step.

Proposition 4.10: Let p be an odd prime. Then
(a) $x^2 \equiv -1 \pmod{p}$ is solvable if and only if $p \equiv 1 \pmod{4}$.
(b) If $p \equiv 1 \pmod{4}$ then $x = \pm \left(\frac{p-1}{2}\right)!$ are the only solutions $\pmod{p}$.

Proof: (a) By Euler's criterion, $x^2 \equiv -1 \pmod{p}$ is solvable if and only if $(-1)^{\frac{p-1}{2}} \equiv 1 \pmod{p}$. But $(-1)^{(p-1)/2}$ is either 1 or -1 depending on whether $(p-1)/2$ is even or odd, respectively. Since $p > 2$, $(-1)^{\frac{p-1}{2}} \equiv 1 \pmod{p}$ if and only if $(p-1)/2$ is even if and only if $p \equiv 1 \pmod{4}$.
(b) If $p \equiv 1 \pmod{4}$, then $(p-1)! = 1 \cdot 2 \cdot \ldots \cdot \frac{p-1}{2} \cdot (p-1) \cdot (p-2) \cdot \ldots \cdot \left(p - \frac{p-1}{2}\right)$. So

$$(p-1)! \equiv \left(\frac{p-1}{2}\right)! \cdot (-1)^{\frac{p-1}{2}} \cdot \left(\frac{p-1}{2}\right)! = [(p-1)!]^2 \pmod{p}$$

since $p \equiv 1 \pmod 4$. By Wilson's theorem, $(p-1)! \equiv -1 \pmod p$. Thus, if $x = \left(\frac{p-1}{2}\right)!$, then $x^2 \equiv -1 \pmod p$.

Clearly $x = -\left(\frac{p-1}{2}\right)!$ is also a solution. Furthermore, the two solutions are distinct since $p \nmid 2 \cdot \left(\frac{p-1}{2}\right)!$ with p larger than any of the factors on the right. By Lagrange's theorem (or Proposition 4.8) there can be no other solutions $\pmod p$. The proposition is established. ∎

Proposition 4.10 was originally proved by Euler in 1749 by other means. The proof given here essentially follows one due to Lagrange in 1773.

Exercise 4.2

1. Let $p > 3$ be prime and suppose $p \nmid a$.
   (a) Show that $x^3 \equiv a \pmod p$ is solvable if $p \equiv 2 \pmod 3$.
   (b) If $p \equiv 1 \pmod 3$, show that $x^3 \equiv a \pmod p$ is solvable if and only if $a^{(p-1)/3} \equiv 1 \pmod p$.
2. Show that if p is an odd prime with $p \mid (n^2 + 1)$ for some n, then $p \equiv 1 \pmod 4$.
3. Use Exercise 2 to show that there are infinitely many primes $p \equiv 1 \pmod 4$.
4. If p is an odd prime and $p \mid (n^2 + 3)$ for some n, must it be the case that $p \equiv 3 \pmod 4$?
5. (a) Verify Theorem 4.8 for $p = 11, n = 3$, and $a = 5$ by finding all solutions.
   (b) Verify Theorem 4.8 for $p = 37, n = 4$, and $a = 33$ by finding all solutions.
6. Determine how many incongruent solutions $\pmod{11}$ there are to the following congruence relations:
   (a) $x^6 \equiv 5 \pmod{11}$
   (b) $x^7 \equiv 5 \pmod{11}$
   (c) $x^6 \equiv 3 \pmod{11}$
7. Determine how many incongruent solutions $\pmod{77}$ there are to the following congruence relations:
   (a) $x^6 \equiv 5 \pmod{77}$
   (b) $x^3 \equiv 6 \pmod{77}$
8. Find all solutions to the following:
   (a) $x^6 \equiv 5 \pmod{11}$
   (b) $3x^6 \equiv 5 \pmod{11}$
   (c) $x^5 \equiv 2 \pmod{13}$
   (d) $7x^5 \equiv 2 \pmod{13}$
9. (a) Find the smallest positive solution to $x^2 \equiv -1 \pmod{37}$.
   (b) Find a solution to the congruence in Example 4.4.

10. Use Theorem 4.9 to determine whether or not the following congruences are solvable:
    (a) $x^4 \equiv 4 \pmod{53}$
    (b) $x^3 \equiv 5 \pmod{29}$
    (c) $x^3 \equiv 18 \pmod{101}$
11. Use Euler's criterion to determine which of the following are solvable:
    (a) $x^2 \equiv 2 \pmod{13}$
    (b) $x^2 \equiv 3 \pmod{13}$
    (c) $x^2 \equiv 3 \pmod{71}$
    (d) $x^2 \equiv 48 \pmod{71}$
12. Use Euler's criterion to determine which of the following are solvable:
    (a) $x^2 \equiv 2 \pmod{19}$
    (b) $x^2 \equiv 11 \pmod{19}$
    (c) $x^2 \equiv 3 \pmod{97}$
    (d) $x^2 \equiv 348 \pmod{97}$
13. Use Proposition 4.10 to determine whether the following are solvable:
    (a) $x^2 \equiv -1 \pmod{19}$
    (b) $x^2 \equiv -1 \pmod{2099}$
    (c) $x^2 \equiv -1 \pmod{7001}$
    (d) $x^2 \equiv -1 \pmod{2^{11213} - 1}$
14. Use Proposition 4.10 to find all solutions to $x^2 \equiv -1 \pmod{17}$.

4.3 The Legendre symbol and Gauss' lemma

In this section quadratic residues are studied in greater detail. We begin with some simple observations.

The problem of solving the general quadratic congruence

$$ax^2 + bx + c \equiv 0 \pmod{n} \text{ with } \gcd(a, n) = 1$$

reduces to solving the quadratic congruences

$$ax^2 + bx + c \equiv 0 \pmod{p} \tag{4.6}$$

where $p \mid n$ by the Chinese remainder theorem and Hensel's lemma. The latter congruence is equivalent to a simpler quadratic congruence as we now demonstrate.

Proposition 4.11: Let p be an odd prime and $p \nmid a$. Congruence (4.6) is solvable if and only if the congruence

$$y^2 \equiv d \pmod{p} \tag{4.7}$$

is solvable where $y \equiv x + (2a)^* b \pmod{p}$ and $d \equiv (2a)^{*2} b^2 - a^* c \pmod{p}$.

Proof: $ax^2 + bx + c \equiv 0 \pmod p$ is solvable if and only if $x^2 + a^*bx + a^*c \equiv 0 \pmod p$ is solvable if and only if $(x^2 + a^*bx + (2^*a^* b)^2) + (a^*c - (2^*a^* b)^2) \equiv 0 \pmod p$ is solvable if and only if $(x + 2^*a^* b)^2 \equiv (2^*a^* b)^2 - a^*c \pmod p$.
But $2^*a^* = (2a)^*$ by Exercise 2.2.14. ∎

It is now apparent that the determination of whether a general quadratic congruence is solvable boils down to determining quadratic residues modulo p.

Proposition 4.12: Let p be an odd prime and $p \nmid a$. Consider the congruence relation

$$x^2 \equiv a \pmod p \tag{4.8}$$

(a) If eq. (4.8) is solvable, then there are exactly two incongruent solutions modulo p.

(b) Equation (4.8) is solvable if and only if a is congruent to one of $1^2, 2^2, \ldots, \left(\frac{p-1}{2}\right)^2$ $(\bmod\, p)$.

(c) The numbers $1^2, 2^2, \ldots, \left(\frac{p-1}{2}\right)^2$ are all incongruent $(\bmod\, p)$.

Proof:
(a) Let $x = b$ be a solution of eq. (4.8). Then $x = -b$ is also a solution. Further, $b \not\equiv -b \pmod p$ since $p \nmid 2b$. By Lagrange's theorem there are no others.
(b) ($\Leftarrow$) If $a \equiv r^2 \pmod p$, then let $x = r$.
($\Rightarrow$) Since any integer x with $p \nmid x$ is congruent to one of $1, 2, \ldots, p-1$; it must be the case that $a \equiv r^2 \pmod p$ for some r where $1 \leq r \leq p-1$. But there is duplication due to the fact that $(p-r)^2 \equiv r^2 \pmod p$, and hence a is congruent to one of $1^2, 2^2, \ldots, \left(\frac{p-1}{2}\right)^2 \pmod p$.
(c) Suppose $a^2 \equiv b^2 \pmod p$ with $1 \leq a < b \leq (p-1)/2$. Then $p \mid (a+b)(b-a)$ and by Euclid's lemma either $p \mid a+b$ or $p \mid b-a$. But this is impossible since $1 \leq b-a < a+b < p-1$. ∎

Corollary 4.12.1: There are precisely $(p-1)/2$ quadratic residues and $(p-1)/2$ quadratic nonresidues modulo p.

The previous observation will prove valuable in our discussion in Chapter 5 concerning integers expressible as the sum of four squares. Next, we introduce an extremely useful symbolism due to Legendre (1798).

Definition 4.4: Let p be an odd prime and $p \nmid a$. The *Legendre symbol* $\left(\frac{a}{p}\right)$ is defined by

$$\left(\frac{a}{p}\right) = \begin{cases} +1 \text{ if } a \text{ is a quadratic residue } (\bmod\, p) \\ -1 \text{ if } a \text{ is a quadratic nonresidue } (\bmod\, p) \end{cases}$$

Proposition 4.13: Let p be an odd prime and $p \nmid a$. Then

(a) $\left(\frac{a^2}{p}\right) = 1$

(b) If $a \equiv b \pmod{p}$, then $\left(\frac{a}{p}\right) = \left(\frac{b}{p}\right)$

(c) $\left(\frac{-1}{p}\right) = (-1)^{(p-1)/2}$

(d) $\left(\frac{a}{p}\right) \equiv a^{(p-1)/2} \pmod{p}$.

Proof: (a) and (b) The proof is immediate.
(c) By Proposition 4.10, $\left(\frac{-1}{p}\right) = 1$ if and only if $p \equiv 1 \pmod 4$. But $(-1)^{(p-1)/2}$ is 1 if $p \equiv 1 \pmod 4$ and -1 if $p \equiv 3 \pmod 4$.
(d) This is a restatement of Euler's criterion since $a^{(p-1)/2} \equiv \pm 1 \pmod p$. ∎

In fact, in the rest of the chapter we will refer to Proposition 4.13(d) as *Euler's criterion*. It has a very important consequence.

Corollary 4.13.1: Let p be an odd prime and suppose $p \nmid ab$. Then

$$\left(\frac{ab}{p}\right) = \left(\frac{a}{p}\right)\left(\frac{b}{p}\right). \tag{4.9}$$

Proof: By Euler's criterion,
$\left(\frac{ab}{p}\right) \equiv (ab)^{(p-1)/2} = a^{(p-1)/2} \cdot b^{(p-1)/2} \equiv \left(\frac{a}{p}\right)\left(\frac{b}{p}\right) \pmod p$. Equality follows since $p > 2$. ∎

By induction, Corollary 4.13.1 can be extended so that if $p \nmid a_1 a_2 \cdots a_n$ then $\left(\frac{a_1 a_2 \cdots a_n}{p}\right) = \left(\frac{a_1}{p}\right) \cdot \left(\frac{a_2}{p}\right) \cdots \left(\frac{a_n}{p}\right)$. Note that if we extend the definition of the Legendre symbol so that $\left(\frac{a}{p}\right) = 0$ if $p \mid a$, then Corollary 4.13.1 says that the function $f(a) = \left(\frac{a}{p}\right)$ is completely multiplicative for a given p. This is a great computational aid since it reduces the computation of $\left(\frac{a}{p}\right)$ to $\left(\frac{q}{p}\right)$ where $q \mid a$ is prime.

Example 4.6: Determine whether $x^2 \equiv 455 \pmod{17}$ is solvable.
Solution: Since computational facility is one of the goals of our discussion, we provide three separate solutions. Knowing a variety of techniques is useful.
(i) Since $455 \equiv 13 \pmod{17}$ we need only determine $\left(\frac{13}{17}\right)$ by Proposition 4.13(b). We easily check $1^2, 2^2, \ldots, 8^2 \pmod{17}$ and find $13 \equiv 8^2 \pmod{17}$ and hence $x^2 \equiv 455 \pmod{17}$ is solvable.

(ii) Factoring yields $455 = 5 \times 7 \times 13$. By formula (4.8), $\left(\frac{455}{17}\right) = \left(\frac{5}{17}\right)\left(\frac{7}{17}\right)\left(\frac{13}{17}\right)$.

Invoking Proposition 4.12(b), the only quadratic residues $(\bmod 17)$ are $1, 2, 4$, $8, 9, 13, 15$, and 16. So $\left(\frac{5}{17}\right) = -1$, $\left(\frac{7}{17}\right) = -1$, and $\left(\frac{13}{17}\right) = 1$. So $\left(\frac{455}{17}\right) = (-1)(-1)(+1) = 1$ and $x^2 \equiv 455 \ (\bmod 17)$ is solvable.

(iii) Begin as in (i), noting $455 \equiv 13 \ (\bmod 17)$. Euler's criterion says $x^2 \equiv 13 \ (\bmod 17)$ is solvable if and only if $13^8 \equiv 1 \ (\bmod 17)$. But $13 \equiv -4 \ (\bmod 17)$ and so $13^2 \equiv 16 \equiv -1 \ (\bmod 17)$. Hence, $13^8 \equiv (13^2)4 \equiv (-1)^4 = 1 \ (\bmod 17)$. So $x^2 \equiv 455 \ (\bmod 17)$ is solvable.

The next theorem is a remarkable result which took Gauss a decade to discover (published in 1808). It will be the key to our proof of the law of quadratic reciprocity in the next section.

Theorem 4.14 (Gauss' Lemma): Let p be an odd prime and $p \nmid a$. List the integers $a, 2a, \ldots, \left(\frac{p-1}{2}\right)a \ (\bmod p)$ so that they are all reduced between $-\left(\frac{p-1}{2}\right)$ and $\left(\frac{p-1}{2}\right)$ inclusive. Let n denote the number of negative integers in the list. Then $\left(\frac{a}{p}\right) = (-1)^n$.

Proof: Note that if $1 \le s < t \le \frac{p-1}{2}$, then $sa \not\equiv ta \ (\bmod p)$. Otherwise, $p \mid (ta - sa) = a(t-s)$ which is impossible.

Similarly, $sa \not\equiv -ta \ (\bmod p)$ since otherwise $p \mid a(s+t)$ which is impossible since $s + t < p$.

Now let $r_s \equiv sa \ (\bmod p)$ with $-(p-1)/2 \le r_s \le (p-1)/2$. By the above, $r_s \neq r_t$ $(\bmod p)$ and $r_s \not\equiv -r_t \ (\bmod p)$ for $1 \le s < t \le \frac{p-1}{2}$. In addition, $r_s \not\equiv 0 \ (\bmod p)$ for all s. Hence the numbers $|r_1|, |r_2|, \ldots, |r_{(p-1)/2}|$ are all distinct positive integers and form a permutation of $1, 2, \ldots, (p-1)/2$. So

$$r_1 \, r_2 \, \cdots \, r_{(p-1)/2} = (-1)^n \, 1 \cdot 2 \, \cdots \, (p-1)/2 = (-1)^n \cdot \left(\frac{p-1}{2}\right)! \tag{4.10}$$

But $r_s \equiv sa \ (\bmod p)$, and so

$$r_1 r_2 \cdots r_{\frac{p-1}{2}} \equiv a(2a) \cdots \left(\frac{p-1}{2}\right)a \equiv a^{\frac{p-1}{2}}\left(\frac{p-1}{2}\right)! \ (\bmod p) \tag{4.11}$$

Equating eqs. (4.10) and (4.11), $(-1)^n \cdot \left(\frac{p-1}{2}\right)! \equiv a^{\frac{p-1}{2}}\left(\frac{p-1}{2}\right)! \ (\bmod p)$.

Hence, $(-1)^n \equiv a^{\frac{p-1}{2}} (\bmod p)$.

Euler's criterion implies $\left(\frac{a}{p}\right) \equiv (-1)^n \ (\bmod p)$. But $p > 2$ and so $\left(\frac{a}{p}\right) = (-1)^n$. ∎

Example 4.7: Let us use Gauss' lemma to calculate $\left(\frac{3}{17}\right)$.

In this case $\frac{p-1}{2} = 8$. We consider the integers $3, 6, 9, 12, 15, 18, 21, 24$ which are congruent to $3, 6, -8, -5, -2, 1, 4, 7 (\bmod 17)$, respectively. Since the number of negative integers in the list is $n = 3$, $\left(\frac{3}{17}\right) = (-1)^3 = -1$. Hence 3 is a quadratic nonresidue modulo 17.

Corollary 4.14.1: Let p be an odd prime. Then $\left(\frac{2}{p}\right) = (-1)^{(p^2-1)/8}$.

Notice that if $p = 8k + r$, then $\frac{p^2-1}{8} = 8k^2 + 2rk + \frac{r^2-1}{8}$. Hence the parity of $(p^2 - 1)/8$ is the same as the parity of $(r^2 - 1)/8$. This simplifies matters greatly. For example, $\left(\frac{2}{163}\right) = -1$ since $163 \equiv 3 \pmod 8$ and $(3^2 - 1)/8$ is odd.

Proof: Consider the even integers $2, 4, \ldots, p - 1$. The n in Gauss' lemma is the number of s for which $(p + 1)/2 \le 2s \le p - 1$. Equivalently, n is the number of integers divisible by 4 between $p + 1$ and $2p - 2$ inclusive.

(i) If $p \equiv 1 \pmod 4$, then the relevant integers are $p + 3, p + 7, \ldots, 2p - 2$.
    So $n = 1 + \frac{(2p-2)-(p+3)}{4} = (p - 1)/4$. If $p \equiv 1 \pmod 8$ then n is even.
    If $p \equiv 5 \pmod 8$ then n is odd.

(ii) If $p \equiv 3 \pmod 4$ then the relevant integers are $p + 1, p + 5, \ldots, 2p - 2$.

So $n = 1 + \frac{(2p-2)-(p+1)}{4} = (p + 1)/4$. If $p \equiv 3 \pmod 8$ then n is odd.
    If $p \equiv 7 \pmod 8$ then n is even.
    Applying Gauss' lemma,

$$\left(\frac{2}{p}\right) = \begin{cases} 1 \text{ if } p \equiv \pm 1 \pmod 8 \\ -1 \text{ if } p \equiv \pm 5 \pmod 8 \end{cases}$$

But $(p^2 - 1)/8$ is even if $p \equiv \pm 1 \pmod 8$ and $(p^2 - 1)/8$ is odd if $p \equiv \pm 5 \pmod 8$. Therefore, $\left(\frac{2}{p}\right) = (-1)^{(p^2-1)/8}$. ∎

Exercise 4.3

1. Verify Corollary 4.12.1 directly for $p = 11, 13$, and 17.
2. Determine whether $x^2 \equiv -1 \pmod p$ is solvable for the following values of p:
    (a) $p = 23$, (b) $p = 449$, (c) $p = 17389$, and (d) $p = 2^{607} - 1$.
3. Determine whether $x^2 \equiv 2 \pmod p$ is solvable for the following values of p:
    (a) $p = 23$, (b) $p = 449$, (c) $p = 17389$, and (d) $p = 2^{607} - 1$.
4. (a) Determine whether $3x^2 + 17x + 14 \equiv 0 \pmod{13}$ is solvable by using Proposition 4.11. If so, find all solutions.
    (b) Determine whether $2x^2 + 11x - 7 \equiv 0 \pmod{11}$ is solvable. If so, find all solutions.
5. Show that $x^2 \equiv -a^2 \pmod p$ is solvable if and only if $p = 2$ or $p \equiv 1 \pmod 4$.
6. Prove Corollary 4.13.1 directly in case
    (a) a and b are quadratic residues $\pmod p$;
    (b) a is a quadratic residue and b is a quadratic nonresidue $\pmod p$.
7. Prove Corollary 4.13.1 using the observation that a is a quadratic residue $\pmod p$ if and only if ind a is even (see Exercise 4.1.21).
8. Show that if p is an odd prime and $p \mid (n^2 - 2)$ for some n, then $p \equiv \pm 1 \pmod 8$.

9. Show that $(x^2 - 11)(x^2 - 13)(x^2 - 143) \equiv 0 \pmod{m}$ is solvable for all positive integers m even though $(x^2 - 11)(x^2 - 13)(x^2 - 143) = 0$ has no integral solution.

10. Determine the following:
    (a) $\left(\frac{7}{11}\right)$; (b) $\left(\frac{11}{7}\right)$; (c) $\left(\frac{7}{13}\right)$; (d) $\left(\frac{13}{7}\right)$ (e) $\left(\frac{11}{13}\right)$; (f) $\left(\frac{13}{11}\right)$; (g) $\left(\frac{5}{13}\right)$; and (h) $\left(\frac{13}{5}\right)$.
    Any conjectures?

11. Apply Gauss' lemma to determine
    (a) $\left(\frac{3}{23}\right)$; (b) $\left(\frac{-2}{23}\right)$; and (c) $\left(\frac{5}{23}\right)$.

12. Apply Gauss' lemma to determine
    (a) $\left(\frac{-2}{31}\right)$; (b) $\left(\frac{3}{31}\right)$; and (c) $\left(\frac{5}{31}\right)$.

13. Apply Gauss' lemma to determine
    (a) $\left(\frac{5}{73}\right)$; (b) $\left(\frac{6}{73}\right)$; and (c) $\left(\frac{7}{73}\right)$

14. Determine $\left(\frac{-2}{p}\right)$ for p an odd prime.

15. Apply Gauss' lemma as in the proof of Corollary 4.14.1 to determine $\left(\frac{3}{p}\right)$ for p an odd prime. (You will have to consider p modulo 12.)

16. (a) Is $x^2 \equiv 220 \pmod{13}$ solvable?
    (b) Is $x^2 \equiv 525 \pmod{13}$ solvable?

17. Does there exist an a for which $x^2 \equiv a + k \pmod{19}$ is solvable for $1 \le k \le 5$? Explain.

18. Give an example to show that for some composite n the product of two quadratic nonresidues $(\bmod\, n)$ can be a quadratic nonresidue $(\bmod\, n)$.

19. (a) Show that if a is a quadratic residue $(\bmod\, p)$ and $ab \equiv 1 \pmod{p}$, then b is a quadratic residue $(\bmod\, p)$.
    (b) Show that the product of the quadratic residues $(\bmod\, p)$ is congruent to $1, -1 \pmod{p}$ depending on whether $p \equiv 3, 1 \pmod{4}$, respectively.
    (c) What is the situation for the product of the quadratic nonresidues?

20. Prove that there are infinitely many primes $p \equiv 7 \pmod{8}$:
    (a) Let $p_1, \ldots, p_r$ be a set of primes $\equiv 7 \pmod{8}$ and let $N = 16(\prod_{i=1}^{r} p_i)^2 - 2$. Show that $N \equiv 6 \pmod{8}$ and $2 \parallel N$ (2 exactly divides N).
    (b) Let p be an odd prime with $p \mid N$. Show that $\left(\frac{2}{p}\right) = 1$ and so $p \equiv \pm 1 \pmod{8}$.
    (c) Argue that there must be at least one $p \mid N$ with $p \equiv 7 \pmod{8}$. Conclude that there are infinitely many primes $p \equiv 7 \pmod{8}$.

4.4 The law of quadratic reciprocity and extensions

One of the high points in elementary number theory is the law of quadratic reciprocity. Both Euler (1744) and Legendre (1785) formulated the law, but neither was able to complete a satisfactory proof. Gauss (not yet 19 years old) rediscovered the law and succeeded in proving it on April 8, 1796, as he dutifully noted in his diary. He called it his "theorema fundamentale". Gauss kept returning to the law of quadratic reciprocity searching for proofs of it that would generalize to higher reciprocity laws.

In all, Gauss published six proofs and no doubt discovered several others. He was fully aware of its central importance and the role it would play in the further development of number theory. Leopold Kronecker (1823–1891) later referred to its proof as the "test of strength of Gauss' genius."

Theorem 4.15 (Law of Quadratic Reciprocity): If p and q are distinct odd primes, then $\left(\frac{p}{q}\right)\left(\frac{q}{p}\right) = (-1)^{(p-1)(q-1)/4}$.

Corollary 4.15.1: (a) $\left(\frac{p}{q}\right) = \left(\frac{q}{p}\right)$ if either $p \equiv 1 \pmod 4$ or $q \equiv 1 \pmod 4$.
(b) $\left(\frac{p}{q}\right) = -\left(\frac{q}{p}\right)$ if $p \equiv q \equiv 3 \pmod 4$.

Proof of Corollary: The proof is left as Exercise 4.4.7.

Example 4.8: Determine the value of the Legendre symbol $\left(\frac{23}{101}\right)$.
Solution: By Corollary 4.15.1, $\left(\frac{23}{101}\right) = \left(\frac{101}{23}\right)$ since $101 \equiv 1 \pmod 4$. But $101 \equiv 9 \pmod{23}$ and 9 is a perfect square. Hence by Proposition 4.13, $\left(\frac{23}{101}\right) = \left(\frac{9}{23}\right) = 1$.

Our proof of Theorem 4.15 mimics Gauss' third proof. We begin with a lemma. Note that the condition that a be odd below does not limit the lemma's generality. If a is even, then simply replace a by $p + a$.

Lemma 4.15.1: If p is an odd prime and a is odd with $p \nmid a$, then $\left(\frac{a}{p}\right) = (-1)^t$ where $t = [a/p] + [2a/p] + \cdots + [(p-1)a/2p]$ and $[x]$ denotes the greatest integer less than or equal to x.

Proof of Lemma: As in Gauss' lemma, reduce the integers $ka \pmod p$ with $1 \le k \le (p-1)/2$ to those of smallest absolute value. Let r_k denote the negative residues $(1 \le k \le n)$ and let s_k denote the positive residues $(1 \le k \le m)$. Here $m + n = (p-1)/2$.

Note that when $ka \equiv s_k \pmod p$, by the division algorithm we can write

$$ka = p \cdot [ka/p] + s_k$$

On the other hand, when $ka \equiv r_k \pmod p$ we can write

$$ka = p \cdot [ka/p] + (p + r_k)$$

It follows that

$$a + 2a + \cdots + (p-1)a/2 = \sum_{k=1}^{(p-1)/2} p \cdot [ka/p] + \sum_{k=1}^{n} (p + r_k) + \sum_{k=1}^{m} s_k. \qquad (4.12)$$

In addition,

$$1 + 2 + \cdots + (p-1)/2 = \sum_{k=1}^{n} -r_k + \sum_{k=1}^{m} s_k \qquad (4.13)$$

Subtracting eq. (4.13) from eq. (4.12),

$$(a-1) \sum_{k=1}^{(p-1)/2} k = pt + \sum_{k=1}^{n} (p + 2r_k)$$

Thus

$$(a-1) \sum_{k=1}^{(p-1)/2} k = p(t+n) + 2 \sum_{k=1}^{n} r_k$$

But a is odd. Hence the left-side of the equation above is even. It follows that $p(t+n)$ is even. But p odd implies that $t \equiv n \pmod 2$. By Gauss' lemma, $\left(\frac{a}{p}\right) = (-1)^t$. ∎

Proof of Theorem 4.15: Consider the rectangle in Figure 4.1 with vertices at $(0,0)$, $(\frac{p}{2}, 0)$, $(\frac{q}{2}, 0)$, and $(\frac{p}{2}, \frac{q}{2})$.

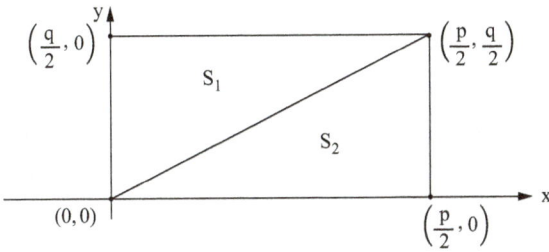

Figure 4.1: Rectangle of dimension $p/2 \times q/2$ divided into upper and lower triangles S_1 and S_2.

The diagonal shown does not pass through any lattice point (x,y) with $x,y \in \mathbb{Z}$ since otherwise $py = qx$ which is impossible. So the diagonal separates the lattice points (x,y) interior to the rectangle into two disjoint sets S_1 and S_2 depending on whether or not $py > qx$ or $py < qx$, respectively. The total number of lattice points within the rectangle is $\frac{p-1}{2} \cdot \frac{q-1}{2} = \frac{(p-1)(q-1)}{4}$. But the lattice points in S_1 are those for which $1 \le x < \frac{py}{q}, 1 \le y \le \frac{q-1}{2}$. Hence the number of lattice points in S_1 is precisely $\sum_{y=1}^{(q-1)/2} [py/q]$. Similarly, the lattice points in S_2 are those for which $1 \le x \le \frac{p-1}{2}, 1 \le y < \frac{qx}{p}$. Hence the number of lattice points in S_2 is $\sum_{x=1}^{(p-1)/2} [qx/p]$.

It follows that

$$\sum_{k=1}^{(q-1)/2} [pk/q] + \sum_{k=1}^{(p-1)/2} [qk/p] = (p-1)(q-1)/4$$

Now apply Lemma 4.15.1 and $\left(\frac{p}{q}\right)\left(\frac{q}{p}\right) = (-1)^{(p-1)(q-1)/4}$. ∎

Example 4.9: Determine whether $x^2 \equiv 356 \pmod{1993}$ is solvable.

Solution: The number 1993 is prime and $356 = 4 \times 89$. Thus, it suffices to find $\left(\frac{89}{1993}\right)$. Since $1993 \equiv 1 \pmod 4$, the quadratic reciprocity law ensures $\left(\frac{89}{1993}\right) = \left(\frac{1993}{89}\right)$. But $\left(\frac{1993}{89}\right) = \left(\frac{35}{89}\right)$ since $1993 \equiv 35 \pmod{89}$. Continuing, $\left(\frac{35}{89}\right) = \left(\frac{5}{89}\right)\left(\frac{7}{89}\right) = \left(\frac{89}{5}\right)\left(\frac{89}{7}\right) = \left(\frac{4}{5}\right)\left(\frac{5}{7}\right) = \left(\frac{5}{7}\right) = \left(\frac{7}{5}\right) = \left(\frac{2}{5}\right) = -1$ since 2 is a quadratic nonresidue $\pmod 5$. Hence, $x^2 \equiv 356 \pmod{1993}$ is not solvable.

Example 4.10: Determine all primes p for which $x^2 \equiv 5 \pmod p$ is solvable.

Solution: Certainly $x^2 \equiv 5 \pmod 2$ is solvable as is $x^2 \equiv 5 \pmod 5$; so assume p is odd and $p \neq 5$. Then $\left(\frac{5}{p}\right) = \left(\frac{p}{5}\right)$ since $5 \equiv 1 \pmod 4$. But then $\left(\frac{p}{5}\right) = 1$ if $p \equiv \pm 1 \pmod 5$ and $\left(\frac{p}{5}\right) = -1$ if $p \equiv \pm 3 \pmod 5$ by Proposition 4.13(b). Hence $x^2 \equiv 5 \pmod p$ is solvable if and only if $p = 2, 5$, or $p \equiv \pm 1 \pmod 5$.

The law of quadratic reciprocity together with Corollary 4.14.1 and Proposition 4.13(c) enables us to readily determine the Legendre symbol $\left(\frac{a}{p}\right)$ in many situations. The key at each step of the calculation is to reduce the numerator (modulo the denominator) and then to factor the numerator before applying Theorem 4.15. One slight limitation is that the denominator must always be prime. In order to deal with the situation where the denominator may not be prime, we must extend the definition of the Legendre symbol itself.

Definition 4.5: Let $n > 1$ be an odd integer with $\gcd(a, n) = 1$. If $n = p_1 p_2 \cdots p_t$ is a product of (not necessarily distinct) primes, then define the *Jacobi symbol* by $\left(\frac{a}{n}\right) = \left(\frac{a}{p_1}\right)\left(\frac{a}{p_2}\right) \cdots \left(\frac{a}{p_t}\right)$ where all factors on the right-hand side are Legendre symbols.

We now state for the Jacobi symbol many of the analogs of our results about the Legendre symbol.

Proposition 4.16: Let m and n be odd integers greater than 1 and let a and b be relatively prime to both m and n:

(a) If $a \equiv a' \pmod n$, then $\left(\frac{a}{n}\right) = \left(\frac{a'}{n}\right)$.

(b) $\left(\frac{ab}{n}\right) = \left(\frac{a}{n}\right)\left(\frac{b}{n}\right)$.

(c) $\left(\frac{a}{m}\right)\left(\frac{a}{n}\right) = \left(\frac{a}{mn}\right)$.

(d) If $x^2 \equiv a \pmod n$ is solvable, then $\left(\frac{a}{n}\right) = 1$.

Proof: The proof is left for Exercise 4.4.8.

The contrapositive of Proposition 4.16(d) is quite useful. For example, $\left(\frac{2}{1155}\right) = \left(\frac{2}{3}\right)\left(\frac{2}{5}\right)\left(\frac{2}{7}\right)\left(\frac{2}{11}\right)$ by Proposition 4.16(c) since $1155 = 3 \times 5 \times 7 \times 11$. Hence by Corollary 4.14.1, $\left(\frac{2}{1155}\right) = (-1)(-1)(+1)(-1) = -1$. So by Proposition 4.16(d), $x^2 \equiv 2 \pmod{1155}$ is not solvable.

We now state and prove the quadratic reciprocity law for the Jacobi symbol.

Proposition 4.17: Let m and n be odd integers greater than 1 and suppose $\gcd(m, n) = 1$:

(a) $\left(\frac{-1}{n}\right) = (-1)^{(n-1)/2}$

(b) $\left(\frac{2}{n}\right) = (-1)^{(n^2-1)/8}$

(c) $\left(\frac{m}{n}\right)\left(\frac{n}{m}\right) = (-1)^{(n-1)(m-1)/4}$

Corollary 4.17.1: (a) $\left(\frac{n}{m}\right) = \left(\frac{m}{n}\right)$ if either $m \equiv 1 \pmod 4$ or $n \equiv 1 \pmod 4$.

(b) $\left(\frac{n}{m}\right) = -\left(\frac{m}{n}\right)$ if $m \equiv n \equiv 3 \pmod 4$.

Proof: Let $m = q_1 q_2 \cdots q_s$ and let $n = p_1 p_2 \cdots p_t$.

(a) By Proposition 4.13(c), $\left(\frac{-1}{n}\right) = \prod_{i=1}^{t}\left(\frac{-1}{p_i}\right) = \prod_{i=1}^{t}(-1)^{(p-1)/2}$.

Let $r = \sum_{i=1}^{t}(p_i - 1)/2$. Then $\left(\frac{-1}{n}\right) = (-1)^r$.
If $a \equiv b \equiv 1 \pmod 4$, then $(a-1)/2$, $(b-1)/2$, and $(ab-1)/2$ are all even.
If $a \equiv b \equiv 3 \pmod 4$, then $(a-1)/2$ and $(b-1)/2$ are odd and $(ab-1)/2$ is even.
In either case, $\frac{a-1}{2} + \frac{b-1}{2} \equiv \frac{ab-1}{2} \pmod 2$.
If $a \not\equiv b \pmod 4$, then $(a-1)/2 + (b-1)/2$ and $(ab-1)/2$ are both odd. So in all
cases, $(a-1)/2 + (b-1)/2 \equiv \frac{ab-1}{2} \pmod 2$.
But $p_1, \ldots, p_t$ are odd and thus by induction, $r \equiv \frac{1}{2}\left(-1 + \prod_{i=1}^{t} p_i\right) \pmod 2$.
Since $n = \prod_{i=1}^{t} p_i$, $\left(\frac{-1}{n}\right) = (-1)^{(n-1)/2}$.

(b) By Corollary 4.14.1, $\left(\frac{2}{n}\right) = \prod_{i=1}^{t}\left(\frac{2}{p_i}\right) = \prod_{i=1}^{t}(-1)^{(p_i^2-1)/8}$.
Let $u = \sum_{i=1}^{t}(p_i^2 - 1)/8$. Then $\left(\frac{2}{n}\right) = (-1)^u$.
If $a \equiv \pm 1 \pmod 8$ and $b \equiv \pm 1 \pmod 8$, then $(a^2 - 1)/8$, $(b^2 - 1)/8$, and $\left((ab)^2 - 1\right)/8$
are all even.
If $a \equiv \pm 5 \pmod 8$ and $b \equiv \pm 5 \pmod 8$, then $(a^2 - 1)/8$ and $(b^2 - 1)/8$ are odd,
whereas $((ab)^2 - 1)/8$ is even.
In either case, $(a^2 - 1)/8 + (b^2 - 1)/8 \equiv \frac{((ab)^2 - 1)}{8} \pmod 2$.
If $a \not\equiv \pm b \pmod 8$, then $(a^2 - 1)/8 + (b^2 - 1)/8$ and $((ab)^2 - 1)/8$ are both odd. In all
instances, $(a^2 - 1)/8 + (b^2 - 1)/8 \equiv \frac{((ab)^2 - 1)}{8} \pmod 2$. By induction,
$u \equiv \frac{1}{8}\left(-1 + \prod_{i=1}^{t} p_i^2\right) \pmod 2$. Hence $u \equiv \frac{(n^2-1)}{8} \pmod 2$ and so

$$\left(\frac{2}{n}\right) = (-1)^{(n^2-1)/8}.$$

(c) By Proposition 4.16(c), $\left(\frac{m}{n}\right) = \prod_{i=1}^{t}\left(\frac{m}{p_i}\right) = \prod_{i=1}^{t}\prod_{j=1}^{s}\left(\frac{q_j}{p_i}\right)$.

Hence $\left(\frac{m}{n}\right) = \prod_{i=1}^{t}\prod_{j=1}^{s}\left(\frac{p_i}{q_j}\right)(-1)^{(p_i-1)(q_j-1)/4}$ by Proposition 4.15.

But by Proposition 4.16(c) again,

$\prod_{j=1}^{s}\left(\frac{p_i}{q_j}\right)(-1)^{(p_i-1)(q_j-1)/4} = \left(\frac{p_i}{m}\right)(-1)^r$ where $r = \frac{(p_i-1)}{4}\sum_{j=1}^{s}(q_j - 1)$.

By the definition of the Jacobi symbol,

$$\left(\tfrac{m}{n}\right) = \left(\tfrac{n}{m}\right)(-1)^u \text{ where } u = \tfrac{1}{4}\sum_{i=1}^{t}\sum_{j=1}^{s}(p_i-1)(q_j-1).$$

But $u = \sum_{i=1}^{t}(p_i-1)/2 \cdot \sum_{j=1}^{s}(q_j-1)/2$. Furthermore, $\sum_{i=1}^{t}(p_i-1)/2 \equiv \tfrac{n-1}{2} \pmod 2$ and $\sum_{j=1}^{s}(q_j-1)/2 \equiv \tfrac{m-1}{2} \pmod 2$ as in the proof of part (a). Therefore, $\left(\tfrac{m}{n}\right) = \left(\tfrac{n}{m}\right)(-1)^{(n-1)(m-1)/4}$. Equivalently, $\left(\tfrac{m}{n}\right)\left(\tfrac{n}{m}\right) = (-1)^{(n-1)(m-1)/4}$. ∎

An important point is that $x^2 \equiv a \pmod n$ is not necessarily solvable just because $\left(\tfrac{a}{n}\right) = 1$. In fact, as long as there is a $p \mid n$ with $\left(\tfrac{a}{p}\right) = -1$, then $x^2 \equiv a \pmod n$ has no solutions. For example, $x^2 \equiv 2 \pmod{15}$ is not solvable, yet $\left(\tfrac{2}{15}\right) = \left(\tfrac{2}{3}\right)\left(\tfrac{2}{5}\right) = (-1)(-1) = 1$. However, if p is prime, then the Legendre symbol $\left(\tfrac{a}{p}\right)$ is identical to the Jacobi symbol $\left(\tfrac{a}{p}\right)$ and Proposition 4.17 is a great computational aid.

Example 4.11: Evaluate $\left(\tfrac{315}{1997}\right)$.

Solution: By Proposition 4.17, $\left(\tfrac{315}{1997}\right) = \left(\tfrac{1997}{315}\right) = \left(\tfrac{107}{315}\right) = -\left(\tfrac{315}{107}\right) = -\left(\tfrac{-6}{107}\right) = -\left(\tfrac{-1}{107}\right)\left(\tfrac{2}{107}\right)\left(\tfrac{3}{107}\right) = -1(-1)(+1)\left(\tfrac{3}{107}\right) = \left(\tfrac{3}{107}\right) = -\left(\tfrac{107}{3}\right) = -\left(\tfrac{2}{3}\right) = -(-1) = 1$. Since 1997 is prime, $\left(\tfrac{315}{1997}\right)$ is a Legendre symbol, and hence $x^2 \equiv 315 \pmod{1997}$ is solvable.

Much research has been done dealing with the distribution of quadratic residues and nonresidues $\pmod p$. One result that bears at least superficial resemblance to Artin's primitive root conjecture is the following concerning quadratic nonresidues. Our proof is largely based on that of Ireland and Rosen [63].

Proposition 4.18: Let a be a nonsquare integer. Then there are infinitely many primes p such that $\left(\tfrac{a}{p}\right) = -1$.

In fact, it can be shown analytically that for a given nonsquare integer a, a is a quadratic residue for half of the primes and a quadratic nonresidue for the remaining half (excepting those primes for which $p \mid a$).

Proof: If $a = r^2 s$ then $\left(\tfrac{a}{p}\right) = \left(\tfrac{r^2}{p}\right)\left(\tfrac{s}{p}\right) = \left(\tfrac{s}{p}\right)$ by formula (4.8). Hence, we may assume that a is square-free and that $a = 2^c p_1 \cdots p_t$ where the p_i are distinct odd primes and $c = 0$ or 1.

We distinguish two cases:

(i) If $a = 2$ then let $q_1, \ldots, q_k$ be a non-empty finite set of primes larger than 3 for which $\left(\tfrac{2}{q_i}\right) = -1$. (Note that $\left(\tfrac{2}{5}\right) = -1$ and hence our argument is not vacuous.) Let $b = 8q_1 \cdots q_k + 3$. Note that b is not divisible by 3 or any of the q_i. But $b \equiv 3 \pmod 8$ and hence $\left(\tfrac{2}{b}\right) = -1$ by Proposition 4.17(b).

Now let $b = r_1 \cdots r_s$ where the r_i are prime. Since $\left(\tfrac{2}{b}\right) = \left(\tfrac{2}{r_1}\right) \cdots \left(\tfrac{2}{r_s}\right)$, there is at least one value of i for which $\left(\tfrac{2}{r_i}\right) = -1$. Hence there is some prime r_i not among the set $\{3, q_1, \ldots, q_k\}$ for which $\left(\tfrac{2}{r_i}\right) = -1$. So no finite list of primes p for which $\left(\tfrac{2}{p}\right) = -1$ can be complete.

(ii) If $a \neq 2$, then $a = 2^c p_1 \cdots p_t$ where $c = 0$ or 1 and $t \geq 1$. Let $q_1, \ldots, q_k$ be a finite set of primes distinct from the p_j. (Unlike case (i), we are not requiring any other conditions on the q_i.) Let n be a nonresidue $(\bmod\, p_t)$ (n is guaranteed to exist by Corollary 4.12.1). By the Chinese remainder theorem we can find a simultaneous solution $x = b$ to the following:

$$\begin{cases} x \equiv 1 \ (\bmod\, q_i) \text{ for } i = 1, \ldots, k \\ x \equiv 1 \ (\bmod\, p_j) \text{ for } j = 1, \ldots, t-1 \\ x \equiv n \ (\bmod\, p_t) \\ x \equiv 1 \ (\bmod\, 8) \end{cases}$$

Since $b \equiv 1 \ (\bmod\, 8)$, b is odd. Let $b = r_1 \cdots r_s$ be its prime decomposition.

$b \equiv 1 \ (\bmod\, 8)$ implies $\left(\frac{2}{b}\right) = 1$ and $\left(\frac{p_j}{b}\right) = \left(\frac{b}{p_j}\right)$ by Proposition 4.17. Hence

$$\left(\frac{a}{b}\right) = \left(\frac{2}{b}\right)^c \left(\frac{p_1}{b}\right) \cdots \left(\frac{p_t}{b}\right) = \left(\frac{b}{p_1}\right) \cdots \left(\frac{b}{p_t}\right) = \left(\frac{1}{p_1}\right) \cdots \left(\frac{1}{p_t - 1}\right) \left(\frac{n}{p_t}\right) = -1$$

since n is a quadratic nonresidue $(\bmod\, p_t)$.

By the definition of the Jacobi symbol, though, $\left(\frac{a}{b}\right) = \left(\frac{a}{r_1}\right) \cdots \left(\frac{a}{r_s}\right)$. So $\left(\frac{a}{r_i}\right) = -1$ for at least one r_i with $1 \leq i \leq s$.

Since $b \equiv 1 \ (\bmod\, q_i)$ for $i = 1, \ldots, k$, it must be the case that r_i is not among the set $\{q_1, \ldots, q_k\}$. Since $q_1, \ldots, q_k$ are arbitrary, it follows that no finite set contains all primes p for which $\left(\frac{a}{p}\right) = -1$. This concludes our proof. ∎

We complete this chapter with a few brief historical comments. Just as the Jacobi symbol generalizes the Legendre symbol, the Jacobi symbol itself admits many vast generalizations. One such extension is the Kronecker symbol, named after Leopold Kronecker, where we allow the denominator to be even (see Exercise 4.4.16.) Gauss' work on quadratic reciprocity bore fruit in his preliminary investigations of what we now call algebraic number theory and the discovery of the biquadratic reciprocity law. The first published proof of the biquadratic reciprocity law was furnished by Gauss' student Ferdinand Gotthold Eisenstein (1823–1852) in 1844. Eisenstein also proved the cubic reciprocity law about the same time. The subsequent work of Dirichlet, Dedekind, Kummer, Hurwitz, Hilbert, Takagi, Artin, and Furtwängler has led to a beautiful edifice known as class field theory. In this setting, the quadratic reciprocity law and Gauss' keen insight can be more fully appreciated.

Exercise 4.4

1. Evaluate the following Legendre symbols:
   (a) $\left(\frac{21}{79}\right)$, (b) $\left(\frac{-5}{87}\right)$, (c) $\left(\frac{71}{547}\right)$, and (d) $\left(\frac{81}{857}\right)$.

2. Evaluate the following Legendre symbols:
   (a) $\left(\frac{1776}{1511}\right)$, (b) $\left(\frac{-65}{1949}\right)$, (c) $\left(\frac{103}{1999}\right)$, and (d) $\left(\frac{15}{10007}\right)$.

3. Determine all odd primes p for which $x^2 \equiv a \pmod{p}$ is solvable:
   a. $a = -3$ $(p \neq 3)$
   b. $a = -5$ $(p \neq 5)$
   c. $a = 7$ $(p \neq 7)$
   d. $a = 11$ $(p \neq 11)$
   e. $a = 6$ $(p \neq 3)$

4. Determine whether $5x^2 + 3x + 29 \equiv 0 \pmod{43}$ is solvable. If so, find all solutions (mod 43).

5. Determine whether the following congruences are solvable:
   (a) $x^2 \equiv 41 \pmod{85}$, (b) $x^2 \equiv 23 \pmod{105}$,
   (c) $x^2 \equiv 31 \pmod{91}$, (d) $x^2 \equiv 11 \pmod{119}$

6. Find x for which $x^2 \equiv 23 \pmod{101}$.

7. Prove Corollary 4.15.1.

8. Prove Proposition 4.16.

9. Euler conjectured that if is any natural number and p and q are primes with $p = 4ns + r$ and $q = 4nt + r'$ where $0 < r < 4n$ and r' is either r or $4n - r$, then the quadratic character of $n \pmod{p}$ and $n \pmod{q}$ are the same. Show that this is equivalent to the law of quadratic reciprocity.

10. What is the smallest prime p for which $1, 2, 3$, and 4 are all quadratic residues $\pmod{p}$?

11. (a) What is the smallest prime p for which $2, 3$, and 5 are all quadratic nonresidues $\pmod{p}$? What about $2, 3, 5$, and 7?
    (b) Is there a prime p for which $2, 3, 5$, and 6 are all quadratic nonresidues?

12. Establish the following for primes p with $p \equiv 1 \pmod{4}$:
    (a) $\sum_{n=1}^{p-1} n = \frac{p(p-1)}{4}$ where we sum over quadratic residues only.
    (b) $\sum_{n=1}^{p-1} n = \frac{p(p-1)}{4}$ where we sum over quadratic nonresidues only.
    (c) $\sum_{n=1}^{p-1} n \left(\frac{n}{p}\right) = 0$.

13. Establish the following for all primes $p > 3$:
    (a) $\sum_{n=1}^{p-1} n \equiv 0 \pmod{p}$ where we sum over quadratic residues only.
    (b) $\sum_{n=1}^{p-1} n \equiv 0 \pmod{p}$ where we sum over quadratic nonresidues only.

14. Evaluate the following Jacobi symbols:
    (a) $\left(\frac{21}{91}\right)$, (b) $\left(\frac{210}{95}\right)$, (c) $\left(\frac{-35}{187}\right)$, and (d) $\left(\frac{73}{325}\right)$.

15. Evaluate the following Jacobi symbols:
    (a) $\left(\frac{147}{2069}\right)$, (b) $\left(\frac{153}{6141}\right)$, (c) $\left(\frac{1331}{12125}\right)$, and (d) $\left(\frac{1003}{371293}\right)$.

16. Define the Kronecker symbol $\left(\frac{a}{n}\right)$ as follows:
    (i) $\left(\frac{a}{n}\right) = 0$ if $\gcd(a, b) > 1$. Else
    (ii) $\left(\frac{a}{n}\right)$ is the Jacobi symbol $\left(\frac{a}{n}\right)$ if n is odd.
    (iii) $\left(\frac{a}{n}\right) = \left(\frac{a}{2}\right)^s \left(\frac{a}{c}\right)$ where $n = 2^s c$ (c odd) and $a \equiv 1 \pmod 4$.
    Here $\left(\frac{a}{2}\right) = \begin{cases} 1 \text{ if } a \equiv 1 \pmod 8 \\ -1 \text{ if } a \equiv 5 \pmod 8 \end{cases}$

Let m, n, k be positive integers and $a \equiv 1 \pmod 4$. Show the following:
    (a) $\left(\frac{a}{2}\right) = \left(\frac{2}{a}\right)$.
    (b) $\left(\frac{a}{2}\right)^k = \left(\frac{a}{2^k}\right)$.
    (c) $\left(\frac{a}{m}\right)\left(\frac{a}{n}\right) = \left(\frac{a}{mn}\right)$.

Chapter 5
Sums of squares

5.1 Fundamentals of Diophantine equations

A *Diophantine equation* is an equation with integral coefficients of the form

$$f(x_1, \ldots, x_m) = n \tag{5.1}$$

for a given function f in which integral solutions are sought for the indeterminates $x_1, \ldots, x_m$.

In our book, f is a polynomial, but there are examples of Diophantine equations in which f involves various exponential functions. Oftentimes, n is given and we seek information concerning the form and number of solutions for $x_1, \ldots, x_m$. In particular, does the given equation have any integral solutions? Recall that we have already determined when a linear Diophantine equation $ax + by = n$ is solvable. By Corollary 2.1.2, it is solvable precisely when $\gcd(a, b)$ divides n. In this case, there were infinitely many solutions and we were able to specify those solutions.

Our study of congruences often proves quite useful in showing a given Diophantine equation is not solvable. If it can be shown that

$$f(x_1, \ldots, x_m) \equiv n \pmod{t} \tag{5.2}$$

is unsolvable for some t, then certainly eq. (5.1) has no solution. For example, $x_1^2 + x_2^2 + x_3^2 \equiv 7 \pmod 8$ has no solutions since all squares are congruent to 0, 1, or 4 (mod 8) and hence $x_1^2 + x_2^2 + x_3^2$ is congruent to 0, 1, 2, 3, 4, 5, or 6 (mod 8). Since $247 \equiv 7 \pmod 8$, it follows that $x_1^2 + x_2^2 + x_3^2 = 247$ is insoluble.

In the other direction, the conditions under which solutions to eq. (5.2) for $t = p^m$ for all primes p and natural numbers m coupled with a real n-tuple $(x_1, \ldots, x_m)$ solution to eq. (5.1) guarantee a solution to the Diophantine equation (5.1) depends deeply on the form of f itself (e.g., see Exercise 4.3.9). The general study of such questions relates to what is known as the Hasse principle named after Helmut Hasse (1898–1979) and is reserved for more specialized works. However, the notion that all the "local" solutions team up to ensure that a "global" solution always works with quadratic forms as was proved by Hasse in 1923 and hence with the equations we will study in this chapter.

If a given Diophantine equation is solvable, then are there infinitely many solutions or just finitely many? If there are finitely many, can we list all solutions? If not, can we at least determine how many different solutions there are? That is, how many ordered m-tuples $(x_1, \ldots, x_m)$ satisfy eq. (5.1)? A related question is the following: if we do not distinguish between solutions which differ only in the order or sign of the terms, then how many essentially distinct solutions are there? That is, how many ordered m-tuples $(x_1, \ldots, x_m)$ with $0 \le x_1 \le \cdots \le x_m$ satisfy eq. (5.1)?

https://doi.org/10.1515/9783111579283-005

For example, $x_1^2 + x_2^2 = 13$ has eight different solutions, namely $(2, 3)$, $(2, -3)$, $(-2, 3)$, $(-2, -3)$, $(3, 2)$, $(3, -2)$, $(-3, 2)$, and $(-3, -2)$. However, $x_1^2 + x_2^2 = 13$ has just one essentially distinct solution $(2, 3)$. We will take up some examples of questions of this sort in Chapter 10.

In other situations, we ask for which n is eq. (5.1) solvable?

That is the sort of question we will commonly consider in this chapter. For example, in Section 5.2, we determine which positive integers n can be expressed as the sum of two squares. We will also determine all Pythagorean triplets, that is, solutions to the Diophantine equation

$$x^2 + y^2 = z^2 \tag{5.3}$$

The study of these questions will lead naturally to other related questions concerning the representation of integers as sums of squares. In particular, in Section 5.3, we determine which squares can be expressed as the sum of three squares. In Section 5.4, it is proven that all natural numbers can be expressed as the sum of at most four squares, the proof of which benefitted from the painstaking labor of several notable mathematicians. In Section 5.5, Legendre's equation

$$ax^2 + by^2 = cz^2 \tag{5.4}$$

is analyzed utilizing some of our previous work on quadratic residues. In addition, in this chapter we will apply our solution of eq. (5.3) to show that the Fermat equation

$$x^4 + y^4 = z^4 \tag{5.5}$$

has no nontrivial solutions.

We conclude this section with a brief survey of a more general nature. Rather than considering a particular Diophantine equation, mathematicians sometimes consider classes of Diophantine equations. For example, Fermat's last theorem states that the class of all Diophantine equations of the form $x^n + y^n = z^n$ is unsolvable for $n \geq 3$.

In 1772, Euler conjectured that all equations of the form

$$x_1^n + x_2^n + \cdots + x_m^n = z^n$$

had no nontrivial solutions provided that $1 < m < n$. Notice that this class of equations subsumes the Fermat equations above. Euler's conjecture was disproved by L.J. Lander and T.R. Parkin in 1966 after an extensive computer search. Their counterexample for $n = 5$ is

$$27^5 + 84^5 + 110^5 + 133^5 = 144^5$$

It is the smallest such example in the sense that any other with $n = 5$ must have $z > 144$.

Several number theorists turned their attention to the case $n = 4$ following Lander and Parkin's success. In 1988, utilizing the theory of elliptic curves, Noam Elkies found such a counter example:

$$2,682,440^4 + 15,365,639^4 + 18,796,760^4 = 20,615,673^4$$

His success spurred others on to find the smallest counter example for $n = 4$. By an extensive computer search on several Connection Machines, Roger Frye (1988) found the following:

$$95,800^4 + 217,519^4 + 414,560^4 = 422,481^4$$

The modern theory of Diophantine equations goes well beyond what we can accomplish here. Generally speaking, Diophantine equations give rise to curves in projective two-space (ordinary "affine" two-space together with a "line at infinity") and can be classified by an invariant of the curve known as its genus. The genus is always a non-negative integer. Consider the equation $f(x, y, z) = 0$, where f is a homogeneous polynomial (all terms of equal degree) of degree n. Points (x, y, z) where all three first partial derivatives vanish are called singular points. If N is the number of singular points, then the genus g can be defined as $g = \frac{1}{2}(n-1)(n-2) - N$. For example, if f is of degree 2, then its genus $g = 0$ and N must be zero.

Diophantine equations are often grouped into three main categories depending on whether their genus is 0, 1, or 2 or more. The equations we consider in this chapter along with Pell's equation considered in Chapter 6 have genus 0. In general, they have infinitely many integral solutions.

Curves of genus 1 passing through at least one rational point (all three coordinates rational) are called elliptic curves. They are of the form $y^2 z = x^3 + axz^2 + bz^3$, where a and b are integers. In this case, in 1929, C. Siegel (1896–1981) proved that there are only finitely many integral solutions (but usually infinitely many rational solutions). For example, if $z = 1$, $a = 0$, and $b = -2$, we get the equation $x^3 = y^2 + 2$ considered by Bachet and Fermat. It so happens that its only integral solution is $(x, y) = (3, \pm 5)$.

It was conjectured by L.J. Mordell (1888–1972) in 1922 that any curve of genus greater than or equal to 2 had at most finitely many rational solutions. This deep theorem was established by Gerd Faltings in 1983 and remains one of the key results in this area. It follows directly that Fermat's equation $x^n + y^n = z^n$ can have only finitely many solutions for any given $n \geq 3$.

Despite many successes in solving Diophantine equations, mathematicians still have plenty of work to keep them occupied for the foreseeable future.

Exercise 5.1

1. (a) Show that $x^2 + 2y^2 = 805$ has no integral solutions. (Hint: work mod 8.)
   (b) Show that $x^2 + 3y^2 = 805$ has no integral solutions.
2. (a) Show that $3x^2 + 7y^2 = 2001$ has no integral solutions.
   (b) Show that $3x^2 - 5y^2 = 10001$ has no integral solutions.
3. (a) Find all solutions to the Diophantine equation $3x^2 = 15x - 18$.
   (b) Find all solutions to the Diophantine equation $x^5 - 16x = 0$.
4. (a) Find two essentially distinct solutions to $x^2 + y^2 = 65$ with $\gcd(x, y) = 1$.
   (b) Find two essentially distinct solutions to $x^2 + y^2 + z^2 = 110$.
5. List all primes less than 100 which can be expressed as the sum of two squares. Can you make a conjecture based on your list?
6. Find three consecutive integers that are expressible as the sum of two squares. Can you find four consecutive integers with this property?
7. List all primes less than 100 which can be expressed as $x^2 + 2y^2$. Any conjectures?
8. Infinitely many triangular numbers are squares.
   (a) Find a solution to $2m^2 + 1 = n^2$ with $m, n \in N$. (This is an example of Pell's equation, treated in Section 6.5.).
   (b) If (m, n) is a solution to $2m^2 + 1 = n^2$, then show that the triangular number $1 + 2 + \cdots + (n^2 - 1)$ is a perfect square.
   (c) Show that if $2m^2 + 1 = n^2$, then $2(2mn)^2 + 1$ is a perfect square. Conclude that infinitely many triangular numbers are squares.
   (d) Construct three squares larger than 1 which are triangular numbers. (Euler also proved that no triangular number larger than 1 is either a cube or a biquadrate.)
9. List all integers less than 100 that can be expressed as the sum of 5 nonzero squares. Any conjectures?
10. List all squares less than 200 that can be expressed as the sum of 3 nonzero squares.
11. (a) What is the smallest positive integer that requires the sum of four nonzero squares?
    (b) What is the smallest positive integer that requires the sum of nine nonzero cubes?
12. (a) Let p and q be odd primes. Show that the Diophantine equation $x^2 = py + q$ is solvable if and only if $(q/p) = 1$.
    (b) Solve $x^2 = 13y + 29$.
    (c) Solve $x^2 = 17y + 43$.
    (d) Show that $x^2 + 2x = 11y + 1$ is unsolvable.
13. Show that between any two positive cubes lies a perfect square.
14. Show that for every n there are infinitely many solutions to the Diophantine equation $x^n + y^n = z^{n+1}$.

15. Let $S = \{1, 3, 8, 120\}$. Show that 1 plus the product of any two elements of S is a perfect square. (Let me know if you can find a similar set with five natural numbers!)

16. Recall the conjecture of Paul Erdös and Ernst Straus which states that the equation $\frac{4}{n} = \frac{1}{x} + \frac{1}{y} + \frac{1}{z}$ is solvable for all $n > 1$ (Exercise 1.1.3). Show that if n is prime, then either one or two of x, y, and z is divisible by n.

17. (a) Verify that $3^3 + 4^3 + 5^3 = 6^3$.

   (b) Verify that $30^4 + 120^4 + 272^4 + 315^4 = 353^4$ (R. Norrie, 1911). (This was the first example given of the sum of four biquadrates equaling a biquadrate.)

   (c) Verify that $167^3 + 436^3 = 228^3 + 423^3 = 255^3 + 414^3$ (J. Leech, 1967).

18. Find all values of $n \leq 10$ for which $n! + 1$ is a perfect square. It is unknown if there are any larger values of n with this property.

19. (a) Prove that the product of two consecutive positive integers cannot be a perfect square.

   (b) Prove that the product of three consecutive positive integers cannot be a perfect square. (Erdös has shown that the product of any k consecutive positive integers with $k > 1$ cannot be a perfect square.)

5.2 Sums of two squares

Consider the Diophantine equation

$$x^2 + y^2 = n \tag{5.6}$$

Many mathematicians dating back to Diophantus himself discussed which integers n could be expressed as the sum of two integral squares. Many conjectures proved to be erroneous. However, in a letter dated December 9, 1632, the Dutch mathematician A. Girard (1595–1632) stated he had determined that all primes of the form $4n + 1$, all squares, all products of such primes and squares, and the double of all of the previous numbers were expressible as the sum of two squares. (Notice that we allow the square 0.) In a letter to Mersenne dated December 25, 1640, Fermat rediscovered Girard's result and added many observations concerning the number of such representations. He also claimed to have an irrefutable proof, but did not include any details. After some considerable effort, the first full proof of Girard's result was published by Euler in 1754. We proceed to a discussion of it here.

The product of two integers, each the sum of two squares, is itself expressible as the sum of two squares. This follows from the following beautiful identity known and effectively used by Leonardo of Pisa (1225):

$$\left(x_1^2 + y_1^2\right)\left(x_2^2 + y_2^2\right) = \left(x_1 x_2 + y_1 y_2\right)^2 + \left(x_1 y_2 - x_2 y_1\right)^2 \tag{5.7}$$

(In fact, the identity was explicitly stated by Abu Jafar al-Khazin in 950.) In modern terminology, eq. (5.7) expresses the fact that the product of the norms of two Gaussian integers is the norm of the product. A *Gaussian integer* is a complex number of the form $a + bi$, where a and b are (rational) integers and $i^2 = -1$. The norm of $a + bi$ is the product $(a + bi)(a - bi) = a^2 + b^2$. (The Gaussian integers used to derive eq. (5.7) are precisely $x_1 - iy_1$ and $x_2 + iy_2$, respectively.) Similarly, if we take the norms of $x_1 + iy_1$ and $x_2 + iy_2$, then we obtain the companion equation

$$\left(x_1^2 + y_1^2\right)\left(x_2^2 + y_2^2\right) = (x_1 x_2 - y_1 y_2)^2 + (x_1 y_2 + x_2 y_1)^2 \tag{5.7'}$$

Since the primes are the multiplicative building blocks of the natural numbers, by eq. (5.7) we need to determine which primes can be expressed as the sum of two squares.

Theorem 5.1: The prime p is expressible as the sum of two squares if and only if $p = 2$ or $p \equiv 1 \pmod 4$. In either case, p has a unique representation as the sum of two squares.

Proof: (Existence) ($\Rightarrow$) The "only if" part of the proof is trivial since all squares are congruent to either 0 or 1(mod 4) and hence their sum can be congruent to only 0, 1, or 2 (mod 4).

($\Leftarrow$) Since $2 = 1^2 + 1^2$, it suffices to show that all primes $p \equiv 1 \pmod 4$ are the sum of two squares. Assume $p \equiv 1 \pmod 4$. By Proposition 4.10(a), $\left(\frac{-1}{p}\right) = 1$, and $x^2 \equiv -1 \pmod p$ is solvable for some $x < p$. Hence there exists an m such that $mp = x^2 + 1$, a sum of two squares. Clearly $1 \le m < p$ since $x \le p - 1$ and $x^2 + 1 < p^2$. Now assume that m_1 is the smallest such m for which $mp = a^2 + b^2$ is solvable. So

$$m_1 p = a_1^2 + b_1^2 \tag{5.8}$$

for some a_1 and b_1, where $1 \le m_1 < p$. We need to demonstrate that, in fact, $m_1 = 1$.

Assume $m_1 > 1$ and let $a_2 \equiv a_1 \pmod{m_1}$ and $b_2 \equiv b_1 \pmod{m_1}$, where $|a_2| \le m_1/2$ and $|b_2| \le m_1/2$. Furthermore, it cannot be the case that both a and b_2 are zero. Otherwise $m_1^2 \mid (a_1^2 + b_1^2)$ which implies $m_1 \mid p$ by eq. (5.8).

But p prime then implies $m_1 = p$, contradicting the fact $1 \le m_1 < p$.

Since $a_1^2 + b_1^2 \equiv a_2^2 + b_2^2 \equiv 0 \pmod{m_1}$, there is an $m_2 > 0$ such that

$$m_1 m_2 = a_2^2 + b_2^2 \tag{5.9}$$

In addition,

$$m_2 = \frac{a_2^2 + b_2^2}{m_1} \le \frac{(m_1/2)^2 + (m_1/2)^2}{m_1} < m_1$$

Multiply eq. (5.8) by eq. (5.9) to obtain

$$m_1^2 m_2 p = \left(a_1^2 + b_1^2\right)\left(a_2^2 + b_2^2\right)$$

Applying identity (5.7),

$$m_1^2 m_2 p = (a_1 a_2 + b_1 b_2)^2 + (a_1 b_2 - a_2 b_1)^2. \text{ But}$$

$$a_1 a_2 + b_1 b_2 \equiv a_1^{\,2} + b_1^{\,2} \equiv 0 \ (\mathrm{mod}\ m_1) \text{ and}$$

$$a_1 b_2 - a_2 b_1 \equiv a_1 b_1 - a_1 b_1 \equiv 0 \ (\mathrm{mod}\ m_1). \text{ Hence,}$$

$$m_2 p = \left(\frac{a_1 a_2 + b_1 b_2}{m_1}\right)^2 + \left(\frac{a_1 b_2 - a_2 b_1}{m_1}\right)^2,$$

is a sum of two integral squares. But $m_2 < m_1$, a contradiction to the minimality of m_1. Thus $m_1 = 1$ and p can be written as the sum of two squares.

(Uniqueness) If $p = 2$, then $p = 1^2 + 1^2$ is the unique representation of p as the sum of two natural numbers. Suppose that $p \equiv 1 \ (\mathrm{mod}\ 4)$ and that

$$p = x^2 + y^2 = u^2 + v^2, \text{ where } 0 < x < y < \sqrt{p} \text{ and } 0 < u < v < \sqrt{p}.$$

It follows that $x^2 \equiv -y^2 (\mathrm{mod}\ p)$ and $u^2 \equiv -v^2 (\mathrm{mod}\ p)$. Hence $x^2 u^2 \equiv y^2 v^2 (\mathrm{mod}\ p)$ and so $p \mid (xu - yv)(xu + yv)$. By Euclid's lemma either (i) $p \mid xu - yv$ or (ii) $p \mid xu + yv$.

In case (i), $xu - yv = 0$ since $-p < xu - yv < p$ from the bounds on $x, y, u,$ and v. It follows that $xu = yv$. But $x < y$ and $u < v$ lead to a contradiction.

In case (ii), $xu + yv = p$ since $0 < xu + yv < 2p$. Thus

$$p^2 + (xv - yu)^2 = (xu + yv)^2 + (xv - yu)^2$$

$$= x^2 u^2 + x^2 v^2 + y^2 v^2 + y^2 u^2$$

$$= \left(x^2 + y^2\right)\left(u^2 + v^2\right) = p^2.$$

So $xv - yu = 0$ and $v = \frac{yu}{x}$. Then

$$p = u^2 + v^2 = u^2 + \frac{y^2 u^2}{x^2} = \frac{u^2}{x^2}\left(x^2 + y^2\right) = \frac{u^2}{x^2} p.$$

So $\frac{u^2}{x^2} = 1$ and $x = u, y = v$. Therefore, p has a unique representation as the sum of two squares. ∎

The technique of assuming a smallest positive integral solution m_1 to a given equation and deriving a smaller positive integral solution m_2 goes back to Fermat and is known as Fermat's method of infinite descent. Hence Fermat's assertion that he had an "irrefutable proof" seems quite credible in this instance.

It follows from Theorem 5.1 that the primes 101, 3037, and 333041 are all expressible as the sum of two squares. On the other hand, the primes 67, 2027, and $2^{756,839} - 1$ are not. The uniqueness of the representation for suitable primes allows us to verify

that 1189 is composite since $1189 = 30^2 + 17^2 = 33^2 + 10^2$. (See if you can factor 1189 by using eqs. (5.7) and (5.7').).
   We now treat the general case.

Theorem 5.2 (Sum of two squares theorem): Equation (5.6) is solvable if and only if all prime divisors p of n with $p \equiv 3 \pmod 4$ occur to an even power.

Proof: ($\Leftarrow$) Let $n = s^2 m$, where m is square-free. Certainly s^2 can be written as the sum of two squares since $s^2 = s^2 + 0^2$. By our hypothesis, m is not divisible by any prime $p \equiv 3 \pmod 4$. By Theorem 5.1, all primes dividing m are expressible as the sum of two squares. Repeated application of identity eq. (5.7) gives the result.
   ($\Rightarrow$) Suppose $n = x^2 + y^2$ with $\gcd(x,y) = d$ and let $p \mid n$ with $p \equiv 3 \pmod 4$. In fact, let $p^{c||n}$ and $p^r || d$. We will show that $c = 2r$, and hence c is even.
   Let $x = da$ and let $y = db$ where necessarily $\gcd(a,b) = 1$. Thus,
   $n = d^2(a^2 + b^2) = d^2 t$ say. So $p^{c-2r} || t$. Assume for now that $c - 2r > 0$.
   Hence $a^2 + b^2 = t$ with $\gcd(a,b) = 1$. If $p \mid t$ and $p \mid a$, then $p \mid b^2$ and $p \mid b$ by Euclid's lemma. Similarly, if $p \mid t$ and $p \mid b$, then $p \mid a$. But $\gcd(a,b) = 1$ and so $p \nmid ab$.
   By Fermat's little theorem, $a^{p-1} \equiv 1 \pmod p$ and hence $ba^{p-1} \equiv b \pmod p$. Let $u = ba^{p-2}$ and so $au \equiv b \pmod p$. Then $a^2(1 + u^2) \equiv a^2 + b^2 = t \equiv 0 \pmod p$. Since $p \nmid a^2$, it follows that $1 + u^2 \equiv 0 \pmod p$ by Proposition 2.4(a). But then $u^2 \equiv -1 \pmod p$ and $\left(\frac{-1}{p}\right) = 1$, contradicting Proposition 4.10(a).
   It follows that $p \nmid t$ and so $c - 2r = 0$, that is, c is even. ∎

Example 5.1: Express 16660 as a sum of 2 squares in 2 different ways.
Solution: $16660 = 2^2 \cdot 5 \cdot 7^2 \cdot 17$ and so is expressible as the sum of two squares by Theorem 5.2. We write $5 = 2^2 + 1^2$ and $17 = 4^2 + 1^2$. Applying eq. (5.7), we have $5 \cdot 17 = 9^2 + 2^2$. Next multiply through by $2^2 \cdot 7^2$ to obtain $16660 = 2^2 \cdot 7^2 \cdot 5 \cdot 17 = 126^2 + 28^2$.
   Another solution is gotten by writing $5 = 1^2 + 2^2$ and $17 = 4^2 + 1^2$. Now eq. (5.7) leads to $5 \cdot 17 = 6^2 + 7^2$ and $16660 = 84^2 + 98^2$.
   The problem of representing an integer n as a sum of two squares reduces to factoring n and then representing each prime p dividing n as a sum of two squares. In most of our examples and exercises n can be easily factored. Furthermore, the primes dividing n will be fairly small and easily expressed as $x^2 + y^2$. We will study factorization methods in Chapter 7. Here we introduce without proof an algorithm for representing appropriate primes p as the sum of two squares.
   Let p be a prime congruent to 1 (mod 4). In 1848 C. Hermite (1822–1901) developed a continued fraction algorithm to find x and y such that $x^2 + y^2 = p$. (Similar but somewhat inferior algorithms were developed by J.A. Serret and G. Cornaccia (1908) as well.) In 1972, Brillhart made some modifications to derive an efficient method based on the Euclidean algorithm. In any event, we call the following the *Hermite–Brillhart* algorithm:

1. Find N such that $p \mid (N^2 + 1)$ with $0 < N < \frac{p}{2}$. If $p = N^2 + 1$, then we are done. If not, then proceed to step 2.
2. Apply the Euclidean algorithm to p and N. Let k be the smallest positive integer for which the remainder $r_k < \sqrt{p}$. Then $p = r_k^2 + r_{k+1}^2$.

Example 5.2: Find integers x and y for which $73 = x^2 + y^2$.
Solution: $p = 73$ is a prime congruent to 1 (mod 4). Hence it is representable as the sum of two squares. We apply the Hermite–Brillhart algorithm:

We find that $p \mid 730$ and $730 = 27^2 + 1$. Now apply the Euclidean algorithm to 73 and 27:

$73 = 2 \cdot 27 + 19$ (Notice that $19 > \sqrt{73}$.)
$27 = 1 \cdot 19 + 8$ (Notice that $8 < \sqrt{73}$, so just go one more step.)
$19 = 2 \cdot 8 + 3$, etc.
Hence $k = 2$ and $73 = r_2^2 + r_3^2 = 8^2 + 3^2$.

Note that the existence of N in the Hermite–Brillhart algorithm is guaranteed by our proof of Theorem 5.1. In fact, by Euler's criterion if a is a quadratic nonresidue $(\bmod\, p)$, then $N \equiv a^{\frac{p-1}{4}}$ $(\bmod\, p)$ suffices. Here is an example with a larger value of p.

Example 5.3: Find integers x and y for which $15101 = x^2 + y^2$.
Solution: The integer $p = 15101$ is prime and is congruent to 1 (mod 4). But $15101 \equiv 5$ (mod 8) and hence $a = 2$ is a quadratic nonresidue $(\bmod\, p)$ by Corollary 4.14.1. Hence $N \equiv 2^{3775}$ $(\bmod\, p)$. By the binary modular exponentiation algorithm (or with appropriate computer software) we find $N = 1943$. Now apply the Euclidean algorithm:

$$15101 = 7 \cdot 1943 + 1500$$

$$1943 = 1 \cdot 1500 + 443$$

$$443 = 2 \cdot 171 + 101 \quad (\text{notice that} \;\; 101 < \sqrt{15101} = 122.88\ldots)$$

$$171 = 1 \cdot 101 + 70, \text{ etc.}$$

Hence, $15101 = 101^2 + 70^2$.

In Section 9.2, we discuss the problem of determining the total number of representations of an integer n as a sum of two squares. This is a much more difficult problem. In fact, we just derive the average number of such representations.

We now turn our attention to finding all Pythagorean triplets, given by the Diophantine equation (5.3). Certainly, you are already familiar with several solutions from a geometry class, such as $(3, 4, 5), (5, 12, 13)$, and $(8, 15, 17)$. We assume x and y are nonzero so that eq. (5.3) does not reduce to the trivial equation $x^2 = z^2$. Notice that we may assume x, y, and $z > 0$ since if (x, y, z) is a solution to eq. (5.3), then so is $(\pm x, \pm y, \pm z)$.

If $\gcd(x, y) = d$, then $d^2 \mid z^2$ and $d \mid z$. Similarly, if $\gcd(x, z) = d$, then $d \mid y$, and if $\gcd(y, z) = d$ then $d \mid x$. So, for example, the Pythagorean triplet $(15, 20, 25)$ can be viewed as being generated by $(3, 4, 5)$. In fact, all Pythagorean triplets (x, y, z) are of the form (dx_0, dy_0, dz_0), where $x_0^2 + y_0^2 = z_0^2$. This leads to the following distinction.

Definition 5.1: A solution to the Pythagorean equation (5.3), where x, y, and z are relatively prime is called a *primitive solution*. Otherwise, the solution is called *imprimitive*.

Hence to find all solutions to eq. (5.3), it suffices to find all primitive solutions. For more general Diophantine equations, a solution $x_1, x_2, \ldots, x_m$ is primitive if $\gcd(x_1, x_2, \ldots, x_m) = 1$. In the case of the Pythagorean equation, we have shown that if x, y, and z are relatively prime, then in fact x, y, and z are pairwise relatively prime. We prove the following result:

Theorem 5.3 (Pythagorean triplets theorem): Every positive primitive solution of the Diophantine equation:
$x^2 + y^2 = z^2$ *is of the form*
$$x = 2ab, y = a^2 - b^2, z = a^2 + b^2 \tag{5.10}$$

where $\gcd(a, b) = 1, a > b$, *and* a *and* b *are of opposite parity.*

Proof: Let (x, y, z) be a positive primitive solution to eq. (5.3). If both x and y are odd, then $x^2 \equiv y^2 \equiv 1 \pmod 4$ and $z^2 \equiv 2 \pmod 4$, a contradiction. So without loss of generality, assume x is even and y odd. If follows that z is odd and hence both $z + y$ and $z - y$ are even. So $x/2$, $(z + y)/2$, and $(z - y)/2$ are all positive integers and we can write

$$\left(\frac{x}{2}\right)^2 = \left(\frac{z+y}{2}\right)\left(\frac{z-y}{2}\right) \tag{5.11}$$

If $d \mid (z + y)/2$ and $d \mid (z - y)/2$, then $d \mid z$ and $d \mid y$ since d must divide the sum and difference of $(z + y)/2$ and $(z - y)/2$. But y and z are relatively prime and hence so are $(z + y)/2$ and $(z - y)/2$. But by eq. (5.11) their product is a perfect square and hence each of them is a perfect square. Hence there are integers a and b with $a > b > 0$ such that

$$\left(\frac{z+y}{2}\right) = a^2 \text{ and } \left(\frac{z-y}{2}\right) = b^2 \tag{5.12}$$

Solving eq. (5.12) for y and z we get $y = a^2 - b^2$ and $z = a^2 + b^2$. From eq. (5.3), it follows that $x = 2ab$. If $\gcd(a, b) > 1$, then x, y, and z would not be pairwise relatively prime. If a and b are both odd, then x, y, and z would all be even.

Finally, note that if $x = 2ab, y = a^2 - b^2$, and $z = a^2 + b^2$, then $x^2 + y^2 = z^2$. Theorem 5.3 follows necessarily. ∎

Formula (5.10) is essentially given by Diophantus, though it should be recalled that many special cases and formulas for subclasses of Pythagorean triplets appeared in many cultures throughout antiquity.

Theorem 5.3 completely characterizes all solutions to eq. (5.3) since any imprimitive solution is simply a multiple of one given by eqs. (5.10). Hence, we have a full accounting of integral-sided right triangles, called Pythagorean triangles. Solving problems like the next example are now easily accomplished.

Example 5.4: Find all Pythagorean triangles having hypotenuse less than or equal to 40.
Solution: We need to find all solutions to $x^2 + y^2 = z^2$ with $z \leq 40$. We will find all primitive solutions with $z \leq 40$ utilizing Theorem 5.3. Then we can find all appropriate imprimitive solutions quite readily.

To find primitive solutions we make a chart of relatively prime a and b of opposite parity with $a > b$ and $z = a^2 + b^2 \leq 40$. The values of x and y can be calculated later:

a	b	z
2	1	5
3	2	13
4	1	17
4	3	25
5	2	29
6	1	37

The primitive Pythagorean triplet with $z = 5$ gives rise to imprimitive Pythagorean triplets with $z = 10$, 15, 20, 25, 30, 35, and 40. Similarly, the primitive Pythagorean triplet with $z = 13$ generates solutions with $z = 26$ and 39. In addition there is a Pythagorean triplet with $z = 34$ formed by doubling $z = 17$. Hence the integral-sided right triangles having hypotenuse less than or equal to 40 are those with sides (3, 4, 5), (6, 8, 10), (9, 12, 15), (12, 16, 20), (15, 20, 25), (18, 24, 30), (21, 28, 35), (24, 32, 40), (5, 12, 13), (10, 24, 26), (15, 36, 39), (8, 15, 17), (16, 30, 34), (7, 24, 25), (20, 21, 29), and (12, 35, 37).

We are now in a position to prove an important special case of Fermat's last theorem, denoted FLT. Recall that Fermat claimed that the Diophantine equation $x^n + y^n = z^n$ had no nontrivial solutions for all $n \geq 3$. To prove FLT it suffices to establish the result for exponents $n = 4$ and $n = p$ for all odd primes p. This follows since if $x^n + y^n = z^n$, where $n = mp$, then $(x^m)^p + (y^m)^p = (z^m)^p$ is a solution with exponent p. In addition, if $n \geq 3$ and there is no odd prime $p \mid n$, then $n = 2^a$ for some $a \geq 2$. But then $(x^b)^4 + (y^b)^4 = (z^b)^4$, where $b = 2^{a-2}$ is a solution with exponent 4. Fermat introduced his method of infinite descent in proving the next result.

Theorem 5.4: The Diophantine equation

$$x^4 + y^4 = z^2 \qquad (5.13)$$

has no nontrivial solutions.

We consider any solution where $xy = 0$ to be trivial. Hence a nontrivial solution means one, where x, y, and z are all positive. Note that Theorem 5.4 says that the sum of two squares of squares cannot be a perfect square. Hence the result fits naturally with the theme of this section. In addition, notice that if there were a nontrivial solution to the Diophantine equation (5.5):

$$X^4 + Y^4 = Z^4$$

then $x = X$, $y = Y$, and $z = Z^2$ would solve eq. (5.13). Hence the important corollary:

Corollary 5.4.1 (Fermat's last theorem for $n = 4$): *The Diophantine equation (5.5) has no nontrivial solutions.*

Proof. Suppose eq. (5.13) is solvable. Then there would be a solution (x, y, z) to eq. (5.13) with x, y, z all positive and z least. If $\gcd(x, y) = d > 1$, then $d^4 \mid z^2$ and by unique factorization $d^2 \mid z$. But then $(x/d, y/d, z/d^2)$ would be another solution of eq. (5.13) contrary to the minimality of z. Hence $\gcd(x, y) = 1$. It follows that x, y, and z are pairwise relatively prime and so (x^2, y^2, z) form a primitive Pythagorean triplet.

Without loss of generality, assume x is even and y is odd. By Theorem 5.3, there exist relatively prime positive integers a and b with $a > b$ of opposite parity such that

$$x^2 = 2ab, y^2 = a^2 - b^2, z = a^2 + b^2 \qquad (5.14)$$

If b were odd and a even, then $y^2 = a^2 - b^2 \equiv 3 \pmod{4}$, a contradiction.

Hence a is odd and b is even. But a and b relatively prime implies a, b, and y are pairwise relatively prime. Hence (b, y, a) form a primitive Pythagorean triplet. So there exist relatively prime positive integers r and s such that

$$b = 2rs, y = r^2 - s^2, a = r^2 + s^2 \qquad (5.15)$$

By eqs. (5.14) and (5.15),

$$x^2 = 4rs\left(r^2 + s^2\right) \qquad (5.16)$$

But $\gcd(r, s) = 1$ and hence $\gcd\left(r, r^2 + s^2\right) = \gcd\left(s, r^2 + s^2\right) = 1$. Formula (5.16) implies that r, s, and $r^2 + s^2$ are all perfect squares. Hence there are positive integers u, v, and w such that

$$r = u^2, s = v^2, \text{ and } r^2 + s^2 = w^2 \qquad (5.17)$$

But then $u^4 + v^4 = w^2$, another solution to eq. (5.13). Notice that $0 < w \le w^2 = r^2 + s^2 = a \le a^2 < a^2 + b^2 = z$. So (u, v, w) is another positive solution to eq. (5.13) with $w < z$. This contradicts the minimality of z and establishes the result. ∎

By way of example, consider the Diophantine equation $x^4 + 81y^4 = 4z^2$. This is of the form $X^4 + Y^4 = Z^2$, where $X = x$, $Y = 3y$, and $Z = 2z$. By Theorem 5.4 the equation has no nontrivial solution. However, it does have "trivial" solutions. In particular, if $y = 0$, then x is even. Letting $x = 2s$ leads to the solutions $(x, y, z) = (2s, 0, 2s^2)$ for $s \in \mathbb{Z}$. Similarly, if $x = 0$, then y is even. Letting $y = 2t$ leads to the solutions $(x, y, z) = (0, 2t, 18t)$ for $t \in \mathbb{Z}$. Together these give all the integral solutions to $x^4 + 81y^4 = 4z^2$.

Exercise 5.2

1. (a) Show that if $n \equiv 3$ (mod 4), then n is not expressible as the sum of two squares.
    (b) Is it true that if $n \equiv 1$ (mod 4), then n is expressible as the sum of two squares?

2. Show that abc is square-free if and only if a, b, and c are square-free and pairwise relatively prime.

3. Use the fact that $3^2 + 4^2 = 5^2$ and $5^2 + 12^2 = 13^2$ to find two right triangles with hypotenuse 65.

4. Which of the numbers 109, 147, 221, 539, 585, 625, and 2754 can be expressed as the sum of two squares? (If so, indicate at least one representation.)

5. Verify that the following integers n are not prime by finding at least two representations of n as a sum of two squares:
    (a) $n = 221$ (b) $n = 493$ (c) $n = 629$ (d) $n = 1073$ (e) $n = 2501$

6. (Match wits with Euler) Determine that $n = 1{,}000{,}009$ is not prime by finding two representations for n as a sum of two squares.

7. How many ways can 245 be written as the sum of two squares? Explain why this does not contradict Theorem 5.1.

8. (a) Let p_1 and p_2 be primes congruent to 1 (mod 4). Use eqs (5.7) and (5.7′) to show that p_1p_2 has two essentially distinct representations as a sum of two squares.
    (b) Find all essentially distinct representations of $1105 = 5 \cdot 13 \cdot 17$ as a sum of two squares.

9. (a) What is the smallest integer having two essentially distinct representations as the sum of two squares?
    (b) What is the smallest integer having three essentially distinct representations as the sum of two squares?
    (c) Show that the number of essentially distinct representations as the sum of two squares can be made arbitrarily large.

10. Gauss proved that if a prime $p = 4k + 1$, then $p = x^2 + y^2$ where

$x \equiv \frac{1}{2}\binom{2k}{k} \pmod{p}, y \equiv (2k)! x \pmod{p}$, and $0 < |x|, |y| < \frac{p}{2}$.

   (a) Use Gauss' result to express 5, 13, and 17 as a sum of two squares.

   (b) Comment on the practicality of this method.

11. Use the Hermite–Brillhart algorithm to find x and y for which

   (a) $41 = x^2 + y^2$ (b) $101 = x^2 + y^2$

   (c) $181 = x^2 + y^2$ (d) $1693 = x^2 + y^2$.

12. Let F_n denote the nth Fibonacci number. Recall that
   $F_{2n+1} = F_n^2 + F_{n+1}^2$ (see Exercise 1.2.9(b)). Use this formula to express the following as
   a sum of two squares: 233, 610, 1597, 10946, and 16724.

13. Let p be a prime congruent to 1 (mod 4) (Fermat 1641).

   (a) Show that p^2 is the hypotenuse for two distinct Pythagorean triangles.

   (b) If r is a positive integer, then show that p^r is the hypotenuse for r distinct
   Pythagorean triangles.

14. (a) Let p_1 and p_2 be distinct primes congruent to 1 (mod 4). Show that $p_1 p_2$ is the
   hypotenuse for two distinct Pythagorean triangles.

   (b) Let $p_1, \ldots, p_r$ be distinct primes congruent to 1 (mod 4). Show that $\prod_{i=1}^{r} p_i$ is
   the hypotenuse for 2^{r-1} distinct Pythagorean triangles.

15. Find two Pythagorean triangles with hypotenuse 85. Are there any others?

16. Find a primitive Pythagorean triplet (x, y, z) with $x = 60$ and $z - y = 18$.

17. Show that if (x, y, z) is a Pythagorean triplet, then $60 \mid xyz$.

18. How many Pythagorean triangles are there with all sides less than or equal to 50?

19. How many Pythagorean triangles are there with perimeter less than or equal
   to 100?

20. (a) Show that $(x, y, z) = (3, 4, 5)$ is the only Pythagorean triplet consisting of three
   consecutive integers.

   (b) Show that there are infinitely many primitive Pythagorean triplets (x, y, z),
   where $z = x + 1$.

   (c) Show that there are infinitely many primitive Pythagorean triplets (x, y, z),
   where $z = y + 2$.

21. Find all Pythagorean triangles whose area and perimeter are equal.

22. (a) Let $s = (a + b + c)/2$ be the semi-perimeter of a triangle with sides a, b, and c.
   Use the Pythagorean triplets theorem to show that if A is the area of an integral-sided right triangle, then $s \mid A$.

   (b) Use Heron's formula to show further that $\frac{(s-a)(s-b)(s-c)}{s}$ is a perfect square.

23. (a) Show that the inradius (radius of the inscribed circle) of a Pythagorean triangle is always an integer. (Hint: Partition the triangle via lines drawn from the
   center of the inscribed circle to the three vertices.)

   (b) Show that for every $r \geq 1$ there is a Pythagorean triangle with inradius r.

   (c) Determine all Pythagorean triangles with inradius $r = 10$.

24. Find all solutions of $x^2 + 9y^2 = z^2$.
25. (a) Find all solutions of $16x^4 + y^4 = z^2$.
    (b) Find all solutions of $16x^4 + 81y^4 = 25z^2$.
26. Use Theorem 5.3 to show that there are infinitely many solutions to the Diophantine equation $x^3 + y^2 = z^2$.
27. Show there are infinitely many solutions to $x^2 + y^2 = z^4$.
28. Prove that $x^4 - y^4 = z^2$ has no nontrivial solutions.
29. (a) Verify Diophantus' result that if $a^2 + b^2 = c^2$, then $(ab)^4 + (bc)^4 + (ca)^4 = (a^4 + a^2b^2 + b^4)^2$.
    (b) Let $a = 3$, $b = 4$, and $c = 5$ to find the smallest sum of three biquadrates equaling a perfect square.
30. Show that if m is expressible as the sum of two squares and n is not, then mn is not expressible as the sum of two squares. What is the situation for the product of two integers neither of which is the sum of two squares?
31. Show that if m is expressible as the sum of two rational squares, then m is expressible as the sum of two integral squares.
32. (a) Verify Frenicle de Bessy's observation (1657) that $9^3 + 10^3 = 1^3 + 12^3$.
    (b) Verify Euler's observation that $133^4 + 134^4 = 59^4 + 158^4$.
    (c) Verify Euler's observation that $2903^4 + 12231^4 = 10203^4 + 10381^4$.
    (d) Can you find distinct integers a, b, c, and d such that $a^5 + b^5 = c^5 + d^5$?
    If so, you are the first person to do so (though no one has shown it cannot be done).
33. Explain why Fermat's method of infinite descent is equivalent to PMI.
34. Use Fermat's method of infinite descent to prove that $\sqrt{2}$ is irrational.
35. Show that if p is prime and $n \geq 3$, then $x^n + py^n = p^2 z^n$ is unsolvable except for $x = y = z = 0$.
36. Show that $x^n + y^n = z^n$ is solvable in integers with $xyz \neq 0$ if and only if $\frac{1}{x^n} + \frac{1}{y^n} = \frac{1}{z^n}$ is solvable.
37. Show that for any prime p and integer k there exists an integer $n \equiv k \pmod{p}$ expressible as the sum of two squares.
38. Show that if $p \equiv 3 \pmod 4$ is prime and $n \equiv kp \pmod{p^2}$ for some k with $1 \leq k < p$, then z is not expressible as the sum of two squares.
39. (a) Find 6 consecutive positive integers (less than 100) none of which is expressible as the sum of two squares.
    (b) Show that there exist arbitrarily many consecutive positive integers not expressible as the sum of two squares.
40. (a) Show that no Pythagorean triangle has square area.
    (b) The congruent number problem is the difficult problem of determining for which n does there exist a rational-sided right triangle with area n. The congruent number problem has been essentially solved by J. Tunnell (1983). Verify that five is a congruent number corresponding to the triangle with sides

$\frac{3}{2}$, $\frac{20}{3}$, and $\frac{41}{6}$. (If the Birch and Swinnerton–Dyer conjecture is true, then all square-free $n \equiv 5$, 6, or $7 \pmod 8$ are congruent numbers among others.)

41. Show that there are infinitely many primitive Pythagorean triangles having square perimeter. What are the two smallest such triangles?

5.3 Sums of three squares

Determining the set of all n for which the Diophantine equation

$$x^2 + y^2 + z^2 = n \tag{5.18}$$

is solvable is a difficult problem first worked out by Gauss (but independently solved and first published by Legendre – 1798.) The result states that n is expressible as the sum of three squares if and only if n is not of the form $4^a(8m+7)$. In his *Disquisitiones*, Gauss went further by determining the number of representations of an integer n as the sum of three squares. The method of proof goes well beyond what we discuss here. Instead, we will consider three other intriguing problems dealing with the sum of three squares. One is, under what conditions is the sum of three squares itself a square. Another is, which integers are representable as the sum of three squares where at least two of the squares are identical. The third is a particular Diophantine equation that has interesting links with other number-theoretic functions. More specifically, we solve the following problems:

Problem 1: Find all solutions to the Diophantine equation

$$x^2 + y^2 + z^2 = w^2 \tag{5.19}$$

Problem 2: Determine necessary and sufficient conditions on n for the solubility of the Diophantine equation

$$x^2 + 2y^2 = n \tag{5.20}$$

Problem 3: Investigate solutions to the Diophantine equation

$$x^2 + y^2 + z^2 = 3xyz \tag{5.21}$$

We begin with problem 1 which was first worked out by the Japanese mathematician Yoshisuke Matsunago in the early nineteenth century. Problem 1 is equivalent to finding all rectangular parallelepipeds (boxes) with integer sides and integer body diagonal. Notice that it suffices to find all primitive solutions to eq. (5.19) since if $d = \gcd(x,y,z,w)$ then (x_0,y_0,z_0,w_0) satisfies eq. (5.19), where $x = dx_0, y = dy_0, z = dz_0$, and $w = dw_0$.

Theorem 5.5: All positive primitive solutions to eq. (5.19) with x odd and $y \leq z$ are given by

$$x = \frac{a^2 + b^2 - c^2}{c}, \quad y = 2a, \quad z = 2b, \quad w = \frac{a^2 + b^2 + c^2}{c} \tag{5.22}$$

for natural numbers a, b, c with $a \leq b$ and $c \mid (a^2 + b^2)$ with $c < \sqrt{a^2 + b^2}$.

Proof: Let (x, y, z, w) be a positive primitive solution to eq. (5.19). If w is even then an odd number of x, y, and z are even. If x is even and y and z odd, say, then $w^2 \equiv 2 \pmod 4$, a contradiction. Similarly, if x, y, and z are all even, then two divides each of x, y, z, and w, contradicting the primitivity of our solution. Hence w is odd and an odd number of x, y, and z must be odd. However, if all x, y, and z are odd, then $w^2 \equiv 3 \pmod 4$, a contradiction. Thus, it must be the case that w is odd and exactly one of x, y, and z is odd. Without loss of generality, let us assume that x is odd and $y \leq z$.

Let $y = 2a$ and $z = 2b$ with $a \leq b$. Since $w > x$ and both x and w are odd, let $w - x = 2c$. Then eq. (5.19) implies $x^2 + 4a^2 + 4b^2 = (x + 2c)^2$. So $c^2 + cx = a^2 + b^2$. Since $c \mid c^2 + cx$, it follows that $c \mid a^2 + b^2$. Also, $c < \sqrt{a^2 + b^2}$ because $c^2 < a^2 + b^2$. But $x = \frac{a^2 + b^2 - c^2}{c}$ and since $w = x + 2c$, we obtain $w = \frac{a^2 + b^2 + c^2}{c}$. In addition, notice that x, y, z, and w are defined uniquely by eq. (5.21) for given a, b, and c.

Finally, we can directly verify that if x, y, z, and w are given as in eqs. (5.22) for appropriate a, b, and c, then they satisfy eq. (5.19). ∎

For example, if $a = 1$, $b = 3$, and $c = 2$, then we get $x = 3, y = 2, z = 6$, and $w = 7$. In fact, $2^2 + 3^2 + 6^2 = 7^2$ is a primitive solution to eq. (5.19). Unfortunately, Theorem 5.5 does not guarantee that all a, b, c satisfying the conditions listed will generate a primitive solution of eq. (5.19). See Exercise 5.3.1(b) in this regard.

We next turn our attention to problem 2. The first step is to determine which primes p are expressible as $x^2 + 2y^2$. Fermat correctly stated the result in a letter to Pascal of 1654. (See if you can discover the result before proceeding.) Our proof will closely imitate that of Theorem 5.1. The proof will again employ Fermat's method of infinite descent. The equivalent of formula (5.7) appropriate in this instance is the following:

$$\left(x_1^2 + 2y_1^2\right)\left(x_2^2 + 2y_2^2\right) = \left(x_1 x_2 + 2y_1 y_2\right)^2 + 2\left(x_1 y_2 - x_2 y_1\right)^2 \tag{5.23}$$

The identity above can be directly verified. To appreciate its genesis involves the study of the number field $\mathbb{Q}(\sqrt{-2})$ just as formula (5.7) derives from the Gaussian field $\mathbb{Q}(\sqrt{-1})$. Our exposition will not require any algebraic number theory, but it will demonstrate once again the importance of quadratic reciprocity.

Theorem 5.6: The prime p is expressible as $x^2 + 2y^2$ if and only if $p = 2$ or $p \equiv 1 \pmod 8$ or $p \equiv 3 \pmod 8$.

Proof: The prime 2 is a special case. Clearly $2 = 0^2 + 2 \cdot 1^2$. In what follows assume all primes are odd.

($\Rightarrow$) Since all squares are congruent to 0, 1, or 4 (mod 8), $x^2 + 2y^2$ is congruent to 0, 1, 2, 3, 4, or 6 (mod 8). But no odd primes are congruent to 0, 2, 4, or 6 (mod 8). Thus p is congruent to 1 or 3 (mod 8).

($\Leftarrow$) Let $p \equiv 1$ or 3 (mod 8). $x^2 \equiv -2 \pmod p$ is solvable if and only if the Legendre symbol $\left(\frac{-2}{p}\right) = 1$. But $\left(\frac{-2}{p}\right) = \left(\frac{2}{p}\right)\left(\frac{-1}{p}\right) = 1$ if and only if $p \equiv 1$ or 3 (mod 8). This follows from combining Proposition 4.10(a) and Corollary 4.14.1 (or from Exercise 4.3.14). So $x^2 \equiv -2 \pmod p$ is solvable and hence there exists an $m < p$ such that $mp = x^2 + 2$, which is a square plus twice a square (namely 1^2). Assume that m_1 is the smallest positive m for which $mp = a^2 + 2b^2$ is solvable. So there exists a_1 and b_1 with $1 \le m_1 < p$ such that

$$m_1 p = a_1^2 + 2b_1^2 \tag{5.24}$$

Assume for the sake of argument that $m_1 > 1$. Choose $a_2 \equiv a_1 \pmod{m_1}$ and $b_2 \equiv b_1 \pmod{m_1}$, where $|a_2| \le m_1/2$ and $|b_2| \le m_1/2$. Note that not both a_2 and b_2 are zero or else $m_1^2 \mid a_1^2 + 2b_1^2$, contrary to eq. (5.24).

Since $a_1^2 + 2b_1^2 \equiv a_2^2 + 2b_2^2 \equiv 0 \pmod{m_1}$, there exists $m_2 > 0$ such that

$$m_1 m_2 = a_2^2 + 2b_2^2 \tag{5.25}$$

In addition,

$$m_2 = \frac{a_2^2 + 2b_2^2}{m_1} \le \frac{(m_1/2)^2 + 2(m_1/2)^2}{m_1} < m_1$$

Multiplying eq. (5.24) by eq. (5.25) we obtain

$$m_1^2 m_2 p = \left(a_1^2 + 2b_1^2\right)\left(a_2^2 + 2b_2^2\right)$$

Applying identity eq. (5.23) gives

$$m_1^2 m_2 p = (a_1 a_2 + 2b_1 b_2)^2 + 2(a_1 b_2 - a_2 b_1)^2$$

But

$$a_1 a_2 + 2b_1 b_2 \equiv a_1^2 + 2b_1^2 \equiv 0 \pmod{m_1} \text{ and}$$

$$a_1 b_2 - a_2 b_1 \equiv a_1 b_1 - a_1 b_1 \equiv 0 \pmod{m_1}$$

Hence

$$m_2 p = \left(\frac{a_1 a_2 + 2b_1 b_2}{m_1}\right)^2 + 2\left(\frac{a_1 b_2 - a_2 b_1}{m_1}\right)^2$$

which is of the form $x^2 + 2y^2$. But $m_2 < m_1$, contradicting the minimality of m_1. Therefore $m_1 = 1$ and the result follows. ∎

Example 5.5: Let us find a representation of $n = 561$ of the form $x^2 + 2y^2$.

Factor 561 to obtain $561 = 3 \cdot 11 \cdot 17$. By Theorem 5.6, all the prime factors of 561 can be expressed in the desired form. In particular, $3 = 1^2 + 2 \cdot 1^2$, $11 = 3^2 + 2 \cdot 1^2$, and $17 = 3^2 + 2 \cdot 2^2$. Utilizing formula (5.23) twice, we obtain $33 = 3 \cdot 11 = 5^2 + 2 \cdot 2^2$ and $561 = 33 \cdot 17 = 23^2 + 2 \cdot 4^2$. Alternatively, by changing the order of our calculations, we can get $561 = 19^2 + 2 \cdot 10^2$.

Next we prove the analog of Theorem 5.2 for the Diophantine equation (5.20).

Theorem 5.7: Equation (5.20) is solvable if and only if all prime divisors p of n with $p \equiv 5 \pmod 8$ or $p \equiv 7 \pmod 8$ occur to an even power.

Proof: ($\Leftarrow$) Let $n = s^2 m$, where m is square-free. The square s^2 is expressible as $x^2 + 2y^2$ by simply letting $x = s$ and $y = 0$. By hypothesis, m is not divisible by any primes p congruent to either 5 or 7 (mod 8). By Theorem 5.6, all primes dividing m are expressible as $x^2 + 2y^2$. Repeated application of identity (5.23) gives the result.

($\Rightarrow$) Suppose $n = x^2 + 2y^2$ for some x and y with $\gcd(x, y) = d$. Let $p \mid n$ where p is a prime congruent to either 5 or 7 (mod 8). Let $p^c \mid\mid n$ and $p^r \mid\mid d$. We will show that $c = 2r$ and hence c is even as desired.

Let $x = da$ and $y = db$, where $\gcd(a, b) = 1$ necessarily. Hence $n = d^2(a^2 + 2b^2) = d^2 t$ say. So p^{c-2r} / t. Assume for the sake of argument that $c - 2r > 0$ (i.e., $p \mid t$). If in addition $p \mid a$, then $p \mid 2b^2$. But $p \neq 2$ and hence $p \mid b$. Similarly, if $p \mid b$, then $p \mid a$. But $\gcd(a, b) = 1$ and hence $p \nmid ab$.

By Fermat's little theorem, $b^{p-1} \equiv 1 \pmod p$ and thus $ab^{p-1} \equiv a \pmod p$. Let $u = ab^{p-2}$. Then $bu \equiv a \pmod p$. So $b^2(2 + u^2) \equiv a^2 + 2b^2 = t \equiv 0 \pmod p$. But $p \nmid b$ and so $p \mid (2 + u^2)$. Hence $u^2 \equiv -2 \pmod p$ and $\left(\frac{-2}{p}\right) = 1$, contradicting our earlier observation that $\left(\frac{-2}{p}\right) = 1$ only if p is congruent to either 1 or 3 (mod 8). Therefore $p \nmid t$ and $c = 2r$. ∎

For example, $294 = 2 \cdot 3 \cdot 7^2$ is expressible as $x^2 + 2y^2$ while $375 = 3 \cdot 5^3$ is not. In fact, $294 = 14^2 + 2 \cdot 7^2$.

Finally, we turn our attention to eq. (5.21): $x^2 + y^2 + z^2 = 3xyz$. We seek solutions in positive integers x, y, and z. This is known as *Markov's equation* since it was extensively studied in 1880 by the Russian mathematician Andrei A. Markov (1856–1922), best known for his work in statistics and stochastic processes. The Markov equation has two somewhat trivial solutions, namely the Markov triples (1, 1, 1) and (1, 1, 2) which are known as the singular solutions. All other solutions involving three distinct integers are called nonsingular. We call a positive integer a *Markov number* if it appears in any Markov triple.

If (x, y, z) is a solution, then so is $(x, y, 3xy - z)$ since $x^2 + y^2 + (3xy - z)^2 = (x^2 + y^2 + z^2) + (9x^2y^2 - 6xyz) = 9x^2y^2 - 3xyz = 3xy (3xy - z)$. In general, we can permute the numbers x, y, z to produce three other solutions of Markov's equation. Only the two singular Markov triples produce less than three other solutions. If $1 \leq x < y < z$, then $x < z < 3xz - y$ and $y < z < 3yz - x$. Hence, the solution (x, y, z) produces two new solutions $(x, z, 3xz - y)$ and $(y, z, 3yz - x)$ with larger Markov numbers $3xz - y$ and $3yz - x$. The other solution $(x, y, 3xy - z)$ must necessarily be a previous Markov triple with $3xy - z < z$ since the process of producing Markov triples can be reversed. In particular, $(x, y, 3xy - (3xy - z)) = (x, y, z)$. Each nonsingular Markov triple produces exactly two new Markov triples. Furthermore, the reverse process necessarily terminates at the triple $(1, 1, 1)$. Hence, the set of Markov triples form an infinite connected tree with each node having two new branches.

Here is the beginning of the Markov tree:

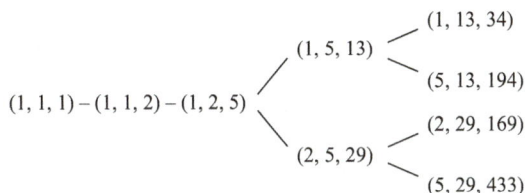

(1, 13, 34)

(1, 5, 13)

(5, 13, 194)

(1, 1, 1) – (1, 1, 2) – (1, 2, 5)

(2, 29, 169)

(2, 5, 29)

(5, 29, 433)

Figure 5.1: Early branches of the Markov tree.

Let us follow a single branch (See Figure 5.1): $(1, 1, 1) \rightarrow (1, 1, 2) \rightarrow (1, 2, 5) \rightarrow (1, 5, 13) \rightarrow (1, 13, 34) \rightarrow (1, 34, 89) \rightarrow (1, 89, 233) \rightarrow (1, 233, 610) \rightarrow (1, 610, 1597) \rightarrow \ldots$

Do these Markov numbers look familiar? I hope so. They are all Fibonacci numbers! Let us see why.

Proposition 5.8: *The triple* $(1, F_{2n-1}, F_{2n+1})$, *where* F_k *is the kth Fibonacci number is a Markov triple.*

Proof: Cassini's identity (Proposition 1.3) states that $F_{k+1} F_{k-1} - F_k^2 = (-1)^k$. Letting $k = 2n$ gives

$$F_{2n}^2 = F_{2n+1}F_{2n-1} - 1.$$

Thus,

$$(F_{2n+1} - F_{2n-1})^2 = (F_{2n} + F_{2n-1} - F_{2n-1})^2 = F_{2n}^2 = F_{2n+1}F_{2n-1} - 1$$

Expanding the left-hand side,

$$F_{2n+1}^2 - 2F_{2n-1}F_{2n+1} + F_{2n-1}^2 = F_{2n+1}F_{2n-1} - 1$$

which implies

$$1^2 + F_{2n-1}^2 + F_{2n+1}^2 = 3 \cdot 1 \cdot F_{2n-1} F_{2n+1}$$ ■

The *Markov unicity conjecture* states that each Markov number appears exactly once as the maximal element of a Markov triple. At present, the proof seems elusive.

Exercise 5.3

1. (a) Use eqs. (5.22) to find solutions to $x^2 + y^2 + z^2 = w^2$ generated from the ordered triples $(a,\ b,\ c) = (1,\ 1,\ 1), (3,\ 5,\ 2)$, and $(3,\ 7,\ 2)$.
    (b) What do the ordered triples $(a, b, c) = (1,\ 2,\ 1)$ and $(3,\ 4,\ 5)$ say about the converse of Theorem 5.5?
2. Find an integral-sided box with body diagonal of length 39.
3. (a) Show that a box with sides of length 44, 117, and 240 has integral face diagonals (Euler). Verify that its body diagonal is irrational (see Exercise 2.3.17).
    (b) Show that a box with sides of length 117, 520, and 756 has two integral face diagonals and an integral body diagonal.
        The existence of a perfect cuboid with integral sides, face diagonals, and body diagonal is unknown!
4. (a) Show that the product of two integers, each expressible as the sum of at most three squares, is not expressible necessarily as the sum of three squares by considering the product of 3 and 5. Hence no identity equivalent to eq. (5.7) exists for the sum of three squares.
    (b) Use Gauss' result to show that the product of two integers, neither expressible as the sum of three squares, is so expressible.
5. Show that if all natural numbers are expressible as the sum of at most three triangular numbers (established by Gauss), then all integers of the form $8k + 3$ can be expressed as the sum of at most three squares. (Hint: Express k as the sum of three triangular numbers).
6. Show that a solution to eq. (5.19) uniquely determines a, b, and c in eqs. (5.22).
7. (a) Verify formula (5.23).
    (b) Derive a formula which shows that, in general, the product of two integers of the form $x^2 + ny^2$ is again of that form.
8. Show that no integer of the form $4^a(8m + 7)$ with $a, m \geq 0$ can be expressed as the sum of three squares. (Hint: First consider the case $a = 0$. If $a > 0$, consider the parity of the three squares required.)
9. Find three essentially distinct solutions to eq. (5.19) with $w = 9$.
10. Which of the integers 66, 200, 493, 525, 637, 1225 can be expressed as $x^2 + 2y^2$?

11. Express the integers 27, 211, 507, 1353, 10450 as $x^2 + 2y^2$.

12. (a) Verify Euler's identity (1750): $a^2 + b^2 + c^2 = (2m - a)^2 + (2m - b)^2 + (2m - c)^2$ where $a + b + c = 3m$.

    (b) Use the above to find an integer expressible in two completely distinct ways as a sum of three squares.

13. (a) Show that $(a^2 - b^2 - 2ab)^2$, $(a^2 + b^2)^2$, and $(a^2 - b^2 + 2ab)^2$ are three squares in arithmetic progression for $0 < a < b$ (see Exercise 1.1.20).

    (b) Show that if (x, y, z) is a primitive Pythagorean triplet, then $(y - x)^2, z^2$, $(y + x)^2$ are in arithmetic progression.

14. (a) Verify the following companion equation to eq. (5.23)

$$(x_1^2 + 2y_1^2)(x_2^2 + 2y_2^2) = (x_1 x_2 - 2y_1 y_2)^2 + 2(x_1 y_2 + x_2 y_1)^2$$

    (b) Find two expressions of the form $x^2 + 2y^2$ for 33, 57, 209, and 473.

15. Recall that n is expressible as the sum of three squares if and only if n is not of the form $4^a(8k + 7)$ with $a, k \geq 0$. What is the largest number of consecutive positive integers expressible as a sum of three squares? What is the largest number of consecutive positive integers not expressible as a sum of three squares?

16. Use Thue's lemma (Exercise 2.5.22) to establish Theorem 5.6.

    (Hint: If $p \equiv 1$ or $3 \pmod 8$, show that there are x and y with $p \mid (x^2 + 2y^2)$ with $1 \leq x, y \leq \sqrt{p}$.)

17. Show that there is a one-to-one correspondence between integral solutions to $x^2 + y^2 + z^2 = 3xyz$ and solutions to $x^2 + y^2 + z^2 = xyz$.

18. Consider the Diophantine equation $x^2 + y^2 + z^2 = 2xyz$.

    (a) Show that if (x_1, y_1, z_1) is a solution, then x_1, y_1, z_1 are all even.

    (b) Show that $(x_2, y_2, z_2) = (x_1/2, y_1/2, z_1/2)$ solves $x^2 + y^2 + z^2 = 4xyz$ with all entries even.

    (c) Inductively, show that there are even integers (x_n, y_n, z_n) solving $x^2 + y^2 + z^2 = 2^n xyz$, where $(x_n, y_n, z_n) = (x_{n-1}/2, y_{n-1}/2, z_{n-1}/2)$ for all $n \geq 2$. Use Fermat's method of infinite descent to show that there are no solutions to $x^2 + y^2 + z^2 = 2xyz$ from the fact that $x_1 > x_2 > \cdots > x_n$.

It can be shown that there are no solutions to $x^2 + y^2 + z^2 = nxyz$ for any n other than 1 or 3.

5.4 Sums of four or more squares

In 1770 Lagrange proved that every positive integer could be expressed by the sum of at the most four squares. In this section we provide its demonstration. Next, we discuss the representation of integers as the sum of k nonzero squares for $k \geq 5$ (i.e., the representation of integers as the sum of precisely k squares). In particular, we will show that there are only a finite set of integers not expressible as the sum of exactly k squares for any $k \geq 5$.

We begin by noting the following remarkable identity due to Euler:

$$\left(w_1^2 + x_1^2 + y_1^2 + z_1^2 \right) \left(w_2^2 + x_2^2 + y_2^2 + z_2^2 \right)$$

$$= \left(w_1 w_2 + x_1 x_2 + y_1 y_2 + z_1 z_2 \right)^2 + \left(-w_1 x_2 + x_1 w_2 + y_1 z_2 - z_1 y_2 \right)^2 \qquad (5.26)$$

In fact, in a note dated May 4, 1748, Euler gave a whole family of such identities of which formula (5.26) is just one special case. In modern terminology, formula (5.26) expresses the fact that the norm of the product of two quaternionic integers is the product of the norms. A quaternionic integer is a number of the form $a + bi + cj + dk$, where a, b, c, d are (rational) integers and $i^2 = j^2 = k^2 = -1$ and $ij = k, jk = i, ki = j$, while $ji = -k, kj = -i$, and $ik = -j$. The quaternionic integers required to derive formula (5.25) are precisely $w_1 + x_1 i - y_1 j + z_1 k$ and $w_2 - x_2 i + y_2 j - z_2 k$.

The importance of formula (5.26) is that it is sufficient to show that all primes are expressible as the sum of four squares in order to prove that all positive integers are so representable. (We allow the square zero in the discussion that follows.) We now state the desired result.

Theorem 5.9 (Sum of Four Squares Theorem): Every positive integer is expressible as the sum of four squares.

Historically, the next significant step taken toward the proof of Theorem 5.9 was the establishment of the following lemma (proved by Euler in 1751). Euler also proved many other partial results relating to Theorem 5.9, but admitted that he was unable to complete a full demonstration.

Lemma 5.9.1: *Let p be prime. Then there are integers a and b such that p divides $a^2 + b^2 + 1$.*

Proof: If $p = 2$ then let $a = 1$ and $b = 0$. So assume that p is odd.

(i) If $p \equiv 1 \pmod 4$, then the Legendre symbol $\left(\frac{-1}{p} \right) = 1$ and so we may let $b = 0$ and then solve $a^2 \equiv -1 \pmod p$ for a.

(ii) If $p \equiv 3 \pmod 4$, then $\left(\frac{-1}{p} \right) = -1$. We need to show that there are a and b such that $a^2 \equiv -(b^2 + 1) \pmod p$. Thus, it suffices to find an integer b for which the Legendre symbol $\left(\frac{-(b^2+1)}{p} \right) = 1$. But $\left(\frac{-(b^2+1)}{p} \right) = \left(\frac{-1}{p} \right)\left(\frac{b^2+1}{p} \right)$. Since $\left(\frac{-1}{p} \right) = -1$, we must find ab such that $\left(\frac{b^2+1}{p} \right) = -1$. But b^2 runs through all quadratic residues $\pmod p$. So it suffices to find a quadratic residue r such that $r + 1$ is a quadratic nonresidue. But that is clearly the case since otherwise $1, 2, 3, \ldots, p-1$ would all be quadratic residues (since 1 is a square), contrary to the fact that there are precisely $(p-1)/2$ quadratic residues $(\bmod\, p)$. ∎

We now turn to the proof of Theorem 5.9 by combining Lagrange's original proof (1770) with some simplifications due to Euler (1773).

Proof: By Euler's identity (5.26), all integers are the sum of four squares if all primes are so representable. Let p be prime. If $p = 2$ then $p = 1^2 + 1^2 + 0^2 + 0^2$ is such a representation. Thus assume p is odd.

By Lemma 5.9.1, there are positive integers a, b, and m such that $a^2 + b^2 + 1 = mp$. In fact, by reducing $(\bmod\, p)$, we may assume that $|a| < p/2$ and $|b| < p/2$. It follows that $m = (a^2 + b^2 + 1)/p < p$. So there exists a, b, c, d, and m with $1 \le m < p$ such that $a^2 + b^2 + c^2 + d^2 = mp$ (let $c = 1, d = 0$).

Now assume that m_1 is the smallest such m for which $mp = a^2 + b^2 + c^2 + d^2$ is solvable. So

$$m_1 p = a_1^2 + b_1^2 + c_1^2 + d_1^2 \tag{5.27}$$

for some a_1, b_1, c_1, and d_1, where $1 \le m_1 < p$. We need to show that, in fact, $m_1 = 1$.

Assume $m_1 > 1$. Choose a_2, b_2, c_2, d_2 such that $a_2 \equiv a_1 (\bmod\, m_1), b_2 \equiv b_1 (\bmod\, m_1)$, $c_2 \equiv c_1 (\bmod\, m_1)$, and $d_2 \equiv d_1 (\bmod\, m_1)$ with

$$|a_2| \le \frac{m_1}{2}, \quad |b_2| \le \frac{m_1}{2}, \quad |c_2| \le \frac{m_1}{2}, \quad |d_2| \le \frac{m_1}{2} \tag{5.28}$$

It cannot be the case that $a_2 = b_2 = c_2 = d_2 = 0$; since then $a_1 \equiv b_1 \equiv c_1 \equiv d_1 \equiv 0 \ (\bmod\, m_1)$ and by (5.27), $m_1 p \equiv 0 \ (\bmod\, m_1^2)$. But then $p \equiv 0 \ (\bmod\, m_1)$ contrary to the fact $1 < m_1 < p$. Similarly, it cannot be the case that $|a_2| = |b_2| = |c_2| = |d_2| = \dfrac{m_1}{2}$, for otherwise eqs. (5.27) and (5.28) imply

$$m_1 p \equiv \frac{m_1^2}{4} + \frac{m_1^2}{4} + \frac{m_1^2}{4} + \frac{m_1^2}{4} = m_1^2 \equiv 0 \ (\bmod\, m_1^2)$$

But then $p \equiv 0 \ (\bmod\, m_1)$, a contradiction as before.

Now $a_2^2 + b_2^2 + c_2^2 + d_2^2 \equiv a_1^2 + b_1^2 + c_1^2 + d_1^2 \equiv 0 \ (\bmod\, m_1)$. Hence there exists an m_2 such that

$$m_1 m_2 = a_2^2 + b_2^2 + c_2^2 + d_2^2 \tag{5.29}$$

In addition, since not all the terms on the right are zero,

$$1 \le m_2 < \frac{(m_1/2)^2 + (m_1/2)^2 + (m_1/2)^2 + (m_1/2)^2}{m_1} = m_1$$

Multiply eq. (5.27) by eq. (5.29) to obtain

$$m_1^2 m_2 p = \left(a_1^2 + b_1^2 + c_1^2 + d_1^2\right)\left(a_2^2 + b_2^2 + c_2^2 + d_2^2\right)$$

Applying identity (5.26),

$$m_1^2 m_2 p = a^2 + b^2 + c^2 + d^2, \quad \text{where } a = a_1 a_2 + b_1 b_2 + c_1 c_2 + d_1 d_2$$

$$b = -a_1 b_2 + b_1 a_2 + c_1 d_2 - d_1 c_2, \quad c = a_1 c_2 - c_1 a_2 + b_1 d_2 - d_1 b_2, \text{ and}$$

$$d = -a_1 d_2 + d_1 a_2 + b_1 c_2 - c_1 b_2$$

But by (5.28), $a \equiv a_1^2 + b_1^2 + c_1^2 + d_1^2 \equiv 0 \pmod{m_1}$. Similarly, $b \equiv c \equiv d \equiv 0 \pmod{m_1}$. Hence $m_2 p = (a/m_1)^2 + (b/m_1)^2 + (c/m_1)^2 + (d/m_1)^2$, a sum of four integral squares.

But $m_2 < m_1$, which contradicts the minimality of m_1. Thus $m_1 = 1$ and p is expressible as the sum of four squares. ∎

We now consider the representation of positive integers as the sum of nonzero squares. As we noted in Section 5.1, if $n \equiv 7 \pmod 8$, then n cannot be expressed as the sum of fewer than four squares. Hence there are infinitely many integers which are not expressible as the sum of k squares for $k \le 3$.

It is also the case that there are infinitely many integers which are not expressible as the sum of four nonzero squares. This follows from the observation that there is a one-to-one correspondence between representations of $2n$ as a sum of four squares and $8n$ as a sum of four squares (including the square zero). If $2n = a^2 + b^2 + c^2 + d^2$, then $8n = (2a)^2 + (2b)^2 + (2c)^2 + (2d)^2$. In the other direction, if $8n = A^2 + B^2 + C^2 + D^2$, then $A^2 + B^2 + C^2 + D^2 \equiv 0 \pmod 8$ and it follows that $A, B, C,$ and D are all even (since odd squares are congruent to 1 (mod 8)). But then $2n = (A/2)^2 + (B/2)^2 + (C/2)^2 + (D/2)^2$, a sum of integral squares. Now notice that $2 = 1^2 + 1^2 + 0^2 + 0^2$ has no representation as the sum of four nonzero squares. It follows that $4^r \cdot 2 = (2^r)^2 + (2^r)^2 + 0^2 + 0^2$ is the only representation of $4^r \cdot 2$ as the sum of four squares. Hence there are infinitely many positive integers not expressible as the sum of four nonzero squares.

The next result shows that the situation is quite different for five or more nonzero squares. It is due to E. Dubouis (1911) who solved a problem of J. Tannery.

Proposition 5.10: *For $k \ge 5$, all but a finite number of positive integers are sums of precisely k nonzero squares.*

Proof: Let $k = 5$. Notice that we can express 169 as the sum of one, two, three, or four nonzero squares (as well as others). In particular, $169 = 13^2 = 12^2 + 5^2 = 12^2 + 3^2 + 4^2 = 11^2 + 4^2 + 4^2 + 4^2$. Let $n \ge 170$. Then $n - 169 = a^2 + b^2 + c^2 + d^2$ by Theorem 5.8 (where some of the squares could be zero). If $abcd \ne 0$, then $n = 13^2 + a^2 + b^2 + c^2 + d^2$, a sum of five nonzero squares. If only one of a, b, c, d is zero (say $= 0$), then $n = 12^2 + 5^2 + a^2 + b^2 + c^2$, a sum of five nonzero squares. Similarly, if a and b are nonzero but $c = d = 0$, then

$n = 12^2 + 3^2 + 4^2 + a^2 + b^2$. Finally, if $n - 169 = a^2$, then $n = 11^2 + 4^2 + 4^2 + 4^2 + a^2$. This establishes the result for $k = 5$ since there are only finitely many positive integers less than 170. (For the sake of completeness check that 1, 2, 3, 4, 6, 7, 9, 10, 12, 15, 18, and 33 are the only positive integers not expressible as the sum of five nonzero squares.) So all integers $n \geq 34$ are expressible as the sum of five nonzero squares.

Now let $k \geq 6$. If $n \geq 29 + k$, then $n + 5 - k \geq 34$ and $n + 5 - k = x_1^2 + x_2^2 + x_3^2 + x_4^2 + x_5^2$, a sum of five nonzero squares. But then $n = 1^2 + \cdots + 1^2 + x_1^2 + x_2^2 + x_3^2 + x_4^2 + x_5^2$ with $k - 5$ terms being 1^2. Hence n is expressible as the sum of k nonzero squares and the proposition follows. ∎

Exercise 5.4

1. Find all essentially distinct representations of $n = 200$ as a sum of at most 4 squares.
2. Verify formula (5.26) directly.
3. (a) Use formula (5.26) to express $420 = 15 \cdot 28$ as a sum of four squares.
   (b) Express $1457 = 31 \cdot 47$ as a sum of four squares.
4. Find appropriate a and b in Lemma 5.9.1 for primes $p = 7$, 11, 19, 23, and 87.
5. Verify that all positive integers less than 170 are expressible as the sum of five nonzero squares save for 1, 2, 3, 4, 6, 7, 9, 10, 12, 15, 18, and 33.
6. List all integers not expressible as the sum of six nonzero squares.
7. Determine all positive integers representable as the sum of four squares with two pairs identical, that is, of the form $2x^2 + 2y^2$.
8. (a) Determine all primes expressible as the sum of four squares three of which are identical, that is, of the form $x^2 + 3y^2$ (Fermat).
   (b) Determine all positive integers of the form $x^2 + 3y^2$.
9. Prove the following theorem of R. Sprague (1949): Every integer greater than 128 is the sum of distinct squares.
10. Determine all n for which 169 can be expressed as the sum of n nonzero squares.
11. Prove the following theorem by filling in the necessary details: Given any integer k, there is an integer m that is expressible as the sum of n nonzero squares for all n with $1 \leq n \leq k$. (a) If x is odd, then $\left(x, \frac{1}{2}(x^2 - 1), \frac{1}{2}(x^2 + 1)\right)$ forms a Pythagorean triplet.
    (b) Given k, let $x_1 = 3$ and $x_i = \frac{x_{i-1}^2 + 1}{2}$ for $2 \leq i \leq k$. Let $m = x_k^2$. Show that m has the desired property.
12. Fill in the details of the following alternate proof of Lemma 5.9.1.
    (a) Let $X = \left\{ x^2 \colon 0 \leq x \leq \frac{p-1}{2} \right\}$ and $Y = \left\{ -y^2 - 1 \colon 0 \leq y \leq \frac{p-1}{2} \right\}$. Show that no two elements of the same set are congruent to one another $(\bmod\, p)$.
    (b) Show that there exists $a^2 \in X$ and $-b^2 - 1 \in Y$ such that $a^2 \equiv -b^2 - 1 \ (\bmod\, p)$. Conclude that $p \mid (a^2 + b^2 + 1)$.

13. Prove the following theorem of H.E. Richert (1949): Every positive integer except 2, 5, 8, 12, 23, 33 is the sum of distinct triangular numbers.

14. Prove the following theorem of V.A. Lebesgue (1872) by filling in the necessary details below: Every odd natural number is the sum of four squares of which two are consecutive.

   (a) By Gauss's result, the number $4n+1$ can be expressed as the sum of 3 squares:

   $$4n+1 = (2a)^2 + (2b)^2 + (2c+1)^2.$$

   (b) Hence $2n+1 = (a+b)^2 + (a-b)^2 + c^2 + (c+1)^2$.

15. (Extension of formula (5.26)).

   (a) Verify that the following identity holds:

   $$\left(w_1^2 + ax_1^2 + by_1^2 + abz_1^2\right)\left(w_2^2 + ax_2^2 + by_2^2 + abz_2^2\right) =$$

   $$(w_1w_2 + ax_1x_2 + by_1y_2 + abz_1z_2)^2 + a(-w_1x_2 + x_1w_2 - b_1z_2 + bz_1y_2)^2$$

   $$+ b(-w_1y_2 + y_1w_2 + ax_1z_2 - az_1x_2)^2 + ab(-w_1z_2 + z_1w_2 - bx_1y_2 + by_1x_2)^2$$

   (b) Express $984 = 24 \cdot 41$ in the form $a^2 + 3b^2 + 5c^2 + 15d^2$.

   (c) Express 451 in the form $2a^2 + b^2 + 3c^2 + 5d^2$.

16. Cannonball problem: Discover something interesting about the sum of the first 24 squares. (It is the only such example discovered by E. Lucas in 1875.)

5.5 Legendre's equation

In this section, we will generalize Theorem 5.3 considerably. First notice that the equation

$$ax^2 + y^2 = z^2 \tag{5.30}$$

with $a > 0$ is solvable by simply letting $x = 2rs, y = ar^2 - s^2$, and $z = ar^2 + s^2$.

Legendre considered the more general equation (5.4)

$$ax^2 + by^2 = cz^2$$

and worked out necessary and sufficient conditions that it be solvable in his two-volume *Essai sur la thèorie des nombres* which was published in 1797 and 1798. This was the first full-length text devoted exclusively to number theory. The solution to eq. (5.4) is rather long and intricate. The crux of the matter is to work out necessary and sufficient conditions for the solvability of

$$ax^2 + by^2 = z^2 \tag{5.31}$$

first and use them to obtain conditions on Legendre's equation (5.4).

Note that for a given a and b, it is conceivable that eq. (5.31) has no solutions (other than the trivial solution $x = y = z = 0$ which we will dismiss.) For example, $5x^2 + 9y^2 = z^2$ has no solution since we may assume without loss of generality that $x, y,$ and z are pairwise relatively prime.

Hence $3 \nmid x$ (else $3 \mid z$) and so $5x^2 \equiv 2 \pmod 3$. But then $z^2 \equiv 2 \pmod 3$ which is impossible.

Let us consider eq. (5.31), where a and b are arbitrary (but fixed) positive integers. Let $\gcd(a, b) = d$ and set $a = a_1 d$ and $b = b_1 d$.

Proposition 5.11: *The Diophantine equation (5.31) is solvable if and only if*

$$\left\{ \begin{array}{l} r^2 \equiv a \pmod b \\ s^2 \equiv b \pmod a \\ t^2 \equiv -a_1 b_1 \pmod d \end{array} \right\} \qquad \begin{array}{l} (5.32), \\ (5.33), \\ (5.34) \end{array}$$

are all solvable.

Notice that if a and b are distinct odd primes, then Proposition 5.10 states that $ax^2 + by^2 = z^2$ is solvable if and only if the Legendre symbols $\left(\frac{a}{b}\right) = 1$ and $\left(\frac{b}{a}\right) = 1$.

Proof: We may assume without loss of generality that in any solution to eq. (5.31), $x, y,$ and z are pairwise relatively prime. Further, since any square could be absorbed by either x^2 or y^2, let us assume throughout that a and b are square-free.

($\Rightarrow$) Equation (5.31) implies $ax^2 \equiv z^2 \pmod b$. If there exists a prime p dividing both b and x, then $p \mid z^2$ and so $p \mid z$. But $\gcd(x, z) = 1$ and hence $\gcd(b, x) = 1$. Now let x^* be such that $xx^* \equiv 1 \pmod b$. Then $a \equiv (x^* z)^2 \pmod b$. If we let $r = x^* z$, then r satisfies eq. (5.32).

In a completely analogous manner, let y^* be an arithmetic inverse of $y \pmod a$. Then $s = y^* z$ is a solution to eq. (5.33).

Next notice that $\gcd(a_1, b_1) = 1$. In fact, $a_1, b_1,$ and d are pairwise relatively prime since a and b are square-free. Equation (5.31) implies $d \mid z^2$. But d is square-free and hence $d^2 \mid z^2$. Thus $d \mid a_1 x^2 + b_1 y^2$ and so $a_1 x^2 \equiv -b_1 y^2 \pmod d$. Let x^* be as above and so $x^* \equiv 1 \pmod d$. Then $(a_1 x^2)(b_1 x^{*2}) \equiv (-b_1 y^2)(b_1 x^{*2}) \pmod d$ and hence $a_1 b_1 \equiv -t^2 \pmod d$, where $t = b_1 y x^*$. that is, $t^2 \equiv -a_1 b_1 \pmod d$ is solvable, satisfying eq. (5.34).

($\Leftarrow$) If $a = 1$ or $b = 1$, then eq. (5.31) is solvable since it reduces to eq. (5.30). If $a = b$, then $d = a$ and the congruence relation equation (5.34) reduces to $t^2 \equiv -1 \pmod a$. But then $a \mid (t^2 + 1)$ and by Exercise 4.2.2, all odd primes $p \mid a$ are congruent to 1 $\pmod 4$. Thus by Theorem 5.2, there are integers m and n such that $a = m^2 + n^2$. Now let $x = m, y = n,$ and $z = a$. Then $ax^2 + b^2 = z^2$; thus eq. (5.31) is solvable.

Now suppose that eqs. (5.32)–(5.34) are satisfied for some a and b with $a > b > 1$. We will find an A with $1 \le A < a$ such that the equivalent congruences for A and b hold and $ax^2 + by^2 = z^2$ is solvable if and only if $AX^2 + bY^2 = Z^2$ is solvable. By the method of infinite descent, this process could be repeated until $A = b$ or $A = 1$. In either case eq. (5.31) is solvable and hence so is the equation with our original a and b.

Choose a solution of eq. (5.33) satisfying $|s| \le a/2$. Since $a \mid (s^2 - b)$, we can write

$$s^2 - b = aAk^2 \tag{5.35}$$

where A is square-free. (A will be the integer previously mentioned.)

Furthermore, $\gcd(k, b) = 1$ since if $p \mid k$ and $p \mid b$, then $p \mid s^2$. But then $p^2 \mid s^2$ and $p^2 \mid k^2$ implies $p^2 \mid b$, contrary to the assumption that b is square-free. A is nonnegative since $aAk^2 = s^2 - b \ge -b > -a$ and there are no negative integers larger than $-a$ divisible by a. But $b \ne s^2$ since b is square-free and $s \ne 1$ (since $a > b > 1$). Thus $A \ge 1$.

In addition,

$$A = (s^2 - b)/ak^2 < s^2/ak^2 \le s^2/a \le a/4 < a \text{ since } |s| \le a/2.$$

Consider the equation

$$AX^2 + bY^2 = Z^2 \tag{5.36}$$

Equation (5.36) is solvable $\Leftrightarrow AX^2 = Z^2 - bY^2$ is solvable

$\Rightarrow ax^2 = aAk^2(Z^2 - bYY^2)$ is solvable where $x = kAX$.
$\Leftrightarrow ax^2 = (s^2 - b)(Z^2 - bYY^2)$ is solvable by eq. (5.34).
$\Leftrightarrow ax^2 = (bY + sZ)^2 - b(sY + Z)^2$ is solvable.
$\Rightarrow ax^2 + by^2 = z^2$ is solvable where $y = sY + Z$ and $z = bY + sZ$.

Thus eq. (5.31) is solvable in x, y, and z if eq. (5.36) is solvable in X, Y, and Z by making the following substitution:

$$x = kAX, y = sY + Z \text{ and } z = bY + sZ \tag{5.37}$$

Finally, we need to verify the three congruence conditions for A and b: Let $\gcd(A, b) = D$ and set $A = A_1 D$ and $b = B_1 D$:

$$\left\{\begin{array}{l} R^2 \equiv A \pmod{b} \\ S^2 \equiv b \pmod{A} \\ T^2 \equiv -A_1 B \pmod{D} \end{array}\right\} \quad \begin{array}{l} (5.32'), \\ (5.33'), \\ (5.34') \end{array}$$

We show individually that each is solvable.

By eq. (5.35), $(s^2 - b)/d = aAk^2/d$ and so $s^2/d - b_1 = a_1 Ak^2$. But $d \mid a$ and $d \mid b$ imply that $d \mid s^2$ and so $d \mid s$ (since d is square-free). Let $s = s_1 d$. Then

$$ds_1^2 - b_1 = a_1 Ak^2 \tag{5.38}$$

Similarly, $r = r_1 d$ and eq. (5.32) is equivalent to $dr_1^2 \equiv a_1 \pmod{b_1}$. Hence $ds_1^2 \equiv a_1 A k k^2 \equiv dA(r_1 k)^2 \pmod{b_1}$. But d, k, a_1 are all relatively prime to b_1 (since d and k are relatively prime to b). Hence $s_1^2 \equiv A(r_1 k)^2 \pmod{b_1}$ and $(s_1 r_1^* k^*)^2 \equiv A \pmod{b_1}$, where r_1^* and k^* are arithmetic inverses of r_1 and $k \pmod{b}$, respectively; thus, $\pmod{b_1}$. Therefore, A is congruent to a square $\pmod{b_1}$.

Also, $-a_1 A k^2 \equiv b_1 \pmod{d}$ by eq. (5.38) and so $-a_1 b_1 A k^2 \equiv b_1^2 \pmod{d}$. From relation (5.34) we have $a_1 b_1 \equiv -t^2 \pmod{d}$ for some t. Hence it follows that $-t^2 A k^2 \equiv b_1^2 \pmod{d}$. But k, a_1, b_1 are all relatively prime to d and so by eq. (5.34) $\gcd(t, d) = 1$. We obtain $A \equiv (b_1 t^* k^*)^2 \pmod{d}$, where t^* and k^* are arithmetic inverses of t and $k \pmod{b}$, and hence $\pmod{d}$. So A is congruent to a square $\pmod{d}$.

Since $\gcd(b_1, d) = 1$, the Chinese remainder theorem (CRT) guarantees that there is an $R \pmod{b}$ such that $R \equiv s_1 r_1 k^* \pmod{b_1}$ and $R \equiv b_1 t^* k^* \pmod{d}$. But then $b_1 \mid (R^2 - A)$ and $d \mid (R^2 - A)$. Since $\gcd(b_1, d) = 1$ and $b = b_1 d$, $R^2 \equiv A \pmod{b}$ are solvable by Corollary 2.1.1(c), eq. (5.32') is satisfied.

By eq. (5.35), $s^2 \equiv b \pmod{A}$ and hence (5.33') is solvable with $S = s$.

Dividing eq. (5.35) by D gives $(s^2 - b)/D = aAk^2/D$. But $D \mid s$ much as $d \mid s$ (again $D \mid b$ and b are square-free). Set $s = S_1 D$. Then $DS_1^2 - B_1 = aA_1 k^2$. Hence $A_1 DS_1^2 - A_1 B_1 = aA_1 k^2$ and $-A_1 B_1 \equiv a(A_1 k)^2 \pmod{D}$. Since $r^2 \equiv a \pmod{D}$ by eq. (5.32) it follows that $-A_1 B_1 \equiv (rAA_1 k)^2 \pmod{D}$, which satisfies eq. (5.34') with $T = rAA_1 k$.

The result now follows by infinite descent. ■

We now deduce necessary and sufficient conditions for eq. (5.4) to be solvable. Let us make the natural supposition that a, b, and c are square-free and pairwise relatively prime.

Theorem 5.12: Legendre's equation (5.4) is solvable if and only if

$$\begin{cases} r^2 \equiv bc \pmod{a} \\ s^2 \equiv ac \pmod{b} \\ t^2 \equiv -ab \pmod{c} \end{cases} \tag{5.39}$$

are all solvable.

Proof: The equation $ax^2 + by^2 = z^2$ is solvable if and only if $acx^2 + bcy^2 = c^2 z^2$ is solvable. Letting $Z = cz$, $ax^2 + by^2 = cz^2$ is solvable if and only if $acx^2 + bcy^2 = Z^2$ is solvable. But a, b, and c are square-free and pairwise relatively prime and so ac and bc are square-free. By Proposition 5.11, $acx^2 + bcy^2 = Z^2$ is solvable if and only if $r^2 \equiv bc \pmod{ac}$, $s^2 \equiv ac \pmod{bc}$, and $t^2 \equiv -ab \pmod{c}$ are all solvable. The last condition follows from the fact $\gcd(a, b) = 1$. But by the CRT, $r^2 \equiv bc \pmod{ac}$ is solvable if and only if $r^2 \equiv bc \pmod{a}$ and $r^2 \equiv bc \pmod{c}$ are solvable since $\gcd(a, c) = 1$. The second congruence holds trivially by letting $r = c$. Thus $r^2 \equiv bc \pmod{ac}$ is solvable if and only if $r^2 \equiv bc \pmod{a}$ is solvable. Similarly, $s^2 \equiv ac \pmod{bc}$ is solvable if and only if $s^2 \equiv ac \pmod{b}$ is solvable. The result follows. ∎

For example, the Diophantine equation $5x^2 + 6y^2 = 7z^2$ has no solution since, in fact, none of the congruences $r^2 \equiv 42 \pmod 5$, $s^2 \equiv 35 \pmod 6$, and $t^2 \equiv -30 \pmod 7$ are solvable.

An example may help clarify the technical details of our proofs.

Example 5.6: Find an integral solution to $17x^2 + 3y^2 = 11z^2$.

Solution: The given equation is solvable if and only if $187x^2 + 33y^2 = z_1^2$ is solvable where $z_1 = 11z$. Proposition 5.11 guarantees that the latter equation is solvable if and only if $r^2 \equiv 187 \pmod{33}$, $s^2 \equiv 33 \pmod{187}$, and $t^2 \equiv -51 \pmod{11}$ are all solvable since $a = 187$, $b = 33$, and $d = 11$.

Since $187 \equiv 22 \pmod{33}$, it suffices to determine the value of the Legendre symbol $\left(\frac{22}{3}\right) = \left(\frac{1}{3}\right) = 1$. (As noted above, $r^2 \equiv 22 \pmod{11}$ is trivially satisfied.)

Hence $r^2 \equiv 187 \pmod{33}$ is solvable.

The congruence $s^2 \equiv 33 \pmod{187}$ is solvable if and only if $s_1^2 \equiv 33 \pmod{11}$ and $s_2^2 \equiv 33 \pmod{17}$ are solvable. Indeed $s_1 \equiv 0 \pmod{11}$ and $s_2 \equiv 4 \pmod{17}$ work. By the CRT this leads to $s = 55$ as a solution to $s^2 \equiv 33 \pmod{187}$.

The congruence $t^2 \equiv -51 \pmod{11}$ is equivalent to $t^2 \equiv 4 \pmod{11}$ and is certainly solvable.

Hence $187x^2 + 33y^2 = z_1{}^2$ is solvable. Notice that $|s| \le a/2$ as required in the proof of Proposition 5.10. Equation (5.35) becomes $3025 - 33 = 2992 = 16 \cdot 187$.

So $A = 1$ and $k = 4$.

Now consider $X^2 + 33Y^2 = Z^2$. This equation is of the form (5.30) and so is solvable by letting $X = 33u^2 - v^2$, $Y = 2uv$, and $Z = 33u^2 + v^2$. In particular, with $u = 2$ and $v = 1$, we obtain the solution $X = 131$, $Y = 4$, and $Z = 133$. Using the substitution eq. (5.37) leads to $x = 524$, $y = 353$, and $z_1 = 7447$. Finally, we obtain $x = 524, y = 353$, and $z = 677$ as a solution to $17x^2 + 3y^2 = 11z^2$.

Please note that there is no guarantee that our method will yield the smallest solution (see Exercise 5.5.13). In fact, by a theorem of L. Holzer (1950), simplified by L.J. Mordell (1968), if Legendre's equation (5.4) is solvable, then there exists a nonnegative solution with $x \le \sqrt{bc}$, $y \le \sqrt{ac}$, and $z \le \sqrt{ab}$. Hence the equation in Example 5.6 must have a solution with $0 \le x \le 5$, $0 \le y \le 13$, and $0 \le z \le 7$. In fact, such solutions are $(1, 3, 2)$ and $(4, 1, 5)$.

Exercise 5.5

1. Verify the given solution to eq. (5.29).
2. (a) In the proof of Proposition 5.10, explain why we could assume x, y, and z were pairwise relatively prime and a and b were square-free.
   (b) In the proof of Theorem 5.11, explain why we could assume a, b, and c were square-free and pairwise relatively prime.
3. (a) Use the solution to eq. (5.29) to find the general solution to $2x^2 + y^2 = z^2$.
   (b) Determine all such solutions with $0 < x \leq 8$.
4. (a) Find the general solution to $x^2 + 3y^2 = z^2$.
   (b) Determine all such solutions with $0 < |x| \leq 12$, $0 < |y| \leq 12$.
5. (a) Use Proposition 5.10 to show that $2x^2 + 7y^2 = z^2$ is solvable.
   (b) Solve $2x^2 + 7y^2 = z^2$.
   (c) Use the solution in part (b) to solve $2x^2 + 12x + 7y^2 - 56y + 130 = z^2$.
6. (a) Rewrite Theorem 5.11 in terms of the Legendre symbol in the case where a, b, and c are odd primes.
   (b) Determine whether $3x^2 + 7y^2 = 11z^2$ is solvable.
7. (a) Verify that $7x^2 + 37y^2 = z^2$ is solvable.
   (b) Solve $7x^2 + 37y^2 = z^2$.
8. (a) Verify that $3x^2 + 5y^2 = 137z^2$ is solvable.
   (b) Solve $3x^2 + 5y^2 = 137z^2$.
9. (a) Determine whether $3x^2 + 13y^2 = z^2$ is solvable.
   (b) If so, solve $3x^2 + 13y^2 = z^2$.
10. (a) Determine whether $5x^2 + 11y^2 = 23z^2$ is solvable.
    (b) If so, solve $5x^2 + 11y^2 = 23z^2$.
11. (a) Determine whether $7x^2 + 11y^2 = 43z^2$ is solvable.
    (b) If so, solve $7x^2 + 11y^2 = 43z^2$.
12. Show if $a \neq b$ and $a \equiv b \equiv 3 \pmod 4$, then $ax^2 + by^2 = z^2$ is not solvable.
13. There are two solutions to the equation in Example 5.6 with x, y, and z all less than or equal to 5. Can you find them?
14. Let a, b, c be square-free and pairwise relatively prime. By a theorem of L.E. Dickson (1929), if $ax^2 + by^2 = cz^2$ is solvable, then every positive integer is expressible in the form $ax^2 + by^2 - cz^2$.
    (a) Find (x, y, z) such that $x^2 + y^2 = z^2 + n$ for $n = 1, \ldots, 20$.
    (b) Find (x, y, z) such that $x^2 + 2y^2 = z^2 + n$ for $n = 1, \ldots, 20$.

Chapter 6
Continued fractions and Farey sequences

6.1 Finite simple continued fractions

In this chapter we shall study fractions of the form

$$a_0 + \cfrac{1}{a_1 + \cfrac{1}{a_2 + \cfrac{1}{a_3 + \dots}}}$$

where a_0 is an integer and a_i is a positive integer for all $i \geq 1$. Such fractions are called *simple continued fractions*. If the sum terminates then the fraction is called *finite*. Clearly a finite simple continued fraction is a rational number. In this chapter it will be shown that all rational numbers have an essentially unique representation as a finite simple continued fraction. Next, we will study successive rational approximations, or convergents, to a finite simple continued fraction. We begin by defining some terms.

Definition 6.1: Consider the expression

$$a_0 + \cfrac{b_1}{a_1 + \cfrac{b_2}{a_2 + \cfrac{b_3}{a_3 + \dots}}}$$

where the a_i's and b_j's are all integers. Such an expression is called a *continued fraction*. If all the b_j's are equal to 1 and $a_i \geq 1$ for all $i \geq 1$ then it is a *simple continued fraction*. Furthermore, if the sum terminates then it is a *finite simple continued fraction*. A simple continued fraction will be denoted simply as $[a_0; a_1, a_2, \dots]$. The a_i is called the *partial quotients* of the associated continued fraction.

For example, $3 + \cfrac{1}{1 + \cfrac{1}{4 + \cfrac{1}{1 + \frac{1}{5}}}} = [3; 1, 4, 1, 5] = 134/35$ as can be verified by simplifying the fraction from the bottom up.

The study of continued fractions has a long and fascinating history dating back at least to the ancient Greeks. In addition, continued fractions have found application to such diverse areas as differential equations, statistics, operations research, electrical networks, and astronomy (see Exercise 6.4.9).

Our first proposition establishes that all rational numbers can be represented by finite simple continued fractions. In fact, there is a simple algorithm to accomplish this, one that you already know.

https://doi.org/10.1515/9783111579283-006

Proposition 6.1: Any rational number can be expressed as a finite simple continued fraction.

Proof: Let $r = a/b$ be a rational number and suppose that

$$a/b = a_0 + r_1/b$$

where $a_0 = [r]$, the greatest integer less than or equal to r. If r is an integer, then $r = a_0$ and we are done. If r is not an integer, then $r_1 \neq 0$ and $1 \leq r_1 < b$ and so

$$b/r_1 = a_1 + r_2/r_1$$

with $a_1 = [b/r_1]$. It necessarily follows that $0 \leq r_2 < r_1$. If $r_2 \neq 0$, then write

$$r_1/r_2 = a_2 + r_3/r_2$$

with $0 \leq r_3 < r_2$, etc.

In general, the ith equation has the form

$$r_{i-2}/r_{i-1} = a_{i-1} + r_i/r_{i-1}$$

Since the r_i's form a decreasing sequence of nonnegative integers, there must be an n for which $r_n > 0$ but $r_{n+1} = 0$. Working backward,

$$r_{n-2}/r_{n-1} = a_{n-1} + 1/a_n = [a_{n-1}; a_n]$$

$$r_{n-3}/r_{n-2} = a_{n-2} + \cfrac{1}{r_{n-2}/r_{n-1}} = [a_{n-2}; a_{n-1}, a_n]$$

and finally $r = [a_0; a_1, \ldots, a_n]$. ■

Notice that if we multiply the ith equation through by r_{i-1} for all i, we obtain

$$r_{i-2} = a_{i-1}r_{i-1} + r_i \text{ where } 0 \leq r_i < r_{i-1}.$$

This is identical to the Euclidean algorithm for determining the greatest common divisor of a and b (Proposition 2.1)! In the proof of Proposition 6.1, we have simply replaced the q_i in the proof of Proposition 2.1 by a_{i-1}. It should now be clear why the a_i are called partial quotients.

Example 6.1: Express 118/35 as a finite simple continued fraction.
Solution: $118 = 3 \cdot 35 + 13$

$$35 = 2 \cdot 13 + 9$$

$$13 = 1 \cdot 9 + 4$$

$$9 = 2 \cdot 4 + 1$$

$$4 = 4 \cdot 1$$

Hence it follows that $118/35 = [3; 2, 1, 2, 4]$.

Alternatively, $\frac{118}{35} = 3 + \frac{13}{35} = 3 + \frac{1}{35/13}$

$$= 3 + \cfrac{1}{2 + 9/13} = 3 + \cfrac{1}{2 + \frac{1}{13/9}}$$

$$= 3 + \cfrac{1}{2 + \cfrac{1}{1 + 4/9}} = 3 + \cfrac{1}{2 + \cfrac{1}{1 + \frac{1}{9/4}}}$$

$$= 3 + \cfrac{1}{2 + \cfrac{1}{1 + \cfrac{1}{2 + \frac{1}{4}}}}$$

Since $4 = 3 + 1$, we can also write $118/35 = [3; 2, 1, 2, 3, 1]$.

The last observation illustrates the fact that if $a_n > 1$, then

$$[a_0; a_1, \ldots, a_n] = [a_0; a_1, \ldots, a_n - 1, 1]$$

Equivalently, if $a_n = 1$, then

$$[a_0; a_1, \ldots, a_n] = [a_0; a_1, \ldots, a_{n-1} + 1]$$

So a finite simple continued fraction representation is not entirely unique due to a rather trivial restructuring of its last partial quotients. The next theorem states that a finite simple continued fraction is unique otherwise. (Note the analogy with decimal representations: for example, $6/5 = 1.2 = 1.199999\overline{9}$ with 9's repeated ad infinitum.)

Proposition 6.2 (Uniqueness of Continued Fraction Expansions over $\mathbb{Q}$):
If $a_n > 1$ and $b_m > 1$ and $[a_0; a_1, \ldots, a_n] = [b_0; b_1, \ldots, b_m]$,
then $n = m$ and $a_i = b_i$ for $1 \leq i \leq n$.

Proof: Let $A_i = [a_i; \ldots, a_n]$ for $0 \leq i \leq n$ and $B_j = [b_j; \ldots, b_m]$ for $0 \leq j \leq m$.
Assume without loss of generality that $n \leq m$. Notice that

$$A_i = a_i + 1/A_{i+1} \text{ for } 0 \leq i \leq n - 1 \tag{6.1}$$

and

$$B_j = b_j + 1/B_{j+1} \text{ for } 0 \leq j \leq m - 1$$

Hence

$$A_{i+1} = \frac{1}{A_i - a_i} \text{ for } 0 \leq i \leq n - 1 \tag{6.2}$$

and

$$B_{j+1} = \frac{1}{B_j - b_j} \text{ for } 0 \leq j \leq m - 1$$

Clearly $A_{i+1} > 1$ for $0 \le i \le n-1$ and $B_{j+1} > 1$ for $0 \le j \le m-1$ (where we use the fact that $a_n > 1$ and $b_m > 1$ to establish the cases $i = n-1$ and $j = m-1$).

So by eq. (6.1) and the fact $a_n = A_n$ and $b_m = B_m$,

$$a_i = [A_i] \text{ for } 0 \le i \le n \text{ and } b_j = [B_j] \text{ for } 0 \le j \le m \qquad (6.3)$$

The assertion can now be proved by induction on i.

For $i = 0$, by hypothesis $A_0 = B_0$ and eq. (6.3) implies $a_0 = b_0$.

Assume that $A_k = B_k$ for some $i = k$, where k satisfies $0 \le k \le n-1$.

By eq. (6.3), $ak_k = b_k$. By eq. (6.2),

$A_{k+1} = \frac{1}{A_k - a_k} = \frac{1}{B_k - b_k} = B_{k+1}$. Again eq. (6.3) implies $a_{k+1} = b_{k+1}$.

Thus $a_i = b_i$ for $0 \le i \le n$. Finally, if $m > n$ then $a_n = A_n = B_n = b_n + 1/B_{n+1}$ with $B_{n+1} > 0$. But $a_n = b_n$ and so $m = n$. ∎

Definition 6.2: Let $r = [a_0; a_1, \ldots, a_n]$ and let $r_i = [a_0; \ldots, a_i]$ for $0 \le i \le n$.

Then the rational number r_i is called the ith *convergent to r*.

Example 6.2: Let $r = [3; 7, 15, 1, 2\ 93]$. Then the convergents to r are

$$r_0 = [3] = 3$$

$$r_1 = [3; 7] = 22/7$$

$$r_2 = [3; 7,15] = 333/106$$

$$r_3 = [3; 7,15,1] = 355/133$$

$$r_4 = [3; 7,15,1,293] = 104348/33215$$

If you convert these to decimals, you will notice no doubt that the convergents are successively better approximations to π. Archimedes (287–212 BCE) demonstrated that $220/71 < \pi < 22/7$ and so $22/7$ is sometimes called the "Archimedean" value of π. Interestingly, the Chinese mathematician Tsu Chung-Chi (430–501 CE) explicitly mentioned 355/113 as being a very accurate approximation to π. We will elaborate on this observation further in Section 6.4.

The next theorem provides a handy technique for readily calculating convergents. In our discussion that follows, p_i represents the numerator of the ith convergent r_i and q_i denotes its denominator.

Theorem 6.3: Let $r = [a_0; a_1, \ldots, a_n]$ and let $r_i = p_i/q_i$ be the ith convergent for $0 \le i \le n$. If we let $p_{-2} = q_{-1} = 0$ and $p_{-1} = q_{-2} = 1$, then

$$p_i = a_i p_{i-1} + p_{i-2} \text{ and } q_i = a_i q_{i-1} + q_{i-2} \text{ for } 0 \le i \le n. \qquad (6.4)$$

Proof: (Induction on i) For $i = 0, r_0 = a_0/1$ and hence $p_0 = a_0 = a_0 \cdot 1 + 0 = a_0 p_{-1} + p_{-2}$ and $q_0 = a_0 \cdot 0 + 1 = a_0 q_{-1} + q_{-2}$.

Now suppose the result is true for $i = k$ for some k where $0 \le k < n$. Then $r_{k+1} = \frac{p_{k+1}}{q_{k+1}} = [a_0; a_1, \ldots, a_k, a_{k+1}] = [a_0; a_1, \ldots, a_{k-1}, a_k + 1/a_{k+1}]$. The latter continued fraction has the same number of partial quotients as the former, though its last partial quotient no longer may be an integer.

Applying the inductive hypothesis to the latter continued fraction, we obtain

$$p_{k+1} = (a_k + 1/a_{k+1}) p_{k-1} + p_{k-2},$$

and

$$q_{k+1} = (a_k + 1/a_{k+1}) q_{k-1} + q_{k-2}$$

It follows that

$$\begin{aligned}
r_{k+1} &= \frac{(a_k + 1/a_{k+1}) p_{k-1} + p_{k-2}}{(a_k + 1/a_{k+1}) q_{k-1} + q_{k-2}} \\
&= \frac{(a_k \cdot a_{k+1} + 1) p_{k-1} + a_{k+1} \cdot p_{k-2}}{(a_k \cdot a_{k+1} + 1) q_{k-1} + a_{k+1} \cdot q_{k-2}} \\
&= \frac{a_{k+1}(a_k p_{k-1} + p_{k-2}) + p_{k-1}}{a_{k+1}(a_k q_{k-1} + q_{k-2}) + q_{k-1}} \\
&= \frac{a_{k+1} p_k + p_{k-1}}{a_{k+1} q_k + q_{k-1}}
\end{aligned}$$

Thus, the case $i = k + 1$ follows and the theorem is proved.

Corollary 6.3.1: Let r be rational and let q_i be the denominator of its ith convergent for $i \ge 0$. Then $q_1 \ge q_0$ and $q_{i+1} > q_i$ for $i \ge 1$. Furthermore, $q_i \ge i$ for $i \ge 0$.

Proof: (Induction on i) By Theorem 6.3,

$$q_2 = a_2 q_1 + q_0 \ge q_1 + 1 > q_1 = a_1 \ge 1 = q_0$$

Assume that $q_i > q_{i-1}$ for some $i \ge 2$. By Theorem 6.3 again,

$$q_{i+1} = a_{i+1} q_i + q_{i-1} \ge q_i + q_{i-1} \ge q_i + 1 > q_i$$

Thus $q_{i+1} > q_i$ for $i \ge 1$. Since $q_1 = 0, q_i \ge i$ for all $i \ge 0$. ∎

Theorem 6.3 suggests a simple method for constructing a table for the successive numerators and denominators of the convergents of a simple continued fraction. Our next example serves as an illustration.

Example 6.3: Let $r = [1; 2, 3, 4, 5]$. After entering all the a_i's and p_i and q_i for negative values of i into the table, formula 6.4 allows us to readily calculate successive values of the p's and q's.

i	-2	-1	0	1	2	3	4
a_i			1	2	3	4	5
p_i	0	1	1	3	10	43	225
q_i	1	0	1	2	7	30	157

So the successive convergents for r are 1, 3/2, 10/7, 43/30, and 225/157.

In the example above, let us calculate the differences $r - r_i$ for $0 \le i \le 3$.

$$225/157 - 1/1 = 68/157$$

$$225/1.57 - 3/2 = -21/314$$

$$225/157 - 10/7 = 5/1099$$

$$225/157 - 43/30 = -1/4710$$

Notice that the convergents are alternately larger and smaller than r. In fact, $r_0 < r_2 < r_4 = r < r_3 < r_1$. Furthermore, the convergents get successively closer to r (in the example the numerators get smaller and the denominators get larger). That these observations hold true in general, we now prove via a sequence of results.

Proposition 6.4: Let $r = [a_0; a_1, \ldots, a_n]$ and p_i and q_i be as in Theorem 6.3 for $0 \le i \le n$. Then

$$p_i q_{i-1} - p_{i-1} q_i = (-1)^{i-1} \tag{6.5}$$

Proof: (Induction on i) For $i = 0$, $p_0 q_{-1} - p_{-1} q_0 = a_0 \cdot 0 - 1 \cdot 1 = (-1)^{0-1}$.

Next assume the assertion is true for $i = k$ for some k with $0 \le k < n$. We now show that the case $i = k+1$ necessarily follows.

$$p_{k+1} q_k - p_k q_{k+1} = (a_{k+1} p_k + p_{k-1}) q_k - p_k (a_{k+1} q_k + q_{k-1}). \text{ by eq} \cdot (6 \cdot 4)$$

$$= -(p_k q_{k-1} - p_{k-1} q_k) = -(-1)^{k-1} = (-1)^k$$

This completes the proof. ■

Corollary 6.4.1:
(a) For $0 \le i \le n$, $\gcd(p_i, q_i) = 1$.
(b) For $1 \le i \le n$, $\gcd(p_i, p_{i-1}) = \gcd(q_i, q_{i-1}) = 1$.

Proof: By formula (6.5), the number 1 can be expressed as a linear combination of p_i and q_i or as p_i and p_{i-1} or as q_i and q_{i-1}. The result follows directly from Corollary 2.1.1 (d). ■

In particular, $\frac{p_i}{q_i}$ is a *reduced* fraction for all i.

Given positive integers p and q, we now have an alternate method to express $\gcd(p,q)$ as a linear combination of p and q (cf. Examples 2.1 and 2.2). The idea is to express the fraction $\frac{p}{q}$ as a finite simple continued fraction $[a_0; a_1, \ldots, a_n]$, use Theorem 6.3 to compute the p_i's and q_i's, and then apply Proposition 6.4 with $i = n$. By way of example, let us redo Examples 2.1 and 2.2 .

Example 6.4: Let us compute $\gcd(54, 231)$ and then find x and y for which $54x + 231y = \gcd(54, 231)$.

$231 = 4 \cdot 54 + 15$, $54 = 3 \cdot 15 + 9$, $15 = 1 \cdot 9 + 6$, $9 = 1 \cdot 6 + 3$, $6 = 2 \cdot 3$. Hence $231/54 = [4; 3, 1, 1, 2]$ and $\gcd(54, 231) = 3$. Notice that we did not have to know $\gcd(231, 54)$ in advance or reduce $231/54$ in order to calculate its finite simple continued fraction. Next, we make a table as in Example 6.3:

i	−2	−1	0	1	2	3	4
a_i			4	3	1	1	2
p_i	0	1	4	13	17	30	77
q_i	1	0	1	3	4	7	18

By formula (6.5), $(-7) \, 77 + 30 \cdot 18 = 1$. Hence $(-7) \cdot 231 + 30 \cdot 54 = 3$.

It is interesting that if we carried out the same process for $231/54$ expressed as the finite simple continued fraction $[4; 3, 1, 1, 1, 1]$, then we obtain a different linear combination. You may wish to verify that we get $11 \cdot 231 + (-47) \cdot 54 = 3$.

Corollary 6.4.2: Let r_i be the ith convergent to $r = [a_0; a_1, \ldots, a_n]$.

(a) $r_i - r_{i-1} = \frac{(-1)^{i-1}}{q_i q_{i-1}}$ for $1 \leq i \leq n$.

(b) $r_i - r_{i-2} = \frac{(-1)^i a_i}{q_i q_{i-2}}$ for $2 \leq i \leq n$.

Proof:

(a) For $1 \leq i \leq n$, $r_i - r_{i-1} = \frac{p_i}{q_i} - \frac{p_{i-1}}{q_{i-1}} = \frac{p_i q_{i-1} - p_{i-1} q_i}{q_i q_{i-1}} = \frac{(-1)^{i-1}}{q_i q_{i-1}}$ by formula 6.5.

(b) For $2 \leq i \leq n$, $r_i - r_{i-2} = (r_i - r_{i-1}) + (r_{i-1} - r_{i-2}) = \frac{(-1)^{i-1}}{q_i q_{i-1}} + \frac{(-1)^{i-2}}{q_{i-1} q_{i-2}}$ by part (a) $= \frac{(-1)^i (q_i - q_{i-2})}{q_i q_{i-1} q_{i-2}}$.

But formula (6.4) implies that $a_i = (q_i - q_{i-2})/q_{i-1}$. The result follows. ∎

We are now prepared to prove the main results of this section.

Theorem 6.5: Let r_i be the ith convergent to $r = [a_0; a_1, \ldots, a_n]$. Then

$$r_0 < r_2 < r_4 < \cdots < r_n = r < \cdots < r_5 < r_3 < r_1.$$

Proof: If i is even, then $r_i - r_{i-2} > 0$ by Corollary 6.4.2(b) and the fact the q's are positive. Hence the even-indexed convergents form a monotonically increasing sequence. If i is odd, then $r_i - r_{i-2} < 0$ by Corollary 6.4.2 (b). Hence the odd-indexed convergents form a monotonically decreasing sequence. If n is even, it follows that r_n is the largest even-numbered convergent; and if n is odd, then r_n is the smallest odd-numbered convergent. The proof is complete if it can be shown that any even-numbered convergent is less than any odd-numbered one.

Let $0 \le h < k \le n$ where h is even and k is odd. The inequality $h < k$ implies $h \le k - 1$. From our observation above, $r_h \le r_{k-1}$.

By Corollary 6.4.2 (a),

$$r_k - r_{k-1} = \frac{(-1)^{k-1}}{q_k q_{k-1}} > 0 \quad \text{and so } r_{k-1} < r_k$$

Combining inequalities, $r_h < r_k$.

If $0 \le k < h \le n$ with h even and k odd, then $k \le h - 1$. From our initial comments, $r_{h-1} \le r_k$. But

$$r_h - r_{h-1} = \frac{(-1)^{h-1}}{q_h q_{h-1}} < 0 \quad \text{and so } r_h < r_{h-1}.$$

Again, by combining inequalities, $r_h < r_k$. The proof is complete. ∎

Corollary 6.5.1: If $r_i = \frac{p_i}{q_i}$ is the ith convergent to r, then $|r - r_i| < \frac{1}{q_i^2}$.

Proof: The proof is left as Exercise 6.1.7. ∎

Theorem 6.6: Let $r = [a_0; a_1, \ldots, a_n]$ with $a_n > 1$. For $1 \le i \le n$,

$$(a) \ |rq_i - p_i| < |rq_{i-1} - p_{i-1}| \tag{6.6}$$

and

$$(b) \ |r - r_i| < |r - r_{i-1}|.$$

Proof:
(a) Let $A_i = [a_i, \ldots, a_n]$ for $0 \le i \le n$ as in the proof of Proposition 6.2. Then $r = [a_0, \ldots, a_i, A_{i+1}] = p_n/q_n$ for $0 \le i \le n - 1$. By formula (6.4),

$$p_n = A_{i+1} p_i + p_{i-1} \text{ and } q_n = A_{i+1} q_i + q_{i-1}. \text{ Thus}$$

$$r = \frac{A_{i+1} p_i + p_{i-1}}{A_{i+1} q_i + q_{i-1}}.$$

So, $A_{i+1}(rq_i - p_i) = -(rq_{i-1} - p_{i-1})$. Taking absolute values, $A_{i+1}|rq_i - p_i| = |rq_{i-1} - p_{i-1}|$. But $A_{i+1} > 1$ for $0 \le i \le n - 1$ (recall that

$A_n = a_n > 1$ by hypothesis). Hence

$$|rq_i - p_i| < |rq_{i-1} - p_{i-1}| \text{ for } 0 \le i \le n-1$$

For $i = n$, $|rq_n - p_n| = 0$ while

$$|rq_{n-1} - p_{n-1}| = \tfrac{1}{q_n}|p_n q_{n-1} - p_{n-1}q_n| = \tfrac{1}{q_n} > 0 \text{ by Proposition 6.4.}$$

(b) By Corollary 6.3.1, $q_i = a_i q_{i-1} + q_{i-2} \ge q_{i-1}$ for $i \ge 1$. So

$$|r - r_i| = \frac{1}{q_i}|rq_i - p_i| < \frac{1}{q_i}|rq_{i-1} - p_{i-1}| \text{ by part (a). But}$$

$$\frac{1}{q_i}|rq_{i-1} - p_{i-1}| = \frac{q_{i-1}}{q_i}|r - r_{i-1}| \le |r - r_{i-1}|$$

The result follows. ∎

Exercise 6.1

1. Derive the simple continued fraction expansions for the following rational numbers: (a) $\frac{55}{89}$ (b) $\frac{89}{55}$ (c) $\frac{89}{144}$ (d) $\frac{-245}{81}$ (e) $\frac{2718}{1000}$ (f) $\frac{1414}{1000}$.
2. Use Theorem 6.3 to determine all the convergents for the fractions in Exercise 1
3. Under what conditions is $p_{i+1} > p_i$ for all i? (cf. Corollary 6.3.1).
4. Find the rational numbers represented by the following simple continued fractions:

   (a) $[2; 1, 2, 1, 2]$, (b) $[-3, 1, 5, 2, 4]$, (c) $[0; 2, 4, 6, 8]$, (d) $[-1, 1, 1, 4, 6, 8]$.
5. Show that every rational number can be represented by a finite simple continued fraction of the form $[a_0; \ldots, a_n]$, where n is even
6. (a) Let $0 < p < q$. If $\frac{p}{q} = [a_0; a_1, \ldots, a_n]$, then what is $\frac{q}{p}$?

   (b) How are the convergents of $\frac{q}{p}$ related to those of $\frac{p}{q}$?
7. Prove Corollary 6.5.1
8. Let $r = [1; 1, \ldots, 1]$, the simple continued fraction with n 1's. Show that $r = \frac{F_{n+1}}{F_n}$ where F_n is the nth Fibonacci number.
9. Let $r = [a_0; a_1, \ldots, a_n]$ have nth convergent $r_n = \frac{p_n}{q_n}$. Show that $p_n \ge F_n$ and $q_n \ge F_n$ where F_n is the nth Fibonacci number.
10. Let $r = [a_0; a_1, \ldots, a_n]$ where $n > 2$ and $a_1 > 1$

    Show that $-r = [-1 - a_0, 1, a_1 - 1, a_2, \ldots, a_n]$.
11. Define the *nearest integer continued fraction* for r by defining a_i to be the integer nearest to $\frac{r_{i-1}}{r_i}$ for $0 \le i \le n$ in the proof of Proposition 6.1 (cf. Exercise 2.1.26). Note that the a_i 's need not be positive integers and hence the nearest integer continued fraction is not a simple continued fraction.

    (a) Determine the nearest integer continued fraction expansions for $\frac{46}{133}$ and $\frac{-56}{27}$. Compare the results with the corresponding simple continued fractions.

(b) Determine the nearest integer continued fraction expansions for $\frac{8}{13}$, $\frac{13}{21}$, $\frac{21}{34}$, and $\frac{55}{89}$. Compare the results with the corresponding simple continued fractions.

(c) Determine the nearest integer continued fraction expansion for $\frac{F_{i-1}}{F_i}$ where F_i is the ith Fibonacci number.

12. Apply the technique of Example 6.4 to express
    (a) the gcd(233, 377) as a linear combination of 233 and 377
    (b) the gcd(627, 3267) as a linear combination of 627 and 3267
    (c) the gcd(2225, 3625) as a linear combination of 2225 and 3625.

13. Verify Proposition 6.4 and all its corollaries for $r = [2; 3, 7, 2, 1, 5]$

14. Verify Theorems 6.5 and 6.6 for $r = [1; 2, 1, 4, 2, 1, 3]$

15. (a) Determine the convergents for $\frac{17}{12}$ and compare with those for $\frac{17}{7}$.
    (b) Determine the convergents for $\frac{107}{74}$ and compare with those for $\frac{107}{13}$.
    (c) If $\frac{p_i}{q_i}$ is the ith convergent to $[a_0; a_1, \ldots, a_n]$ where $a_0 \geq 1$, then show that

$$[a_n; \ldots, a_1, a_0] = \frac{p_n}{p_{n-1}}$$

16. Show that $[a_0; a_1, \ldots, a_n]$ and $[a_0; a_n, \ldots, a_1]$ have the same denominator

17. Define the *negative continued fraction* for r by requiring that the b_j 's are all equal to –1 in Definition 6.1.
    (a) Determine the negative continued fractions for $r = \frac{5}{7}$, $\frac{17}{11}$, $\frac{55}{89}$.
    (b) Determine the negative continued fraction for $\frac{n+1}{n}$ for $n > 1$.
    (c) Investigate the analogs of the propositions in this section for negative continued fractions (especially Theorems 6.3, 6.4, and 6.5).

18. (a) Show that for all $n \leq 12$, there are positive integers a and b with $n = a + b$ and a/b has all its partial quotients equal to 1 or 2.
    (b) Show that the above is false for $n = 23$.

6.2 Farey fractions

In this section we introduce the notion of Farey fractions and develop some of their important properties. We then apply those properties to obtain initial results on the rational approximation of irrationals. Though easy to understand and enjoyable to investigate, perhaps surprisingly Farey fractions find deep application in the areas of Diophantine approximation and analytic number theory.

Definition 6.3: Let $\mathcal{F}_n$ denote the set of rational numbers $\frac{a}{b}$ with $0 \leq a \leq b \leq n$ and $\gcd(a, b) = 1$ arranged in increasing order. We call $\mathcal{F}_n$ the sequence of *Farey fractions* of order n.

Example 6.5: Listed below are the Farey fractions of orders 1 through 6:

$$\mathcal{F}_1: \frac{0}{1}, \frac{1}{1}$$

$$\mathcal{F}_2: \frac{0}{1}, \frac{1}{2}, \frac{1}{1}$$

$$\mathcal{F}_3: \frac{0}{1}, \frac{1}{3}, \frac{1}{2}, \frac{2}{3}, \frac{1}{1}$$

$$\mathcal{F}_4: \frac{0}{1}, \frac{1}{4}, \frac{1}{3}, \frac{1}{2}, \frac{2}{3}, \frac{3}{4}, \frac{1}{1}$$

$$\mathcal{F}_5: \frac{0}{1}, \frac{1}{5}, \frac{1}{4}, \frac{1}{3}, \frac{2}{5}, \frac{1}{2}, \frac{3}{5}, \frac{2}{3}, \frac{3}{4}, \frac{4}{5}, \frac{1}{1}$$

$$\mathcal{F}_6: \frac{0}{1}, \frac{1}{6}, \frac{1}{5}, \frac{1}{4}, \frac{1}{3}, \frac{2}{5}, \frac{1}{2}, \frac{3}{5}, \frac{2}{3}, \frac{3}{4}, \frac{4}{5}, \frac{5}{6}, \frac{1}{1}$$

Notice that if a/b is a member of $\mathcal{F}_n$, then a/b is a member of $\mathcal{F}_m$ for all $m \geq n$. Since there are two terms in $\mathcal{F}_1$ and $\phi(n)$ new Farey fractions of order n that were not Farey fractions of order $n-1$, the number of elements in $\mathcal{F}_n$ is $1 + \sum_{k=1}^{n} \phi(k)$. In Section 8.3 we will study this sum in greater detail to derive the average order of $\phi(n)$.

The Farey fractions are named after John Farey, a British geologist and surveyor, who rediscovered the property described in Theorem 6.8 and stated it without proof in the *Philosophical Magazine* in 1816. Later that year, A.L. Cauchy (1789–1857) noticed Farey's observation and supplied a proof. Cauchy named the sequence of fractions after Farey and his name has been attached ever since. In fact, C. Haros had proved the same result in 1802.

Theorem 6.7: Let $\frac{a}{b}$ and $\frac{c}{d}$ be successive terms in $\mathcal{F}_n$. Then
(a) $b + d \geq n + 1$
(b) $ad - bc = -1$

Proof: (a) Consider the fraction $\frac{a+c}{b+d}$. On the one hand,

$$a(b+d) - b(a+c) = ad - bc < 0 \text{ since } \frac{a}{b} < \frac{c}{d}. \text{ So } \frac{a}{b} < \frac{a+c}{b+d}$$

On the other hand,

$$(a+c)d - (b+d)c = ad - bc < 0. \text{ So } \frac{a+c}{b+d} < \frac{c}{d}$$

Hence $\frac{a+c}{b+d}$ lies between $\frac{a}{b}$ and $\frac{c}{d}$. Since $\frac{a}{b}$ and $\frac{c}{d}$ are successive terms in $\mathcal{F}_n$, the fraction $\frac{a+c}{b+d}$ does not appear in $\mathcal{F}_n$ and so its denominator $b + d \geq n + 1$.
(b) Since $\gcd(a, b) = 1$, the equation

$$ay - bx = -1 \tag{6.8}$$

is solvable for some integers $x = x_0, y = y_0$. In addition, $a(y_0 + mb) - b(x_0 + ma) = ay_0 - bx_0 = -1$. So for any integer m, eq. (6.8) is solvable with $x = x_0 + ma, y = y_0 + mb$. Now choose M such that

$$n - b < y_0 + Mb \leq n \text{ and then let } y = y_0 + Mb, x = x_0 + Ma$$

By eqs. (6.8), $\gcd(x, y) = 1$ and $n - b < y \leq n$.

The above implies that $\frac{x}{y}$ is a member of $\mathcal{F}_n$.

Applying eq. (6.8), $\frac{x}{y} = \frac{1 + ay}{by} = \frac{a}{b} + \frac{1}{by} > \frac{a}{b}$.

If it were true that $\frac{x}{y} > \frac{c}{d}$, then $\frac{x}{y} - \frac{c}{d} = \frac{dx - cy}{dy} \geq \frac{1}{dy}$. Hence

$$\frac{1}{by} = \frac{bx - ay}{by} = \frac{x}{y} - \frac{a}{b} = \left(\frac{x}{y} - \frac{c}{d} \right) + \left(\frac{c}{d} - \frac{a}{b} \right)$$

$$\geq \frac{1}{dy} + \frac{bc - ad}{bd} \geq \frac{1}{dy} + \frac{1}{bd} = \frac{b + y}{bdy}$$

But $b + y > n$ because y was chosen so that $n - b < y$. Thus $\frac{1}{by} \geq \frac{b+y}{bdy} > \frac{n}{bdy} \geq \frac{1}{by}$ since $d \leq n$. This is a contradiction. Hence $\frac{x}{y} = \frac{c}{d}$ and so $ad - bc = -1$. ∎

Notice that the proof of Theorem 6.7(b) provides an algorithm for determining the successor of a term in $\mathcal{F}_n$.

Example 6.6: Find the successor of $\frac{2}{7}$ in $\mathcal{F}_{11}$.
Solution: The appropriate equation, $2y - 7x = -1$, is solvable for $x = 1$ and $y = 3$. The general solution is then $x = 1 + 2m$ and $y = 3 + 7m$ for integral m (review Corollary 2.1.2 if necessary). Choose y so that $11 - 7 < y \leq 11$. For $m = 1, y = 10$ and $x = 3$. Thus $\frac{x}{y} = \frac{3}{10}$ is the successor of $\frac{2}{7}$ in $\mathcal{F}_{11}$. (By the next theorem, $\frac{3}{10}$ is the successor of $\frac{2}{7}$ in $\mathcal{F}_m$ for $10 \leq m \leq 16$.)

Definition 6.4: If $\frac{a}{b} < \frac{c}{d}$ are consecutive terms in $\mathcal{F}_n$, then their *mediant* is the rational number $\frac{a+c}{b+d}$.

Theorem 6.8: If $\frac{a}{b} < \frac{c}{d} < \frac{f}{g}$ are three consecutive terms in $\mathcal{F}_n$, then $\frac{c}{d} = \frac{a+f}{b+g}$.

Proof: By Theorem 6.7(b), $ad - bc = -1 = cg - df$. So $(a + f)d = c(b + g)$ and hence

$$\frac{c}{d} = \frac{a + f}{b + g}$$

∎

Given $\mathcal{F}_{n-1}$, note that Theorem 6.8 is especially useful for determining where to place the $\phi(n)$ new fractions in $\mathcal{F}_n$ without converting any fractions to decimals. For example, given $\mathcal{F}_6$ in example 6.4, take the mediants of successive terms and add in all entries with denominator of 7. For instance, $\frac{4}{7} = \frac{1}{2} + \frac{3}{5}$ must appear between $\frac{1}{2}$ and $\frac{3}{5}$. We get

$$\mathcal{F}_7: \frac{0}{1}, \frac{1}{7}, \frac{1}{6}, \frac{1}{5}, \frac{1}{4}, \frac{2}{7}, \frac{1}{3}, \frac{2}{5}, \frac{3}{7}, \frac{1}{2}, \frac{4}{7}, \frac{3}{5}, \frac{2}{3}, \frac{5}{7}, \frac{3}{4}, \frac{4}{5}, \frac{5}{6}, \frac{6}{7}, \frac{1}{1}$$

We can state now our first theorem involving the rational approximation of real numbers.

Theorem 6.9: If x is any real number and n a positive integer, then there is a reduced fraction $\frac{p}{q}$ such that $0 < q \le n$ and

$$\left| x - \frac{p}{q} \right| < \frac{1}{q(n+1)} \tag{6.9}$$

Proof: Assume that $0 \le x < 1$. Let $\frac{a}{b} \le x < \frac{c}{d}$ where $\frac{a}{b}$ and $\frac{c}{d}$ are successive terms of $\mathcal{F}_n$. Form the mediant $\frac{a+b}{c+d}$. Then either

(i) $x \in \left[\frac{a}{b}, \frac{a+c}{b+d} \right)$ or (ii) $x \in \left[\frac{a+c}{b+d}, \frac{c}{d} \right)$.

On the one hand, $\frac{a+c}{b+d} - \frac{a}{b} = \frac{bc-ad}{b(b+d)} = \frac{1}{b(b+d)}$ by Theorem 6.7(b).
But $b + d \ge n + 1$ by Theorem 6.7(a). So in case (i), let $p = a$ and $q = b$ and

$$\left| x - \frac{p}{q} \right| < \frac{1}{b(b+d)} < \frac{1}{q(n+1)}.$$

On the other hand, $\frac{c}{d} + \frac{a+c}{b+d} = \frac{bc-ad}{d(b+d)}$. So in case (ii), let $p = c$ and $q = d$ and

$$\left| x - \frac{p}{q} \right| < \frac{1}{b(b+d)} < \frac{1}{q(n+1)}.$$

Finally, if $m \le x < m + 1$ for some integer m, then $x = m + x_0$ where $0 \le x_0 < 1$. From above, there is a reduced fraction $\frac{p_0}{q_0}$ such that

$$\left| x_0 - \frac{p_0}{q_0} \right| < \frac{1}{q(n+1)}$$

Since $\gcd(p_0, q_0) = 1$, it follows that $\gcd(mq_0 + p_0, q_0) = 1$. So let $p = mq_0 + p_0$ and $q = q_0$ and the proof is complete. ∎

Example 6.7: Find a fraction $\frac{p}{q}$ such that $\left| e - \frac{p}{q} \right| < \frac{1}{8q}$ where $1 \le q \le 7$. Here $e = 2.71828$ ... is the base of the natural logarithm function.

Solution: Let $x = e$. Since $2 \le e < 3$, we let $x_0 = e - 2 = .71828$ ... as in the proof of Theorem 6.9. In this case, $n = 7$ and so we locate where x_0 falls in $\mathcal{F}_7$. We readily find that $5/7 = .71428 ... < x_0 < 3/4 = .75$. The mediant of $5/7$ and $3/4$ is $8/11 = .727272$ .... Since x_0 lies between $5/7$ and $8/11$, we let $p_0 = 5$ and $q_0 = 7$. Finally, let $p = mq_0 + p_0 = 2(7) + 5 = 19$ and $q = q_0 = 7$. As a check we see $\left| e - \frac{19}{7} \right| < .004001 < \frac{1}{7(8)}$.

Exercise 6.2

1. Display the Farey sequences $\mathcal{F}_8$ and $\mathcal{F}_9$
2. (a) Verify Theorem 6.7 on the terms of $\mathcal{F}_6$.
   (b) Verify Theorem 6.8 on the terms of $\mathcal{F}_6$.
3. Verify Theorem 6.9 for $n = 7$ on the following real numbers:
   (a) $\frac{12}{17}$, (b) 3.14, (c) $\sqrt{2}$, (d) $(10)^{1/3}$, and (e) e^2.
4. (a) Find the successor of $5/9$ in $\mathcal{F}_{14}$.
   (b) Find the successor of $3/4$ in $\mathcal{F}_{16}$.
5. (a) Show if $n > 1$, then no two consecutive terms of $\mathcal{F}_n$ have the same denominator.
   (b) Show that there is no n for which all sets of three consecutive terms in $\mathcal{F}_n$ have different denominators.
6. (a) Show that the sum of all the fractions in $\mathcal{F}_n$ is half the number of elements in $\mathcal{F}_n$.
   (b) Show that the sum of the denominators for all the fractions in $\mathcal{F}_n$ is twice the sum of all the numerators.
7. Let $\frac{a}{b} \in \mathcal{F}_n$ have simple continued fraction expansion $[a_0; a_1, \dots, a_m]$.
   Show that $a_0 + \dots + a_m \le n$.
8. Show that if $x \in \mathbb{R}$, then there exist infinitely many pairs of integers p and q such that

$$|qx - p| < \frac{1}{q}$$

9. Find a fraction $\frac{p}{q}$ with $1 \le q \le 7$ such that $\left| \sqrt{2} - \frac{p}{q} \right| < \frac{1}{8q}$ as guaranteed by Theorem 6.9.
10. Find a fraction $\frac{p}{q}$ with $1 \le q \le 9$ such that $\left| \pi - \frac{p}{q} \right| < \frac{1}{10q}$.
11. Find a reduced fraction $\frac{p}{q}$ with $1 \le q \le 9$ such that $\left| \frac{31}{47} - \frac{p}{q} \right| < \frac{1}{10q}$.
12. Show that if $a/b < c/d$ are consecutive terms in $\mathcal{F}_n$ and $b + d = n + 1$, then $\frac{a+b}{c+d}$ is the only term in $\mathcal{F}_{n+1}$ between a/b and c/d.

6.3 Infinite simple continued fractions

In this section we extend the notion of simple continued fractions to include those with an infinite number of partial quotients $a_i, i \geq 0$. Our first order of business is to show that such a fraction is well-defined, i.e., represents a real number. Please note that all our results in Section 6.1 still hold true with the upper limits on the index i removed.

Proposition 6.10: If $[a_0; a_1, \ldots]$ is an infinite simple continued fraction with convergents $r_{2n}, n \geq 0$, then there is a real number $x = \lim\limits_{n \to \infty} r_{2n}$.

Proof. By Theorem 6.5, the even-indexed convergents form a monotonically increasing sequence bounded above by r_1 (or any other odd-indexed convergent). By the bounded convergence theorem from calculus, $\lim_{n \to \infty} r_{2n}$ exists; call it L. Similarly, the odd-indexed convergents form a monotonically decreasing sequence bounded below by r_0 say.

Let $M = \lim_{n \to \infty} r_{2n-1}$. We need only show that $L = M$.

For all $n \geq 1$,

$$0 \leq |r_{2n} - r_{2n-1}| = \frac{1}{q_{2n} q_{2n-1}} < \frac{1}{2n(2n-1)}$$

by Corollary 6.4.2 and the fact $q_k \geq k$ (Corollary 6.3.1). So

$$0 \leq \lim_{n \to \infty} |r_{2n} - r_{2n-1}| = |L - M| \leq \lim_{n \to \infty} \frac{1}{2n(2n-1)} = 0$$

Hence $L = M$ and the proof is complete by defining $x = L$. ∎

As one consequence of this proposition, Theorem 6.5 can now be revised. If r_n is the nth convergent to a real number $x = [a_0; a_1, \ldots]$, then $r_0 < r_2 < r_4 < \cdots < x < \cdots < r_5 < r_3 < r_1$.

We now prove the analog of Proposition 6.2 for irrational numbers.

Theorem 6.11 (Uniqueness of Continued Fraction Expansions over $\mathbb{R}$):

(a) If x is represented as an infinite simple continued fraction, then x is irrational.
(b) If x is irrational, then x is expressible in a unique way as an infinite simple continued fraction.

Proof: (a) Let $x = [a_0; a_1, \ldots]$ with convergents r_0, r_1, etc. By Theorem 6.5, for every $n \geq 1$, x lies between r_{n-1} and r_n. In fact, by Corollary 6.5.1,

$$0 < |x - r_n| < \frac{1}{q_n q_{n-1}}$$

If x were rational, then $x = \frac{a}{b}$ for some integers a and b. It would follow that

$$0 < |x - r_n| = \left| \frac{a}{b} - \frac{p_n}{q_n} \right| < \frac{1}{q_n q_{n-1}} \tag{6.10}$$

Now choose n such that $q_{n-1} > |b|$ (e.g., $n = |b| + 2$). Formula (6.10) implies

$$0 < |a \cdot q_n - b \cdot p_n| < \frac{|b|}{q_{n-1}} < 1$$

But this is a contradiction since a, b, p_n, and q_n are all integers. So x is irrational.

(b) (Existence): Let $x = A_0$ and set $a_0 = [A_0]$. Now let

$$A_1 = \frac{1}{A_0 - a_0} \quad \text{and } a_1 = [A_1]$$

Similarly, define a_n recursively for all $n \geq 1$ by

$$A_n = \frac{1}{A_{n-1} - a_{n-1}} \quad \text{and } a_n = [A_n] \tag{6.11}$$

Notice that A_0 being irrational implies that A_1 is irrational and hence $A_2, A_3, \ldots$ are all irrational in turn. So the sequence $a_0, a_1, \ldots$ does not terminate. Furthermore, $A_0 - a_0 < 1$ and so $A_1 > 1$ and $a_1 \geq 1$. Analogously, by eq. (6.11), $a_n \geq 1$ for all $n \geq 1$. It follows from eq. (6.11) that

$$x = a_0 + \frac{1}{A_1} = a_0 + \frac{1}{a_1 + \frac{1}{A_2}} = \cdots$$

$$= [a_0; a_1, \ldots, a_n, A_{n+1}] \quad \text{for all } n \geq 1$$

Let $r_n = \frac{p_n}{q_n}$ be the nth convergent of $[a_0; a_1, \ldots, a_n, A_{n+1}]$ and then $x = \frac{p_{n+1}}{q_{n+1}}$ is the $(n+1)^{th}$ convergent. Hence

$$0 \leq |x - r_n| = \left| \frac{A_{n+1} p_n + p_{n-1}}{A_{n+1} q_n + q_{n-1}} - \frac{p_n}{q_n} \right| \quad \text{by eq.(6.4)}$$

$$= \left| \frac{q_n p_{n-1} - p_n q_{n-1}}{q_n (A_{n+1} q_n + q_{n-1})} \right| = \left| \frac{1}{q_n (A_{n+1} q_n + q_{n-1})} \right| < \frac{1}{q_n q_{n-1}} \leq \frac{1}{n(n-1)}$$

So $\lim_{h \to \infty} |x - r_n| = 0$ and $x = [a_0; a_1, \ldots]$.

(Uniqueness): Suppose $x = [a_0; a_1, \ldots] = [b_0; b_1, \ldots]$. Let r_n be the nth convergent to $[a_0; a_1, \ldots]$ and R_n be the nth convergent to $[b_0; b_1, \ldots]$. In addition, set

$$A_n = [a_n; a_{n+1}, \ldots] \text{ and } B_n = [b_n; b_{n+1}, \ldots]$$

As in the proof of Proposition 6.2,

$$a_n = [A_n], \quad A_{n+1} = \frac{1}{A_n - a_n}, \quad b_n = [B_n], \quad B_{n+1} = \frac{1}{B_n - b_n} \text{ for } n \geq 0$$

We will show that $a_n = b_n$ for all $n \geq 0$ by induction.

Since $x = A_0 = B_0$, $a_0 = [A_0] = [B_0] = b_0$.

Now suppose that $A_n = B_n$ for some $n \geq 0$. Then $a_n = b_n$ as above.

But $A_{n+1} = \frac{1}{A_n - a_n} = \frac{1}{B_n - b_n} = B_{n+1}$ and hence $a_{n+1} = [A_{n+1}] = [B_{n+1}] = b_{n+1}$.

Thus $a_n = b_n$ for all n and the continued fraction expansion of x is unique. ∎

The proof of Theorem 6.11 provides a method for determining the continued fraction expansions for a given irrational number. In the following example, we apply formula (6.11) until we encounter a repetitive pattern. Our next two theorems will address under what conditions an irrational number eventually has such a repetitive pattern.

Example 6.8: Find the simple continued fraction expansion for $x = \sqrt{7}$.

Solution: Since $2 \leq \sqrt{7} < 3$, $a_0 = \left[\sqrt{7}\right] = 2$.

$$\text{Hence } \sqrt{7} = 2 + \left(\sqrt{7} - 2\right)] \text{ where } 0 < \sqrt{7} - 2 < 1.$$

$$\frac{1}{\left(\sqrt{7}-2\right)} = \frac{\sqrt{7}+2}{3}. \text{ But } 1 \leq \frac{\sqrt{7}+2}{3} < 2. \text{ So } a_1 = 1.$$

$$\text{Hence } \frac{\sqrt{7}+2}{3} = 1 + \frac{\sqrt{7}-1}{3} \cdot \frac{3}{\sqrt{7}-1} = \frac{\sqrt{7}+1}{2} \text{ and } 1 \leq \frac{\sqrt{7}+1}{2} < 2. \text{ So } a_2 = 1.$$

$$\text{Hence } \frac{\sqrt{7}+1}{2} = 1 + \frac{\sqrt{7}-1}{2} \cdot \frac{2}{\sqrt{7}-1} = \frac{\sqrt{7}+1}{3} \text{ and } 1 \leq \frac{\sqrt{7}+1}{3} < 2. \text{ So } a_3 = 1.$$

$$\frac{\sqrt{7}+1}{3} = 1 + \frac{\sqrt{7}-2}{3} \cdot \frac{3}{\sqrt{7}-2} = \sqrt{7}+2 \text{ and } 4 \leq \sqrt{7}+2 < 5. \text{ So } a_4 = 4.$$

$$\text{Hence, } \sqrt{7} + 2 = 4 + \left(\sqrt{7} - 2\right).$$

Thus, $A_5 = A_1$ and $a_5 = a_1 = 1$. Similarly, $a_{n+4} = a_n$ for all $n \geq 1$.

Therefore, $\sqrt{7} = \left[2; \overline{1,1,1,4}\right]$ where the bar over the digits means that the pattern 1114 is repeated ad infinitum.

The next example suggests one important application of the continued fraction expansion for an irrational number.

Example 6.9: Find a rational approximation to $\sqrt{7}$ with an error less than 0.0001.

Solution: By Example 6.8, $\sqrt{7} = \left[2; \overline{1,1,1,4}\right]$. Recall Corollary 6.5.1 which states that $\left|\sqrt{7} - r_n\right| < \frac{1}{q_n^2}$ where $r_n = \frac{p_n}{q_n}$ is the nth convergent to $\sqrt{7}$. We proceed by making a chart for p_n and q_n as shown in Example 6.3.

n	-2	-1	0	1	2	3	4	5	6	7	8
a_n			2	1	1	1	4	1	1	1	4
p_n	0	1	2	3	5	8	37	45	82	127	590
q_n	1	0	1	1	2	3	14	17	31	48	223

By Corollary 6.5.1, it suffices to find q_n such that $q_n^2 > 10000$. But $q_8^2 > 10000$ and hence $\sqrt{7}$ is approximately $\frac{590}{223}$ with an error less than 0.0001. In fact, $\sqrt{7} - \frac{590}{223} < 0.0000115$. (We will return to similar questions in Section 6.4.)

Given an infinite simple continued fraction with a repeating pattern, it is a straight-forward matter to determine what irrational number it represents. Such infinite continued fractions are called *periodic continued fractions*.

Definition 6.5: The infinite simple continued fraction $[a_0; a_1, \ldots]$ is called *periodic* if there are positive integers r and n such that $a_{m+r} = a_m$ for all $m \geq n$. The smallest such integer r is called the *length* of the period.

For example, $[2; 4, 3, 7, \overline{6, 2, 4}]$ is a periodic continued fraction with period length 3 (see Exercise 6.3.1(e)).

Example 6.10: Determine the irrational number represented by $x = [2; 1, 3, \overline{1, 2}]$.
Solution: $x = [2; 1, 3, a_3]$ where $a_3 = [1; 2, a_3]$.

$$a_3 = 1 + \frac{1}{2 + \frac{1}{a_3}} = \frac{3a_3 + 1}{2a_3 + 1}$$

So $\dfrac{3a_3 + 1}{2a_3 + 1}$ which implies that

$$2a_3^2 - 2a_3 - 1 = 0$$

By the quadratic formula, $a_3 = \frac{1}{2} \pm \frac{\sqrt{3}}{2}$. But $a_3 > 0$ implies $a_3 = \frac{1}{2} \pm \frac{\sqrt{3}}{2}$. So

$$x = \left[2; 1, 3, \frac{1}{2} + \frac{\sqrt{3}}{2}\right] = 2 + \cfrac{1}{1 + \cfrac{1}{3 + \cfrac{2}{1 + \sqrt{3}}}}$$

$$= 2 + \frac{5 + 3\sqrt{3}}{6 + 4\sqrt{3}}$$

$$= \frac{17 + 11\sqrt{3}}{6 + 4\sqrt{3}}$$

$$= \frac{15 + \sqrt{3}}{6}$$

Definition 6.6: An irrational number x is a *quadratic surd* if x is a root of a quadratic polynomial with rational coefficients.

Lemma 6.12.1: The number x is a quadratic surd if and only if $x = a + b\sqrt{c}$ where a and b are rationals and c is a square-free positive integer.

Proof: ($\Rightarrow$) If x is a quadratic surd, then $x^2 + Ax + B = 0$ for some $A, B \in \mathbb{Q}$.

By the quadratic formula, $x = \frac{-A}{2} \pm \frac{1}{2}\sqrt{A^2 - 4B}$ where necessarily $A^2 - 4B > 0$ and $A^2 - 4B$ is not a perfect rational square. Thus $\sqrt{A^2 - 4B} = r\sqrt{c}$ for some rational r and square-free integer $c > 0$. So $x = \frac{-A}{2} \pm \frac{r}{2}\sqrt{c}$. Let $a = \frac{-A}{2}$ and $b = \pm\frac{r}{2}$ and $x = a + b\sqrt{c}$.

($\Leftarrow$) Let $x = a + b\sqrt{c}$ where a and b are rationals and c is a square-free positive integer. Then $x^2 = (a^2 + b^2 c) + 2ab\sqrt{c}$. So $x^2 - 2ax + (a^2 - b^2 c) = 0$. Let $A = -2a \in \mathbb{Q}$ and $B = a^2 - b^2 c \in \mathbb{Q}$. Hence x is a quadratic surd. ∎

Theorem 6.12: If x has a periodic simple continued fraction expansion, then x is a quadratic surd.

Proof: Let $x = [a_0; a_1, \ldots, a_{r-1}, \overline{a_r, \ldots, a_{r+k}}]$ and set $y = [\overline{a_r, \ldots, a_{r+k}}]$.
By Theorem 6.11, y is irrational and $y = [a_r; a_{r+1}, \ldots, a_{r+k}, y]$.
By Proposition 6.3,

$$y = \frac{y \cdot P_k + P_{k-1}}{y \cdot Q_k + Q_{k-1}}$$

where P_i/Q_i is the ith convergent of y. Hence

$$Q_k \cdot y^2 + (Q_{k-1} - P_k) \cdot y - P_{k-1} = 0$$

So y is a quadratic surd (just divide through by Q_k).
    By Lemma 6.12.1, we can write $y = a + b\sqrt{c}$ where $a, b \in \mathbb{Q}$ and c is a positive square-free integer. But by Proposition 6.3 again,

$$x = [a_0; a_1, \ldots, a_{r-1}, y] = \frac{y \cdot p_{r-1} + p_{r-2}}{y \cdot q_{r-1} + q_{r-2}}$$

where p_i/q_i is the ith convergent to x. Cross multiplying we obtain,

$$x(aq_{r-1} + q_{r-2} + bq_{r-1}\sqrt{c}) = ap_{r-1} + p_{r-2} + bp_{r-1}\sqrt{c}$$

Let $a_1 = ap_{r-1} + p_{r-2}, a_2 = aq_{r-1} + q_{r-2}, b_1 = bp_{r-1}$, and $b_2 = bq_{r-1}$. Then

$$x = \frac{a_1 + b_1\sqrt{c}}{a_2 + b_2\sqrt{c}} = a + b\sqrt{c}$$

for

$$a = \frac{a_1 a_2 - b_1 b_2 c}{a_2^2 - b_2^2 c} \quad \text{and} \quad b = \frac{a_2 b_1 - a_1 b_2}{a_2^2 - b_2^2 c} \qquad ∎$$

The converse of Theorem 6.12 is also true as was proved by Lagrange in 1770. This significant result completes our discussion of periodic simple continued fractions for this section. However, we will make a detailed analysis of a special subclass of quadratic surds in Section 6.5.

Theorem 6.13 (Periodic Continued Fraction Theorem): If x is a quadratic surd, then x has a periodic simple continued fraction expansion.

Proof: Since x is a quadratic surd, x satisfies a quadratic equation

$$x^2 + Ax + B = 0 \tag{6.12}$$

where A, B are rational and $A^2 - 4B > 0$ with $A^2 - 4B$ not a rational square.

Let $x = [a_0; a_1, \ldots]$ be the infinite simple continued fraction representation of x. Define $A_n = [a_n; a_{n+1}, \ldots]$ for all $n \geq 0$.

Then $x = [a_0; a_1, \ldots, a_{n-1}, A_n]$ for all $n \geq 1$ and by Theorem 6.3,

$$x = \frac{A_n p_{n-1} + p_{n-2}}{A_n q_{n-1} + q_{n-2}} \tag{6.13}$$

Substituting eq. (6.13) into eq. (6.12) and clearing the denominator gives us
$$\left(A_n p_{n-1} + p_{n-2}\right)^2 + A\left(A_n p_{n-1} + p_{n-2}\right)\left(A_n q_{n-1} + q_{n-2}\right) + B\left(A_n q_{n-1} + q_{n-2}\right)^2 = 0.$$

Expanding and collecting like terms, we obtain

$$a_n A_n^2 + b_n A_n + c_n = 0 \tag{6.14}$$

where

$$\begin{cases} a_n = p_{n-1}{}^2 + A p_{n-1} q_{n-1} + B q_{n-1}{}^2 \\ b_n = 2 p_{n-1} p_{n-2} + A p_{n-1} q_{n-2} + A p_{n-2} q_{n-1} + 2 B q_{n-1} q_{n-2} \\ c_n = p_{n-2}^2 + A p_{n-2} q_{n-2} + B q_{n-2}^2 \end{cases}$$

We now calculate the discriminant $b_n^2 - 4 a_n c_n$ of the polynomial in eq. (6.14):

$$b_n^2 - 4 a_n c_n = \left(2 p_{n-1} p_{n-2} + A p_{n-1} q_{n-2} + A p_{n-2} q_{n-1} + 2 B q_{n-1} q_{n-2}\right)^2$$

$$- 4\left(p_{n-1}{}^2 + A p_{n-1} q_{n-1} + B q_{n-1}\right)\left(p_{n-2} + A p_{n-2} q_{n-2} + B q_{n-2}\right)$$

$$= \left(A^2 - 4B\right)\left(p_{n-1} q_{n-2} q_{n-1} + p_{n-2} q_{n-1} - 2 p_{n-1} p_{n-2} q_{n-1} q_{n-2}\right)$$

$$= \left(A^2 - 4B\right)\left(p_{n-1} q_{n-2} - p_{n-2} q_{n-1}\right)^2$$

$$= A^2 - 4B$$

by Proposition 6.4.

But $A^2 - 4B > 0$ and $A^2 - 4B$ is not a rational square from the hypothesis on x. Hence A_n is a quadratic surd for all $n \geq 1$.

We next apply Corollary 6.5.1 to x and its $(n-1)$th convergent r_{n-1}:

$$\left| x - \frac{p_{n-1}}{q_{n-1}} \right| < \frac{1}{q_{n-1}^2} \quad \text{for } n \geq 1.$$

It follows that there exists a number ε_{n-1} such that $|\varepsilon_{n-1}| < 1$ for which

$$x = \frac{p_{n-1}}{q_{n-1}} - \frac{\varepsilon_{n-1}}{q_{n-1}^2}$$

Hence, substituting formula (6.15) into the expression above for a_n yields

$$a_n = \left(q_{n-1}x + \frac{\varepsilon_{n-1}}{q_{n-1}} \right)^2 + A \left(q_{n-1}x + \frac{\varepsilon_{n-1}}{q_{n-1}} \right) q_{n-1} + B q_{n-1}^2$$

$$= \left(x^2 + Ax + B \right) q_{n-1}^2 + \varepsilon_{n-1}(2x + A) + \frac{\varepsilon_{n-1}}{q_{n-1}^2}$$

$$= \varepsilon_{n-1}(2x + A) + \frac{\varepsilon_{n-1}}{q_{n-1}^2}$$

by formula (6.12).

Hence $|a_n| < 2|x| + |A| + 1$.

Similarly, since $c_n = a_{n-1}$, $|c_n| < 2|x| + |A| + 1$.

So a_n and c_n are bounded for all $n \geq 1$.

But $b_n^2 = 4 a_n c_n + A^2 - 4B$. Hence $|b_n| < \sqrt{4|a_n c_n| + A^2 + 4|B|}$. Thus b_n is also bounded for all $n \geq 1$.

It necessarily follows that there are only finitely many different sets of coefficients in the quadratic equations $a_n A_n^2 + b_n A_n + c_n = 0$. So there are indices j, k, and l such that $a_j = a_k = a_l$, $b_j = b_k = b_l$, and $c_j = c_k = c_l$.

Since A_j, A_k, and A_l all satisfy the same same quadratic equation and a quadratic equation has at most two roots, there must be two values, say A_j and A_k, which are equal (pigeonhole principle). But then

$a_j = [A_j] = [A_k] = a_k$, $a_{j+1} = \left[\frac{1}{A_j - a_j} \right] = \left[\frac{1}{A_k - a_k} \right] = a_{k+1}$, and so on. Hence, the simple continued fraction expansion for x is periodic. ∎

Of course, there is much interest in the continued fraction expansions of irrational numbers other than quadratic surds. For example, the English mathematician and astronomer Roger Cotes (1682–1716) corresponded regularly with his teacher Isaac Newton as he proofread and edited the second edition of the *Principia*. Newton discovered that 25/21 was a good approximation to $\sqrt[3]{2}$ to which Cotes made the slight improvement 44/37. If you determine the beginning of the continued fraction for $\sqrt[4]{2}$, you can verify their work and even improve upon it with the next convergent, 1785/1501 (accurate to eight decimal places). In the exercises, you will get plenty of practice.

Exercise 6.3

1. Determine which quadratic surds are represented by the following simple continued fractions:
   (a) $[1; \overline{1}]$
   (b) $[0; \overline{1,2}]$
   (c) $[0; \overline{1,2,1}]$
   (d) $[-1; 2, \overline{3,2}]$
   (e) $[2; 4, 3, 7, \overline{6, 2, 4}]$

2. Determine the periodic simple continued fraction expansions for the following quadratic surds:
   (a) $\sqrt{2}$ (b) $\sqrt{3}$ (c) $\sqrt{11}$ (d) $2 + 3\sqrt{7}$ (e) $1 + 4\sqrt{3}$

3. (a) Show that if x is irrational, then there are infinitely many reduced rationals a/b such that $\left|x - \frac{a}{b}\right| < \frac{1}{b^2}$
   (b) Show that the conclusion above is false if x is rational.
   (c) Find five reduced rationals a/b such that $\left|\sqrt{2} - \frac{a}{b}\right| < \frac{1}{b^2}$

4. Find the simple continued fraction expansions for the following:
   (a) $\sqrt{5}$
   (b) $\sqrt{10}$
   (c) $\sqrt{17}$
   (d) $\sqrt{k^2 + 1}$ where k is an integer

5. Find the simple continued fraction expansions for the following:
   (a) $\sqrt{8}$
   (b) $\sqrt{15}$
   (c) $\sqrt{24}$
   (d) $\sqrt{k^2 - 1}$ where k is an integer > 1

6. For $r \geq 1$, what rational number is represented by the continued fraction $[r; r, r, \overline{r}]$?

7. (a) If p_n/q_n is the nth convergent to $\sqrt{2}$, then what is $\lim\limits_{n \to \infty} p_{n+1}/p_n$? What about $\lim\limits_{n \to \infty} q_{n+1}/q_n$?
   (b) If p_n/q_n is the nth convergent to $\sqrt{k^2 + 1}$, then what is $\lim\limits_{n \to \infty} p_{n+1}/p_n$?

8. (a) Use the continued fraction expansion for $\sqrt{2}$ to find a rational approximation to $\sqrt{2}$ with an error less than 0.0001.
   (b) Find a rational approximation to $\sqrt{3}$ with an error less than 0.0001.
   (c) Find a rational approximation to $\sqrt{10}$ with an error less than 0.00001.

9. Find the first five partial quotients of the following irrational numbers:
   (a) $\sqrt[3]{2}$
   (b) $\sqrt[3]{3}$
   (c) $2 + 3 \cdot 5^{1/4}$
   (d) $\sqrt{2} + \sqrt{3}$

10. (a) Prove that $\log_{10} 2$ is irrational.
    (b) Prove that if a and b are positive integers with b not a rational power of a, then $\log_a b$ is irrational.
11. Find the first five partial quotients of the following irrational numbers:
    (a) $\log_{10} 2$ (b) $\log_2 3$ (c) $\log_7 10$ (d) $\log_3 5$
12. Find the first 10 partial quotients of $\frac{e-1}{e+1}$.

6.4 Rational approximations of irrationals

Our first theorem is a modest improvement of Corollary 6.5.1.

Theorem 6.14: If x is irrational and $r_n = \frac{p_n}{q_n}$ and $r_{n+1} = \frac{p_{n+1}}{q_{n+1}}$ are two consecutive convergents of x, then either

$$\left| x - \frac{p_n}{q_n} \right| < \frac{1}{2q_n^2} \quad \text{or} \quad \left| x - \frac{p_{n+1}}{q_{n+1}} \right| < \frac{1}{2q_{n+1}^2} \tag{6.16}$$

Proof: By Theorem 6.5, x lies between r_n and r_{n+1} for any n. Hence if both inequalities (6.16) were false, then

$$\left| \frac{p_{n+1}}{q_{n+1}} - \frac{p_n}{q_n} \right| > \frac{1}{2q_n^2} + \frac{1}{2q_{n+1}^2} = \frac{q_n^2 + q_{n+1}^2}{2q_n^2 q_{n+1}^2}$$

The strict inequality above is justified since x is irrational. Since $q_n q_{n+1} > 0$,

$$|p_{n+1}q_n - p_n q_{n+1}| > \frac{q_n^2 + q_{n+1}^2}{2q_n q_{n+1}}$$

But $|p_{n+1}q_n - p_n q_{n+1}| = 1$ by Proposition 6.4. This leads to the inequality $q_n^2 + q_{n+1}^2 < 2q_n q_{n+1}$ which is equivalent to $(q_n - q_{n+1})^2 < 0$, a clear contradiction. The result follows. ∎

We now have an immediate corollary.

Corollary 6.14.1: If x is irrational, then there are infinitely many reduced rationals $\frac{p}{q}$ such that $\left| x - \frac{p}{q} \right| < \frac{1}{2q^2}$.

One immediate consequence of Corollary 6.14.1 is that if x is irrational, then then there are infinitely many reduced rationals $\frac{p}{q}$ for which $|p - xq| < \frac{1}{2q}$. Hence given any $\varepsilon > 0$, we can find infinitely many rationals $\frac{p}{q}$ for which $|p - xq| < \varepsilon$ by simply choosing $2q > 1/\varepsilon$. So given any irrational x, there are infinitely many reduced rationals p/q for which qx differs from p by as little as we please. This result goes back to Dirichlet. Furthermore, Dirichlet's theorem can be readily generalized to show that given any x irrational, r any real, and N and ε positive reals, then then there exist integers m and n such that $n > N$ and $|nx - m - r| < \varepsilon$. Dirichlet's theorem is the special case with $r = 0$. If $0 < r < 1$ and $\varepsilon > 0$ is arbitrarily small, then the fractional part of nx is arbritrarily close to r. Hence the fractional parts of nx for $n = 1, 2, 3, \ldots$ are everywhere dense in the real interval $[0, 1]$. Further extensions dealing with the uniform distribution of numbers were made by Hermann Weyl in 1916. We briefly discuss modern applications of this to random number generators in Chapter 8.

It is of general interest to determine what is the largest value of k for which there are infinitely many reduced fractions p/q such that

$$\left| x - \frac{p}{q} \right| < \frac{1}{kq^2} \tag{6.17}$$

for any irrational number x. By Theorem 6.14, k is at least 2. We now show that k is at most $\sqrt{5}$ for $x = \frac{1}{2} + \frac{1}{2}\sqrt{5}$.

Assume that there are infinitely many reduced rationals $\frac{p_n}{q_n}$ such that $\left| \frac{1}{2} + \frac{1}{2}\sqrt{5} - \frac{p_n}{q_n} \right| < \frac{1}{kq^2}$ where $k > \sqrt{5}$.

Define t_n by $\frac{1}{2} + \frac{1}{2}\sqrt{5} - \frac{p_n}{q_n} = \frac{1}{t_n q_n^2}$ for $n \geq 1$. So $t_n > k > \sqrt{5}$ for all n.

Multiplying through by q, $\frac{1}{t_n q} - \frac{\sqrt{5}q_n}{2} = \frac{q_n}{2} - p_n$. So

$$\left(\frac{1}{t_n q_n} - \frac{\sqrt{5}q_n}{2} \right)^2 = \left(\frac{q_n}{2} - p_n \right)^2$$

which implies that

$$\frac{1}{t_n^2 q_n^2} - \frac{\sqrt{5}}{t_n} = p_n^2 - p_n q_n - q_n^2$$

But the right-hand side is an integer for all n and hence so is the left-hand side:

$$\left| \frac{1}{t_n^2 q_n^2} - \frac{\sqrt{5}}{t_n} \right| \leq \left| \frac{1}{t_n^2 q_n^2} \right| + \left| \frac{\sqrt{5}}{t_n} \right| < \frac{1}{q_n^2} + \frac{\sqrt{5}}{t_n} < \frac{1}{n^2} + \frac{\sqrt{5}}{k}$$

Since $k > \sqrt{5}$, there is an N such that $\frac{1}{N^2} + \frac{\sqrt{5}}{k} < 1$. Then

$$\frac{1}{t_n^2 q_n^2} - \frac{\sqrt{5}}{t_n} = 0 \text{ for } n \geq N$$

This implies

$$\sqrt{5}\, t_n = \frac{1}{q_n^2}$$

But $\sqrt{5}\, t_n > 5$, a contradiction.

The next theorem and its corollary are due to the German mathematician Adolf Hurwitz (1859–1919). Hurwitz studied under Felix Klein. He held professorships at the Universities of Göttingen and Königsberg and was one of David Hilbert's most influential teachers. In 1898 Hurwitz proved that identities like eqs. (5.7) and (5.25) for the sum of n

squares exist only for $n = 2$, 4, and 8. Hurwitz inspired several generations of students in number theory and functional analysis. His attempts at proving Waring's conjecture failed, but they laid the groundwork for Hilbert's brilliant triumph (see Chapter 9).

Theorem 6.15 (Hurwitz–1891): If x is irrational and $r_n = \frac{p_n}{q_n}$, $r_{n+1} = \frac{p_{n+1}}{q_{n+1}}$, and $r_{n+2} = \frac{p_{n+2}}{q_{n+2}}$ are three consecutive convergents, then

$$\left| x - \frac{p_n}{q_n} \right| < \frac{1}{\sqrt{5}q_n^2} \quad \text{or} \quad \left| x - \frac{p_{n+1}}{q_{n+1}} \right| < \frac{1}{\sqrt{5}q_{n+1}^2} \quad \text{or} \quad \left| x - \frac{p_{n+2}}{q_{n+2}} \right| < \frac{1}{\sqrt{5}q_{n+2}^2}$$

Proof: For $n \geq 1$, let $s = \frac{q_{n+1}}{q_n}$ and $t = \frac{q_{n+2}}{q_{n+1}}$. If the conclusion of Theorem 6.15 is false, then, since x lies between r_n and r_{n+1},

$$\left| \frac{p_{n+1}}{q_{n+1}} - \frac{p_n}{q_n} \right| \geq \frac{1}{\sqrt{5}q_n^2} + \frac{1}{\sqrt{5}q_{n+1}^2}$$

As in the proof of Theorem 6.14, this implies

$$q_n^2 + q_{n+1}^2 \leq \sqrt{5}q_n q_{n+1}$$

and so $s + 1/s \leq \sqrt{5}$. Similarly, $t + 1/t \leq \sqrt{5}$.

By the quadratic formula and Corollary 6.3.1, it follows that

$$1 < s < \tfrac{1}{2} + \tfrac{1}{2}\sqrt{5} \text{ and } 1 < t < \tfrac{1}{2} + \tfrac{1}{2}\sqrt{5}$$

The strict inequality on the right-hand side is due to the fact that s and t are rational. But by Theorem 6.3,

$$q_{n+2} = a_{n+1}q_{n+1} + q_n \geq q_{n+1} + q_n$$

Dividing through by q_{n+1}, $t \geq 1 + 1/s$. But $s < \tfrac{1}{2} + \tfrac{1}{2}\sqrt{5}$ implies $1/s > -\tfrac{1}{2} + \tfrac{1}{2}\sqrt{5}$.
Thus $t > \tfrac{1}{2} + \tfrac{1}{2}\sqrt{5}$, a contradiction. The result is now proven. ∎

Corollary 6.15.1: If x is irrational, then there are infinitely many reduced rationals $\frac{p}{q}$ such that $\left| x - \frac{p}{q} \right| < \frac{1}{\sqrt{5}q^2}$.

There is a very broad and deep theory regarding the rational approximation of algebraic numbers with significant applications to transcendence theory. For example, if x is an algebraic number of degree $n > 1$ (i.e. x is a root of a polynomial of degree n and no lower degree), then a remarkable theorem due to K.F. Roth (1955) asserts that for any $\varepsilon > 0$ there are at most finitely many reduced rationals $\frac{p}{q}$ such that

$$\left| x - \frac{p}{q} \right| < \frac{1}{q^{2+\varepsilon}}$$

By our results above, the exponent in Roth's theorem is the best possible. The methods involved supersede what we can present here.

Our final theorem further demonstrates the utility of the convergents of a real number. We begin with a definition.

Definition 6.7: Let x be irrational. The reduced rational r/s is called a *best rational approximation* to x if $|qx - p| > |sx - r|$ for all fractions p/q with $q < s$.

Theorem 6.16 (Best Rational Approximation Theorem): Let x be a real number with $x = [a_0; a_1, \ldots]$ and convergents $r_n = \frac{p_n}{q_n}$. Let p be an integer and q be a positive integer. Then for all ≥ 1:

(a) If $\left|x - \frac{p}{q}\right| < |x - r_n|$, then $q > q_n$ and

(b) If $|qx - p| < |q_n x - p_n|$, then $q \geq q_{n+1}$.

So the convergents of x are its best rational approximations.

Proof: Suppose (b) holds. Let $p \in \mathbb{Z}$ and $q \in \mathbb{Z}^+$ such that $\left|x - \frac{p}{q}\right| < |x - r_n|$ with $q \leq q_n$. Then

$$|qx - p| = q\left|x - \frac{p}{q}\right| < q_n|x - r_n| = |xq_n - p_n|.$$

But $q \leq q_n$ implies $q < q_{n+1}$. This contradicts (b).

Hence to establish the theorem, it suffices to prove (b) since (a) must follow.

Suppose $|qx - p| < |q_n x - p_n|$ and yet $q < q_{n+1}$. Consider

$$\begin{cases} p_n u + p_{n+1} v = p \\ q_n u + q_{n+1} v = q \end{cases} \tag{6.18}$$

This is of the form $AX = B$ where $A = \begin{bmatrix} p_n p_{n+1} \\ q_n q_{n+1} \end{bmatrix}$, $X = \begin{bmatrix} u \\ v \end{bmatrix}$, and $B = \begin{bmatrix} p \\ q \end{bmatrix}$. By Proposition 6.4, $\det A = \pm 1 \neq 0$. By Cramer's rule, eq. (6.18) is solvable and has a unique integral solution (u_0, v_0).

The integer $u_0 \neq 0$ since otherwise $p_{n+1} v_0$ and $q_{n+1} v_0 = q$ and so $\frac{p}{q} = \frac{p_{n+1}}{q_{n+1}}$.

But $q < q_{n+1}$ and $\frac{p_{n+1}}{q_{n+1}}$ is in reduced terms, a contradiction. Similarly, $v_0 \neq 0$ since otherwise $p_n u_0 = p$ and $q_n u_0 = q$ and so $\frac{p}{q} = \frac{p_n}{q_n}$. But then $\frac{p}{q} = \frac{p_n}{q_n}$ which contradicts the assumption that $|qx - p| < |q_n x - p_n|$. Hence $u_0 v_0 \neq 0$.

Suppose $v_0 > 0$. Then since $q < q_{n+1}$ and $q_n u_0 = q - q_{n+1} v_0$, it follows that $q_n u_0 < 0$ and $u_0 < 0$.

Suppose $v_0 < 0$. Then by the same reasoning $q_n u_0 > 0$ and $u_0 > 0$.

Thus $u_0 v_0 < 0$. Since x lies between r_n and r_{n+1}, the quantities $q_n x - p_n = q_n(x - r_n)$ and $q_{n+1} x - p_{n+1} = q_{n+1}(x - r_{n+1})$ have opposite signs.

It follows that

$u_0(q_n x - p_n)$ and $v_0(q_{n+1} x - p_{n+1})$ have the same sign. Since (u_0, v_0) satisfies eq. (6.18), it follows that

$$|qx - p| = |x(q_n u_0 + q_{n+1} v_0) - (p_n u_0 + p_{n+1} v_0)|$$

$$= |u_0(q_n x - p_n) + v_0(q_{n+1} x - p_{n+1})|$$

$$= |u_0(q_n x - p_n)| + |v_0(q_{n+1} x - p_{n+1})|$$

$$\geq |u_0(q_n x - p_n)| \geq |q_n x - p_n|, \text{ a contradiction. Therefore, } q \geq q_{n+1}. \quad \blacksquare$$

Corollary 6.16.1: If x is irrational and

$$\left| x - \frac{a}{b} \right| < \frac{1}{2b^2}$$

then $\frac{a}{b}$ equals one of the convergents of x.

Proof: Let the convergents to x be $\frac{p_k}{q_k}$ and suppose $\frac{a}{b} \neq \frac{p_k}{q_k}$ for all k. Define n by $q_n \leq b < q_{n+1}$. Then $|bx - a| \geq |q_n x - p_n|$ by the best rational approximation theorem. But by the hypothesis, $|bx - a| < \frac{1}{2b}$. Hence $\left| x - \frac{p_n}{q_n} \right| < \frac{1}{2bqq_n}$.
But $\frac{a}{b} \neq \frac{p_n}{q_n}$ implies that $bp_n - aq_n$ is a nonzero integer. Hence

$$\frac{1}{bq_n} \leq \frac{|bp_n - aq_n|}{bq_n} = \left| \frac{p_n}{q_n} - \frac{a}{b} \right|$$

$$\leq \left| x - \frac{p_n}{q_n} \right| + \left| x - \frac{a}{b} \right|$$

$$< \frac{1}{2bq_n} + \frac{1}{2b^2}.$$

Hence,

$$\frac{1}{2bq_n} < \frac{1}{2b^2} \text{ and so } q_n > b, \text{ contradicting the definition of } n. \quad \blacksquare$$

At this point, we can discover another interesting aspect of Markov's eq. (5.21). We begin by displaying the infinite simple continued fraction for $\sqrt{2}$. Since $(\sqrt{2} - 1)(\sqrt{2} + 1) = 1$,

$$\sqrt{2} = 1 + \frac{1}{1 + \sqrt{2}}$$

Substituting this expression for $\sqrt{2}$ on the right-side above repeatedly leads to $\sqrt{2} = [1; 2, 2, \bar{2}]$.

We obtain the following chart of convergents to $\sqrt{2}$

n	-2	-1	0	1	2	3	4	5	6	7	8	9	10
a_n			1	2	2	2	2	2	2	2	2	2	2
p_n	0	1	1	3	7	17	41	99	239	577	1393	3,363	8,119
q_n	1	0	1	2	5	12	29	70	169	408	985	2,378	5,741

By Theorem 6.16, the convergents above are the best rational approximations to $\sqrt{2}$. Notice that $1^2 - 2(1)^2 = -1$, $3^2 - 2(2)^2 = 1$, $7^2 - 2(5)^2 = -1$, $17^2 - 2(12)^2 = 1$ and so on. In general, if $\frac{p_n}{q_n}$ is the nth convergent to $\sqrt{2}$, then $p_n{}^2 - 2q_n{}^2 = (-1)^{n+1}$. We will have much more to say about this in the next section when we discuss Pell's equation. The denominators (with indices shifted over one) are called Pell numbers.

Definition 6.8: The nth *Pell number* is defined by $P_0 = 0$, $P_1 = 1$, and $P_{n+1} = 2P_n + P_{n-1}$ for $n \geq 1$.

Hence, the sequence of Pell numbers begins 0, 1, 2, 5, 12, 29, 70, 169, etc. By the way we calculated the convergents to $\sqrt{2}$, $P_n = q_{n-1}$, the denominator of the $(n-1)^{\text{th}}$ convergent to $\sqrt{2}$.

Next, we look at another branch of the Markov tree, namely those containing 2 as one of their elements:

$$(1, 2, 5) \rightarrow (2, 5, 29) \rightarrow (2, 29, 169) \rightarrow (2, 169, 985) \rightarrow (2, 985, 5,741) \rightarrow \cdots$$

Notice odd Pell numbers appearing as Markov numbers! We state this as a proposition and prove it similarly to how we treated Fibonacci numbers appearing as Markov numbers.

Proposition 6.17: If P_k is the kth Pell number, then $(2, P_{2n-1}, P_{2n+1})$ forms a Markov triple for all $n \geq 1$.

We first establish the equivalent of Cassini's equations for Pell numbers.

Lemma 6.17.1: If P_k is the kth Pell number, then $P_{k+1}P_{k-1} - P_k{}^2 = (-1)^k$.

We leave the proof of Lemma 6.17.1 as an inductive proof (Exercise 6.4.10).

Proof of Proposition 6.17: Let $n = 2k$ in Lemma 6.17.1. Then

$$P_{2n}{}^2 + 1 = P_{2n+1}P_{2n-1}$$

Hence,

$$(P_{2n+1} - P_{2n-1})^2 = (2P_{2n} + P_{2n-1} - P_{2n-1})^2$$

$$= 4P_{2n}{}^2$$

$$= 4P_{2n+1}P_{2n-1} - 4$$

by Lemma 6.17.1.

Hence,

$$P_{2n+1}{}^2 - 2P_{2n+1}P_{2n-1} + P_{2n-1}{}^2 = 4P_{2n+1}P_{2n-1} - 4$$

which implies

$$4 + P_{2n-1}{}^2 + P_{2n+1}{}^2 = 6P_{2n-1}P_{2n+1}$$

Thus, $2^2 + P_{2n-1}{}^2 + P_{2n+1}{}^2 = 3 \cdot 2\, P_{2n-1}P_{2n+1}$ as desired. ∎

Before proceeding to our next example, let us prove that π is irrational. The first proof of this result was due to Johann Heinrich Lambert (1728–1777).

Lambert came from an impoverished background and was almost entirely self-taught after dropping out of school at age 12 to help support his large family. Lambert was a true polymath who made significant discoveries in astronomy, the nature of heat, philosophy, religion, history, meteorology, and acoustics. Lambert's map projections are still an important contribution to cartography. In over 150 scholarly works, Lambert wrote on many areas of mathematics including hyperbolic functions, infinite series, conic sections, and the theory of continued fractions. In retrospect, several results were harbingers of non-Euclidean geometry. Lambert even developed a theory of tetragometry which has an analogous relationship to plane quadrilaterals as does trigonometry to triangles.

Lambert was a nonconformist whose peripatetic lifestyle and unusual dress and behavior caused him some difficulties. His appointment at the Prussian Academy at Berlin was temporarily held up by Frederick the Great despite the warm reception afforded him by Euler.

Proposition 6.18 (Johann Lambert – 1761): The constant π is irrational.

The proof given here is due to Ivan Niven (1947). We begin with a useful lemma.

Lemma 6.18.1: Let a and b be positive integers and define

$$f(x) = \frac{1}{n!}x^n(a - bx)^n \tag{6.19}$$

For all integers $m \geq 0, f^{(m)}(0)$ and $f^{(m)}(a/b)$ are integers.
Recall that $f^{(0)}(x)$ is $f(x)$ and $f^{(m)}(x)$ is the mth derivative of $x(x)$ for $m \geq 1$.

Proof of Lemma: By the binomial theorem (Theorem 1.8),

$$f(x) = \frac{1}{n!}x^n(a - bx)^n = \frac{1}{n!}x^n \sum_{k=0}^{n}\binom{n}{k}a^k(-bx)^{n-k}$$

$$= \frac{1}{n!}\sum_{k=0}^{n}\binom{n}{k}(-1)^{n-k}a^k b^{n-k}x^{2n-k}$$

Thus, for $m \geq 0$,

$$f^{(m)}(x) = \frac{1}{n!}\sum_{k=0}^{n}\binom{n}{k}(-1)^{n-k}a^k b^{n-k}(2n-k)\cdots(2n-k-m+1)x^{2n-k-m}$$

$$= \frac{m!}{n!}\sum_{k=0}^{n}\binom{n}{k}\binom{2n-k}{m}(-1)^{n-k}a^k b^{n-k}x^{2n-k-m} \qquad (6.20)$$

Recall that $\binom{i}{j} = 0$ if $j > i$. We now show that $f^{(m)}(0) \in \mathbb{Z}$ for all $m \geq 0$.

If $m \leq n-1$, then $2n-k-m > 0$ for $0 \leq k \leq n$, and hence $f^{(m)}(0) = 0$.

If $m \geq 2n+1$, then $f^{(m)}(x) = 0$ for all, and so $f^{(m)}(0) = 0$.

If $n \leq m \leq 2n$, then we partition the $n+1$ terms in eq. (6.20) depending on whether
(i) $k < 2n - m$, (ii) $k = 2n - m$, or (iii) $k > 2n - m$.

(i) If $k < 2n - m$, then the exponent of x, $2n - k - m$, is positive.

(ii) If $k = 2n - m$, then $\binom{2n-k}{m}x^{2n-k-m} = 1$.

(iii) If $k > 2n - m$, then $\binom{2n-k}{m} = 0$.

Hence

$$f^{(m)}(0) = \frac{m!}{n!}\binom{n}{2n-m}(-1)^{m-n}a^{2n-m}b^{m-n} \in \mathbb{Z}$$

Next, we make the analogous calculations for $f^{(m)}(a/b)$.

If $m \leq n-1$, then $f^{(m)}(a/b) = 0$ by Leibniz's product formula which states that

$$(f_1 \cdot f_2)^{(m)} = \sum_{k=0}^{m}\binom{m}{k}f_1(k)\cdot f_2(m-k)$$

(Just let $f_1(x) = \frac{1}{n!}x^n$ and let $f_2(x) = (a - bx)^n$).

If $m \geq 2n+1$, then $f^{(m)}(x) = 0$ for all x, and so $f^{(m)}(a/b) = 0$.

If $n \le m \le 2n$, then $f^{(m)}(a/b) = \frac{m!}{n!} \sum_{k=0}^{n} \binom{n}{k} \binom{2n-k}{m} (-1)^{n-k} a^{2n-m} b^{m-n}$ which is certainly an integer since $2n - m \ge 0$ and $m - n \ge 0$. ∎

Recall that if $m < F(x) < M$ for $a \le x \le b$, then $m(b-a) < \int_a^b F(x)dx < M(b-a)$.

Proof of Proposition 6.18: Assume that π is rational, say $\pi = a/b$. With $f(x)$ defined as in eq. (6.19), let

$$g(x) = f(x) - f^{(2)}(x) + \cdots + (-1)^n f^{(2n)}(x) \tag{6.21}$$

By Lemma 6.18.1, $g(0)$ and $g(\pi)$ are integers. Now

$$\frac{d}{dx}[g'(x)\sin x - g(x)\cos x] = [g''(x) + g(x)]\sin x.$$

But $g''(x) + g(x) = f(x) + (-1)^n f^{(2n+2)}(x) = f(x)$ since $f^{(2n+2)}(x) = 0$.
By the fundamental theorem of calculus,

$$\int_0^\pi f(x)\sin x\,dx = [g'(x)\sin x - g(x)\cos x] = g(0) + g(\pi) \in \mathbb{Z}$$

But $0 \le f(x)\sin x \le \frac{\pi^n a^n}{n!}$ for $0 \le x \le \pi$.

Since $\sum_{n=1}^{\infty} \frac{\pi^n a^n}{n!}$ converges by the ratio test, it follows that $\lim_{n\to\infty} \frac{\pi^n a^n}{n!} = 0$.

Hence there is an integer N such that $\frac{\pi^n a^n}{n!} < 1/\pi$ for all $n \ge N$.

By our discussion preceding the proof, $0 < \int_0^\pi f(x)\sin x\,dx < 1$. This is a contradiction since there are no integers strictly between 0 and 1. ∎

Example 6.11: Find the best rational approximation to π with denominator less than 10,000.
Solution: Let us determine the beginning of the simple continued fraction for π by the method outlined in the proof of Theorem 6.11:

$$A_0 = \pi = 3.14159265359\ldots, a_0 = [A_0] = 3$$
$$A_1 = 1/(A_0 - a_0) = 7.06251330592\ldots, a_1 = [A_1] = 7$$
$$A_2 = 1/(A_1 - a_1) = 15.996594095\ldots, a_2 = [A_2] = 15$$
$$A_3 = 1/(A_2 - a_2) = 1.00341722818\ldots, a_3 = [A_3] = 1$$
$$A_4 = 1/(A_3 - a_3) = 292.63483365\ldots, a_4 = [A_4] = 292$$

$$A_5 = 1/(A_4 - a_4) = 1.57521580653\ldots, a_5 = [A_5] = 1$$
$$A_6 = 1/(A_5 - a_5) = 1.7384779567\ldots, a_6 = [A_6] = 1$$
$$A_7 = 1/(A_6 - a_6) = 1.35413656011\ldots, a_7 = [A_7] = 1$$

Thus $\pi = [3; 7, 15, 1, 292, 1, 1, 1, \ldots]$.

Next apply Theorem 6.3 as in Example 6.3 to determine the numerators and denominators of the convergents for π (cf. Example 6.2).

i	-2	-1	0	1	2	3	4	5	6	7
a_i			3	7	15	1	292	1	1	1
p_i	0	1	3	22	333	355	103993	104348	208341	312689
q_i	1	0	1	7	106	113	33,102	33,215	66,317	99,532

By the best rational approximation theorem (we will avoid acronyms here), the best rational approximation to π with denominator less than 10,000 is $355/113$. In fact, $355/113 - \pi < 0.00000026677$.

Exercise 6.4

1. Fill in the details to prove that the number e is irrational (Euler – 1737):
   (a) Use the Taylor expansion for $f(x) = e^x$ to show that $e = \sum_{k=0}^{\infty} \frac{1}{k!}$.
   (b) Let $s_n = \sum_{k=0}^{n} \frac{1}{k!}$ for $n \geq 1$.
      Show that $0 < e - s_n < \frac{1}{(n+1)!}\left[1 + \frac{1}{n+1} + \frac{1}{(n+1)^2} + \cdots\right] = \frac{1}{n!n}$.
   (c) Suppose $e = a/b$ where a and b are positive integers with $b > 1$.
      Then $0 < b!\,(e - s_b) < 1/b$.
   (d) Show that $b!\,(e - sb)$ is an integer, hence obtaining a contradiction.
2. (a) Use formula 6.11 to calculate the first twelve convergents in the continued fraction expansion of e.
   (b) Find the best rational approximation to e with denominator less than 1000.
3. (a) Use Corollary 6.16.1 to show that $13/3$ is a convergent to $\sqrt{19}$.
   (b) Show that $41/29$ is a convergent to $\sqrt{2}$.
   (c) Is $17/12$ a convergent to $\sqrt{2}$?
4. (a) Find the best rational approximation to $\sqrt{3}$ with denominator less than 200.
   (b) Find the best rational approximation to $\sqrt{5}$ with denominator less than 200.
5. (a) Verify Hurwitz's theorem on $\sqrt{2}$ for $1 \leq n \leq 10$.
   (b) Verify Hurwitz's theorem on $\sqrt{3}$ for $1 \leq n \leq 10$.
6. Find the best rational approximation to $\sqrt[3]{2}$ with denominator less than 100
7. Find the best rational approximation to π with denominator less than 1,000,000

8. (a) The length of a (tropical) year is approximately 365.2422 days (accurate to four decimal places). Find the best rational approximation to 365.2422 with denominator less than 100.

   (b) The Gregorian calendar has a leap year every 4 years except for century years (ending in 00) which must be divisible by 400. Explain why a perpetual cycle of 33 years consisting of leap years every 4 years for 28 years followed by a leap year in 5 years is as accurate as the Gregorian calendar.

9. (a) Christiaan Huygens (1629–1695) built a cogwheeled planetarium based on the observations that in 365 days the earth covers $359°\,45'40''31'''$ and Saturn $12°\,13'34''18'''$ of its orbit. Determine that the ratio $r = \frac{77708431}{2640858}$.

   (b) Determine the first seven partial quotients of the continued fraction for r.

   (c) Explain why Huygens used a gear ratio of 206/7 for his planetarium. What is the approximate error in degrees each century?

10. Establish Lemma 6.17.1

11. (a) Mimic our construction of Binet's formula for Fibonacci numbers to show that the nth Pell number $P_n = \frac{1}{2\sqrt{2}}\left[\left(1+\sqrt{2}\right)^n - \left(1-\sqrt{2}\right)^n\right]$.

    (b) Define the n-step *Fibonacci number* by the formula $Z_0 = 0$, $Z_1 = 1$, and $Z_{k+1} = nZ_k + Z_{k-1}$ for $k \geq 1$. (The Fibonacci numbers are 1-step and Pell numbers 2-step). Let $M = \frac{n+\sqrt{n^2+4}}{2}$ and $M' = \frac{n-\sqrt{n^2+4}}{2}$. Show that $Z_k = \frac{M^k - M'^k}{\sqrt{n^2+4}}$ for $k \geq 1$.

12. Determine the simple continued fraction for the number $M = \frac{n+\sqrt{n^2+4}}{2}$.

6.5 Pell's equation

For a given integer d, the Diophantine equation

$$x^2 - dy^2 = 1 \qquad (6.22)$$

is called Pell's equation. Notice that if $d = 0$, then $x = \pm 1$. If $d = -1$, then either $(x,y) = (\pm 1,\ 0)$ or $(x,y) = (0,\ \pm 1)$. If $d < -1$, then eq. (6.22) has only the trivial solution $(x,y) = (\pm 1, 0)$. Similarly, if d is a perfect square, say $d = s^2$, then eq. (6.22) becomes $(x+sy)(x-sy) = 1$ which also has only the trivial solution $(x,y) = (\pm 1,\ 0)$. In fact, $(x,y) = (\pm 1,\ 0)$ is a solution for any d, but will be discounted in any further discussion.

The interesting problem is to determine for d a positive non-square integer whether Pell's equation is solvable and, if so, a full description of those solutions. We will show that in the case where d is a positive non-square integer, Pell's equation has infinitely many solutions which may be explicitly described. Some additional mathematical background is necessary including a fuller discussion of the infinite continued fraction expansion for $\sqrt{d}$.

Special cases of eq. (6.22) have been extensively studied dating back over two millennia. Archimedes' infamous "cattle problem" calls for the numbers of eight classes of cattle (cows and bulls from white, yellow, black, and spotted herds) with seven specific relations. The problem can be solved with some considerable effort to obtain an infinite number of solutions, the smallest solution consisting of a total herd of over 50 million cattle. Some have interpreted Archimedes' original challenge to also stipulate that the total number of black and white bulls must be a perfect square and the total number of yellow and spotted bulls must be a triangular number. With these additional conditions the problem boils down to solving eq. (6.22) with $d = 4,729,494$. Despite the relatively small size of d, the problem was not fully solved until 1965 when a computerassisted solution was given by H.C. Williams, R.A. German, and C.R. Zarnke. The smallest total number of cattle consists of 206, 545 digits!

The mathematicians Brahmagupta (fl. 628), Bhaskara (1114–1185), and other Indian mathematicians studied special cases of eq. (6.22). Oftentimes all solutions were given for a particular value of d. In particular, Bhaskara solved the equation for $d = 61$. In 1657, Fermat challenged the mathematical community to show there were infinitely many solutions to eq. (6.22) for d a positive non-square integer. The British mathematician and first president of the Royal Society, William Brounker (1620–1684), provided such a solution later that same year. Brounker's solution was written up by John Wallis (1616–1703) who held the prestigious Savilian chair at Oxford. In 1732 Euler attributed the solution to John Pell (1611–1685) and his name has stuck ever since.

Pell taught mathematics in the Netherlands, is mentioned in some of Wallis's writings, and copied over much of Fermat's correspondence. In addition, he most likely introduced the symbol ÷ for division together with his contemporary J.H. Rahn. However, he seems never to have actually contributed anything to the solution of the equation for which he is honored. Unfortunately, this state of affairs is not unusual in mathematics, whose practitioners are often less scrupulous in historical details than they are in mathematical ones.

Euler solved Pell's equation for several particular values of d and noted that even for some relatively small values of d that the smallest positive solutions were quite large. For example, he showed that the smallest positive solution to Pell's equation with $d = 109$ is $x = 158,070,671,986,249$ and $y = 15,140,424,455,100$. In 1767 Lagrange published the first complete proof showing that Pell's equation always has non-trivial solutions for positive non-square integers d. Lagrange pointed out some errors and obscurities in previous attempted proofs.

If (x, y) is a solution to Pell's equation, then $(\pm x, \pm y)$ are solutions. Hence it suffices to find all positive solutions to Pell's equation. Presently we show the connection between Pell's equation and continued fractions.

Proposition 6.19: Let d be a positive non-square integer. If (x, y) is a positive solution to Pell's equation, then x/y is a convergent of the continued fraction expansion for $\sqrt{d}$.

Proof: Suppose (x, y) is a positive solution to Pell's equation. Then $x^2 - dy^2 = 1$ if and only if $\left(x + y\sqrt{d}\right)\left(x - y\sqrt{d}\right) = 1$. But

$\left(x + y\sqrt{d}\right) > 0$ implies $\left(x - y\sqrt{d}\right) > 0$ and so $x > y\sqrt{d}$. Hence

$$0 < \frac{x}{y} - \sqrt{d} = \frac{x - y\sqrt{d}}{y} = \frac{x^2 - dy^2}{y\left(x + y\sqrt{d}\right)} = \frac{1}{y\left(x + y\sqrt{d}\right)}.$$

So

$$\frac{x}{y} - \sqrt{d} < \frac{1}{y\left(y\sqrt{d} + y\sqrt{d}\right)} = \frac{1}{2y^2\sqrt{d}} < \frac{\sqrt{d}}{2y^2\sqrt{d}} = \frac{1}{2y^2}.$$

Thus

$$\left|\sqrt{d} - \frac{x}{y}\right| < \frac{1}{2y^2}$$

and the result follows from Corollary 6.16.1. ∎

To continue our study of Pell's equation we need a deeper understanding of the continued fraction for $\sqrt{d}$. We begin with some algebraic preliminaries.

Definition 6.9: Let $x = a + b\sqrt{c}$ be a quadratic surd. Then $x' = a - b\sqrt{c}$ is called the *conjugate* of x.

Lemma 6.20.1: If $x = a_1 + b_1\sqrt{c}$ and $y = a_2 + b_2\sqrt{c}$ are quadratic surds, then
(a) $(x + y)' = x' + y'$
(b) $(xy)' = x'y'$
(c) $(x/y)' = x'/y'$.

Proof: The proof is left as Exercise 6.5.2. ∎

Definition 6.10: Let x be a quadratic surd. If x has a periodic continued fraction expansion of the form $x = \overline{[a_0; a_1, \ldots, a_{r-1}]}$, then we say x has a *purely periodic* infinite simple continued fraction expansion.

The next theorem and its converse were proved by the great French prodigy Evariste Galois (1811–1832) in 1828. This result appeared in his first published paper. Fortunately, it did not get overlooked by his peers.

Theorem 6.20: Let $x = a + b\sqrt{c}$ be a quadratic surd with $x > 1$ and $-1 < x' < 0$. Then x has a purely periodic continued fraction expansion.

Conversely, if x has a purely periodic continued fraction expansion, then $x = a + b\sqrt{c}$ is a quadratic surd with $x > 1$ and $-1 < x' < 0$.

Proof: Let $x = [a_0; a_1, \ldots]$ and $A_n = [a_n; a_{n+1}, \ldots]$ for $n \geq 0$.

Recall that $A_n = \frac{1}{A_{n-1} - a_{n-1}}$ and $a_n = [A_n]$ from eqs. (6.11).

Notice, by induction if necessary, that $A_n = a_n + b_n \sqrt{c}$ for some rationals a_n, b_n for $n \geq 0$. By Lemma 6.19.1 and the fact $a_{n-1}' = a_{n-1}$,

$$A'_n = \frac{1}{A'_{n-1} - a_{n-1}} \tag{6.23}$$

If $A_{n-1}' < 0$, then $-1 < A_n' < 0$ since $a_{n-1} \geq 1$ for all $n \geq 1$. But

$-1 < x' = A_0' < 0$ and so $-1 < A'_n < 0$ for all $n \geq 0$.

By eq. (6.23),

$$a_n = A'_n - \frac{1}{A'_{n+1}} \quad \text{for } n \geq 0$$

Since a_n is an integer and $-1 < A_n' < 0$, in fact

$$a_n = [-1/A'_{n+1}]$$

Now x has a periodic continued fraction expansion with period r, say, by the periodic continued fraction theorem. So $A_{j+r} = A_j$ for all $j \geq N$ for some sufficiently large integer N. We now show that we may choose $N = 0$.

$$\text{If } A_{j+r} = A_j, \text{ then } A'_{j+r} = A'_j.$$

So

$$a_{j-1} = [-1/A'_j] = [-1/A'_{j+r}] = a_{j+r-1}$$

Hence

$$A_{j-1} = a_{j-1} + \frac{1}{A_j} = a_{j+r-1} + \frac{1}{A_{j+r}} = A_{j+r-1}$$

Apply this j times to obtain $A_0 = A_r$ and $a_0 = [A_0] = [A_r] = a_r$. So N can be taken to be 0. In general, $a_{nr+k} = a_k$ for all $n \geq 0, 0 \leq k \leq r-1$. It follows that $x = \overline{[a_0, a_1, \ldots, a_{r-1}]}$ and x has a pure periodic continued fraction expansion.

Conversely, assume that x has a purely periodic continued fraction. Then x is a quadratic surd and $x = \overline{[a_0, a_1, \ldots, a_{r-1}]}$ where the a_i 's are positive integers. So $x > a_0 \geq 1$. By Theorem 6.3,

$$x = [a_0; a_1, \ldots, a_{r-1}, x] = \frac{xp_{r-1} + p_{r-2}}{xq_{r-1} + q_{r-2}}.$$

Hence

$$q_{r-1}x^2 + (q_{r-2} - p_{r-1})x - p_{r-2} = 0$$

Let $f(x) = q_{r-1}x^2 + (q_{r-2} - p_{r-1})x - p_{r-2}$. The two roots of f are x and x'. We know $x > 1$. We need only verify that $-1 < x' < 0$. But $f(-1) = (q_{r-1} - q_{r-2}) + (p_{r-1} - p_{r-2})$. By Theorem 6.3 and the fact that the a_i's are all positive, $q_{r-1} - q_{r-2} > 0$ for any $r \geq 1$ and $p_{r-1} - p_{r-2} \geq 0$ for any $r \geq 1$. Hence $f(-1) > 0$. Additionally, $f(0) = -p_{r-2} < 0$ for any $r \geq 1$. Since f is a continuous function of x, by the intermediate value theorem f has a root, x', between -1 and 0 as desired. ∎

Proposition 6.21: Let d be a positive non-square integer. Then

$$\sqrt{d} = \left[D; \overline{a_1, \ldots, a_{r-1}, 2D} \right] \qquad (6.24)$$

where $D = \left\lfloor \sqrt{d} \right\rfloor$. In particular, $\sqrt{d}$ is periodic from a_1 on.

Proof: Let $x = D + \sqrt{d}$ where D is as above. Then $x > 1$ and $x' = D - \sqrt{d}$ satisfies $-1 < x' < 0$. By Theorem 6.20, x has a purely periodic continued fraction. Suppose $x = \left[\overline{a_0, a_1, \ldots, a_{r-1}} \right]$ where r is the period length. In particular, $A_0 = A_r$ and $a_0 = a_r$. Clearly

$$a_0 = [x] = \left[D + \sqrt{d} \right] = 2D$$

Now

$$\sqrt{d} = x - D \text{ and } x = \left[2D; \overline{a_1, \ldots, a_r} \right] = \left[2D; \overline{a_1, \ldots, a_{r-1}, 2D} \right]$$

So

$$\sqrt{d} = \left[D; \overline{a_1, \ldots, a_{r-1}, 2D} \right] \qquad ∎$$

Lemma 6.22.1: Let d be a positive non-square integer, $A_0 = \sqrt{d}$, and $a_n = [A_n], A_{n+1} = \frac{1}{A_n - a_n}$ for $n \geq 0$. Let

$$s_0 = 0, \quad t_0 = 1, \quad s_{n+1} = a_n t_n - s_n, \quad \text{and} \quad t_{n+1} = \frac{d - s_{n+1}^2}{t_n} \text{ for } n \geq 0 \qquad (6.26)$$

Then s_n and t_n are integers, $t_n \neq 0$, and $A_n = \frac{s_n + \sqrt{d}}{t_n}$ for $n \geq 0$.

Proof (Induction on n): For $n = 0$, $s_0 = 0, t_0 = 1$, and $A_0 = \frac{s_0 + \sqrt{d}}{t_0}$.
Now assume for arbitrary n that s_n and t_n are integers with $t_n \neq 0$ and $A_n = \frac{s_n + \sqrt{d}}{t_n}$. Clearly $s_{n+1} = a_n t_n - s_n$ is an integer. Now

$$t_{n+1} = \frac{d - s_{n+1}^2}{t_n}$$

$$= \frac{d - (a_n t_n - s_n)^2}{t_n}$$

$$= \frac{d - s_n^2}{t_n} + 2a_n s_n - a_n 2 t_n$$

$$= t_{n-1} + 2a_n s_n - a_n 2 t_n \in \mathbb{Z}$$

Moreover, $t_{n+1} \neq 0$, since otherwise $d - s_{n+1}^2 = 0$, contradicting the assumption d is not a perfect square

$$A_n - a_n = \frac{s_n + \sqrt{d}}{t_n} - a_n$$

$$= \frac{\sqrt{d} - (a_n t_n - s_n)}{t_n}$$

$$= \frac{\sqrt{d} - s_{n+1}}{t_n}$$

$$= \frac{d - s_{n+1}^2}{t_n \left(\sqrt{d} + s_{n+1} \right)}$$

$$= \frac{t_{n+1}}{\sqrt{d} + s_{n+1}}$$

Thus $A_{n+1} = \frac{1}{A_n - a_n} = \frac{s_{n+1} + \sqrt{d}}{t_{n+1}}$ as desired. ∎

Theorem 6.22: If d is a positive non-square integer and $\frac{p_n}{q_n}$ is the nth convergent to $\sqrt{d}$, then

$$p_n^2 - dq_n^2 = (-1)^{n+1} t_{n+1} \tag{6.26}$$

where t_n is as in eq. (6.25).

Proof: Let $\sqrt{d} = [a_0; a_1, \ldots, a_n, A_{n+1}]$. Then

$$\sqrt{d} = A_0 = \frac{A_{n+1} p_n + p_{n-1}}{A_{n+1} q_n + q_{n-1}}$$

by Theorem 6.3

$$= \frac{\frac{s_{n+1} + \sqrt{d}}{t_{n+1}} p_n + p_{n-1}}{\frac{s_{n+1} + \sqrt{d}}{t_{n+1}} q_n + q_{n-1}}$$

by Lemma 6.22.1

$$= \frac{\left(s_{n+1} + \sqrt{d}\right)p_n + t_{n+1}p_{n-1}}{\left(s_{n+1} + \sqrt{d}\right)q_n + t_{n+1}q_{n-1}}$$

Clearing the denominator,

$$\sqrt{d}\left[\left(s_{n+1} + \sqrt{d}\right)q_n + t_{n+1}q_{n-1}\right] = \left(s_{n+1} + \sqrt{d}\right)p_n + t_{n+1}p_{n-1}$$

and so

$$(s_{n+1}q_n + t_{n+1}q_{n-1})\sqrt{d} + dq_n = p_n\sqrt{d} + (s_{n+1}p_n + t_{n+1}p_{n-1})$$

The fact that $\sqrt{d}$ is irrational implies that

$$s_{n+1}q_n + t_{n+1}q_{n-1} = p_n$$

and

$$dq_n = s_{n+1}p_n + t_{n+1}p_{n-1}$$

Solving for s_{n+1} in the two equations above gives

$$\frac{dq_n - t_{n+1}p_{n-1}}{p_n} = s_{n+1} = \frac{p_n - t_{n+1}q_{n-1}}{q_n}. \text{ So}$$

$$p_n(p_n - t_{n+1}q_{n-1}) = q_n(dq_n - t_{n+1}p_{n-1}) \text{ and hence}$$

$$p_n^2 - dq_n^2 = t_{n+1}(p_nq_{n-1} - p_{n-1}q_n)$$

$$= (-1)^{n-1}t_{n+1} = (-1)^{n+1}t_{n+1} \text{ by Proposition 6.4.} \qquad\blacksquare$$

Corollary 6.22.1: Let r be the period length of the continued fraction expansion for $\sqrt{d}$ where d is a positive nonsquare integer. Then for $k \geq 0$,

$$p_{kr-1}^2 - dq_{kr-1}^2 = (-1)^{kr} \qquad\qquad (6.27)$$

Additionally, all positive solutions of $x^2 - dy^2 = 1$ are given by (p_n, q_n) where $n = kr - 1$ for all $k \geq 1$ if r is even and for all even $k \geq 2$ if r is odd. In particular, there are infinitely many positive solutions to Pell's equation.

Proof: Recall that $t_0 = 1$ and by Lemma 6.21.1 that $A_n = \frac{s_n + \sqrt{d}}{t_n}$. Since $A_0 = A_{kr}$ for all $k \geq 0$, it follows that $\frac{s_0 + \sqrt{d}}{t_0} = \frac{s_{kr} + \sqrt{d}}{t_{kr}}$. Hence $(s_{kr}t_0 - s_0t_{kr}) + (t_0 - t_{kr})\sqrt{d} = 0$. The irrationality of $\sqrt{d}$ implies that $t_0 = 1 = t_{kr}$ for $k \geq 0$. Theorem 6.22 implies that eq. (6.27) holds.

Next, we show that t_n never equals one when n is not a multiple of r.

If $t_n = 1$, then $A_n = s_n + \sqrt{d}$. But A_n has a purely periodic continued fraction expansion. By Theorem 6.20, $-1 < s_n - \sqrt{d} < 0$ and so $\sqrt{d} - 1 < s_n < \sqrt{d}$. Since s_n is an an integer, $s_n = \lceil \sqrt{d} \rceil = D$. Hence $A_n = D + \sqrt{d}$. But then $A_n = A_{kr}$ for all k and the minimality of r implies that n is a multiple of r itself.

Finally, we need to establish that t_n never equals -1. If $t_n = -1$ for some n, then $A_n = -s_n - \sqrt{d}$. But A_n has a purely periodic continued fraction. By Theorem 6.20, $-s_n - \sqrt{d} > 1$ and $-1 < -s_n + \sqrt{d} < 0$. So $\sqrt{d} < s_n < -1 - \sqrt{d}$, an absurdity. The result follows. ■

Notice that if (x_1, y_1) and (x_2, y_2) are two positive solutions to Pell's equation $x^2 - dy^2 = 1$ and $y_2 > y_1$, then necessarily $x_2 > x_1$. We have now proven that if the continued fraction expansion of $\sqrt{d}$ has even period r, then the smallest positive solution of Pell's equation is (p_{r-1}, q_{r-1}) since $q_{n+1} > q_n$ for $n \geq 1$. If r is odd, then (p_{2r-1}, q_{2r-1}) is the smallest positive solution. The smallest positive solution is traditionally called the *fundamental solution* of Pell's equation.

Example 6.12: Find the two smallest positive solutions to $x^2 - 14y^2 = 1$.
Solution: Verify that the continued fraction expansion for $\sqrt{14} = [3; \overline{1, 2, 1, 6}]$ (exercise 6.5.1). Here the period $r = 4$ is even. By Corollary 6.21.1, all positive solutions of $x^2 - 14y^2 = 1$ are given by (p_n, q_n) where $n = 4k - 1$. Let $S(n) = p_n^2 - 14q_n^2$.

We set up the following chart as in Example 6.3:

n	-2	-1	0	1	2	3	4	5	6	7	8
a_n			3	1	2	1	6	1	2	1	6
p_n	0	1	3	4	11	15	101	116	333	449	3027
q_n	1	0	1	1	3	4	27	31	89	120	809
$S(n)$	-14	1	-5	2	-5	1	-5	2	-5	1	-5

So $(x, y) = (15, 4)$ and $(x, y) = (449, 120)$ are the two smallest solutions to $x^2 - 14y^2 = 1$. Notice that $(x, y) = (p - 1, q - 1) = (1, 0)$ gives the trivial solution.

Example 6.13: Find the fundamental solution to $x^2 - 41y^2 = 1$.
Solution: Verify that the continued fraction expansion for $\sqrt{41} = [6; \overline{2, 2, 1, 2}]$ (Exercise 6.5.1). This time the period $r = 3$ is odd. By Corollary 6.22.1, all positive solutions of $x^2 - 41y^2 = 1$ are given by (p_n, q_n) where $n = 6k - 1$. Let $S(n) = p_n^2 - 41q_n^2$.

n	-2	-1	0	1	2	3	4	5	6
a_n			6	2	2	12	2	2	12
p_n	0	1	6	13	32	397	826	2049	25414
q_n	1	0	1	2	5	62	129	320	3969
$S(n)$	-41	1	-5	5	-1	5	-5	1	-5

So $(x, y) = (2049, 320)$ is the smallest positive solution to $x^2 - 41y^2 = 1$.

In practice there is a quicker way to find additional solutions to $x^2 - dy^2 = 1$ once an initial positive solution (x_1, y_1) is found rather than calculating ever larger convergents of $\sqrt{d}$.

Theorem 6.23: If (x_1, y_1) is the fundamental solution to $x^2 - dy^2 = 1$ for positive nonsquare d, then all positive solutions (x_n, y_n) are given by

$$x_n + y_n\sqrt{d} = \left(x_1 + y_1\sqrt{d}\right)^n \text{ for all } n \geq 1$$

Proof: Define x_n and y_n by $x_n + y_n\sqrt{d} =: \left(x_1 + y_1\sqrt{d}\right)^n$.

By Lemma 6.20.1(b),

$$x_n - y_n\sqrt{d} = \left(x_1 - y_1\sqrt{d}\right)^n. \text{ Now}$$

$$x_n^2 - y_n 2\sqrt{d} = \left(x_n - y_n\sqrt{d}\right)\left(x_n + y_n\sqrt{d}\right)$$

$$= \left(x_1 - y_1\sqrt{d}\right)^n \left(x_1 + y_1\sqrt{d}\right)^n$$

$$= \left(x_1^2 - y_1 2\sqrt{d}\right)^n = 1^n = 1$$

So (x_n, y_n) is a solution to Pell's equation.

We now show that there are no other positive solutions to Pell's equation. Suppose there is an ordered pair of positive integers $(s, t) \neq (x_n, y_n)$ for all n such that $s^2 - dt^2 = 1$. Define m by

$$\left(x_1 + y_1\sqrt{d}\right)^m < s + t\sqrt{d} < \left(x_1 + y_1\sqrt{d}\right)^{m+1} \tag{6.28}$$

Clearly $m \geq 1$ by the definition of (x_1, y_1). Now multiply all expressions in eq. (6.28) by the positive number $\left(x_1 - y_1\sqrt{d}\right)^m$ to obtain

$$1 < \left(s + t\sqrt{d}\right)\left(x_1 - y_1\sqrt{d}\right)^m < x_1 + y_1\sqrt{d}$$

Let a and b be such that $a + b\sqrt{d} = \left(s + t\sqrt{d}\right)\left(x_1 - y_1\sqrt{d}\right)^m$. Then

$$a^2 - b^2 d = \left(a + b\sqrt{d}\right)\left(a - b\sqrt{d}\right) \text{ which by Lemma 6.20.1}$$

$$= \left[\left(s + t\sqrt{d}\right)\left(x_1 - y_1\sqrt{d}\right)^m\right]\left[\left(s - t\sqrt{d}\right)\left(x_1 + y_1\sqrt{d}\right)^m\right]$$

$$= \left(s^2 - dt^2\right)\left(x_1^2 - dy_1^2\right)^m = 1$$

Hence (a, b) is a solution of Pell's equation with $1 < a + b\sqrt{d} < x_1 + y_1\sqrt{d}$. The minimality of the positive solution (x_1, y_1) implies that one of a or b must be negative. However, $1 < a + b\sqrt{d}$ implies that $0 < a - b\sqrt{d} < 1$, so

$$a = \frac{1}{2}\left(a + b\sqrt{d}\right) + \frac{1}{2}\left(a - b\sqrt{d}\right) > \frac{1}{2} + 0 > 0$$

and

$$b\sqrt{d} = \frac{1}{2}\left(a + b\sqrt{d}\right) - \frac{1}{2}\left(a - b\sqrt{d}\right) > \frac{1}{2} - \frac{1}{2} = 0.$$

Thus $b > 0$. Hence, all positive solutions (x_n, y_n) to Pell's equation are given by

$$x_n + y_n\sqrt{d} = \left(x_1 + y_1\sqrt{d}\right)^n$$

for all $n \geq 1$. ∎

Example 6.14: Find the second smallest positive solution to $x^2 - 41y^2 = 1$.
Solution: By Example 6.13, we know that the smallest positive solution is $(x_1, y_1) = (2049, 320)$. By Theorem 6.23, $x_2 + \sqrt{41}y_2 = \left(x_1 + \sqrt{41}y_1\right)^2 = \left(2049 + 320\sqrt{41}\right)^2$. But $\left(2049 + 320\sqrt{41}\right)^2 = 8396801 + 1311360\sqrt{41}$ and so $(x_2, y_2) = (8396801, \ 1311360)$.

Exercise 6.5

1. Verify the continued fraction expansions for $\sqrt{14}$ and $\sqrt{41}$ described in Examples 6.12 and 6.13, respectively

2. Prove Lemma 6.20.1.

3. (a) Explain why solutions to $x^2 - 2y^2 = 1$ are useful in approximating $\sqrt{2}$.
   (b) Find three positive solutions to $x^2 - 2y^2 = 1$.

4. (a) Find the two smallest positive solutions for the equation $x^2 - 3y^2 = 1$.
   (b) Find the two smallest positive solutions for the equation $x^2 - 12y^2 = 1$.
   (c) Find the two smallest positive solutions for the equation $x^2 + 4x = 3y^2 - 6y$.

5. (a) Find the two smallest positive solutions for the equation $x^2 - 5y^2 = 1$.
   (b) Find the two smallest positive solutions for the equation $x^2 - 45y^2 = 1$.
   (c) Find the smallest positive solution for the equation $x^2 + 18x = 5y^2 - 40y$.

6. Show that if (x_n, y_n) is a positive solution to Pell's equation $x^2 - dy^2 = 1$, then $(x_{n+1}, y_{n+1}) = \left(x_n^2 + dy_n^2, 2x_ny_n\right)$ is another solution. Explain how this ensures an infinite number of positive solutions. (Does this method give all positive solutions?)

7. Let d be a positive non-square integer. Show that if n is any positive integer, then the equation $x^2 - dy^2 = n^2$ has infinitely many solutions.

8. Show that for all n there are infinitely many solutions to Pell's equation $x^2 - dy^2 = 1$ with $n|y$.

9. Let d be a positive non-square integer and consider the associated Pell's equation $x^2 - dy^2 = -1$. Let r be the period length of the continued fraction expansion for $\sqrt{d}$.

    (a) Show that if r is even, then there are no solutions to the associated Pell's equation.

    (b) Show that if r is odd, then there are infinitely many positive solutions given by $(x_n, y_n) = (p_{nr-1}, q_{nr-1})$ for positive odd integers n.

10. Show that the associated Pell's equation $x^2 - dy^2 = -1$ is unsolvable if $4 \mid d$. (Necessary and sufficient conditions on d for which the associated Pell's equation is solvable are unknown).

11. (a) Prove the following extension of Proposition 6.19:

    Let d be a positive non-square integer and $0 < n < \sqrt{d}$. If (x, y) is a positive solution to the equation $x^2 - dy^2 = n$, then x/y is a convergent of the continued fraction expansion for $\sqrt{d}$. (In fact, the result is true for $0 < |n| < \sqrt{d}$). (b) Show that the upper bound on n in part (a) cannot be extended by considering the equation $x^2 - 6y^2 = 3$.

12. Determine whether the Diophantine equation $x^2 - 3y^2 = -1$ is solvable. If so, determine the fundamental solution.

13. Determine whether the Diophantine equation $x^2 - 5y^2 = n$ is solvable for the following n. If so, determine the fundamental solution:

    (a) $n = -1$ (b) $n = -2$ (c) $n = 2$

    (d) $n = 4$ (e) $n = 9$

14. Determine whether the Diophantine equation $x^2 - 11y^2 = n$ is solvable for the following n. If so, determine the fundamental solution:

    (a) $n = -1$ (b) $n = -2$ (c) $n = 2$

    (d) $n = -3$ (e) $n = 3$

15. Determine whether the Diophantine equation $x^2 - 13y^2 = n$ is solvable for the following n. If so, determine the fundamental solution:

    (a) $n = -1$ (b) $n = -2$ (c) $n = 2$

    (d) $n = -3$ (e) $n = 3$

16. (a) Show that there are infinitely many n for which both $n + 1$ and $2n + 1$ are perfect squares.

    (b) Show that if $n_1 < n_2 < \cdots$ are all such values of n, then $n_k n_{k+1} + 4$ is a perfect square for $k \geq 1$.

17. (a) Verify the identity

$$(x_1^2 - dy_1^2)(x_2^2 - dy_2^2) = (x_1 x_2 + y_1 y_2 d)^2 - d(x_1 y_2 + x_2 y_1)^2$$

    (b) Show that if d is a positive nonsquare integer, then there are infinitely many solutions to $x^2 - dy^2 = 1 - d$. It follows that there are infinitely many n for which both $n + 1$ and $dn + 1$ are perfect squares.

18. Verify Brounker's solution to Pell's equation with $d = 313$; namely
    $x = 32188120829134849$, $y = 1819380158564160$.

6.6 The continued fraction for *e*

We have seen that rational numbers have finite simple continued fractions, while quadratic irrationals have infinite simple continued fractions with a repeating pattern. Most other irrationals have continued fractions with no discernible pattern. One spectacular exception is the number *e* which has a great deal of regularity in its continued fraction expansion. In this section, we will prove the following expounded in Euler's 1737 paper *De fractionibus continuis dissertation* submitted to the St. Petersburg Academy. Though for the sake of historical accuracy, it should be noted that Roger Cotes made note of this very expansion for e in his article *Logometria* which dates its discovery back to 1714.

Theorem 6.24: The simple continued fraction for *e* is given by

$$e = [2; \ 1, \ 2, \ 1, \ 1, \ 4, \ 1, \ 1, \ 6, \ 1, \ 1, \ 8, \ 1, \ 1, \ 10, \ 1, \ 1, \ \ldots] \qquad (6.29)$$

Proof: Note that $2 = 1 + \frac{1}{0+\frac{1}{1}}$. It will be convenient to rewrite formula 6.29. We will establish

$$e = [1; \ 0, \ 1, \ 1, \ 2, \ 1, \ 1, \ 4, \ 1, \ 1, \ 6, \ 1, \ 1, \ 8, \ 1, \ 1, \ 10, \ 1, \ 1, \ \ldots]$$

Let $N = [1; 0, 1, 1, 2, 1, 1, 4, 1, 1, 6, 1, 1, 8, 1, 1, 10, 1, 1, \ldots]$ of value not known a priori. The partial quotients for N satisfy $a_{3i} = a_{3i+2} = 1$ and $a_{3i+1} = 2i$ for all $i \geq 0$.

Since $p_i = a_i p_{i-1} + p_{i-2}$ and $q_i = a_i q_{i-1} + q_{i-2}$, we have

$$p_{3n} = p_{3n-1} + p_{3n-2}$$

$$p_{3n+1} = 2n \, p_{3n} + p_{3n-1}$$

$$p_{3n+2} = p_{3n+1} + p_{3n}$$

and

$$q_{3n} = q_{3n-1} + q_{3n-2}$$

$$q_{3n+1} = 2n \, q_{3n} + q_{3n-1}$$

$$q_{3n+2} = q_{3n+1} + q_{3n}$$

We need to show that $e = \lim\limits_{i \to \infty} \frac{p_i}{q_i}$. To aid in our demonstration, for $n \geq 0$, let

$$A_n = \int_0^1 \frac{x^n(x-1)^n}{n!} e^x dx$$

$$B_n = \int_0^1 \frac{x^{n+1}(x-1)^n}{n!} e^x dx$$

$$C_n = \int_0^1 \frac{x^n(x-1)^{n+1}}{n!} e^x dx.$$

Lemma 6.24.1: For $n \geq 0$:

(a) $A_n = q_{3n} e - p_{3n}$

(b) $B_n = p_{3n+1} - q_{3n+1} e$

(c) $C_n = p_{3n+2} - q_{3n+2} e$

For the sake of continuity, we will defer the proof of the lemma for now.

For all $x \in [0, 1]$, as $n \to \infty$, $x^n (x-1)^n \to 0$ since the maximum of $|x^n (x-1)^n|$ occurs at $x = \frac{1}{2}$. Thus, $\lim_{n \to \infty} \int_0^1 \frac{x^n(x-1)^n}{n!} e^x dx = 0$. That is, $\lim_{n \to \infty} A_n = 0$. Similarly, $\lim_{n \to \infty} B_n = \lim_{n \to \infty} C_n = 0$.

By Lemma 6.24.1,

$$q_{3n} e - p_{3n} = A_n, \quad q_{3n+1} e - p_{3n+1} = -B_n, \text{ and } q_{3n+2} e - p_{3n+2} = -C_n.$$

Thus, $\lim_{i \to \infty} (q_i e - p_i) = 0$. But $q_i \geq 1$ for all $i \geq 2$. Thus, $e = \lim_{i \to \infty} \frac{p_i}{q_i}$ ∎

Proof of Lemma 6.24.1 (Induction on n)

For $n = 0$, $A_0 = \int_0^1 e^x dx = e - 1 = q_0 e - p_0$:

$$B_0 = \int_0^1 xe^x \, dx = (xe^x - e^x)||_0^1 = 1 = p_1 - q_1 e$$

$$C_0 = \int_0^1 (x-1)e^x \, dx = B_0 - A_0 = 2 - e = p_2 - q_2 e$$

Now assume that formulas (a), (b), and (c) hold for some value $n - 1$ for n a positive integer.

To establish part (a), we need to show that $A_n = -B_{n-1} - C_{n-1}$ since

$$- B_{n-1} - C_{n-1} = -(p_{3n-2} - q_{3n-2} e) - (p_{3n-1} - q_{3n-1} e) = -p_{3n} + q_{3n} e = q_{3n} e - p_{3n}$$

Then, for part (b), we need to show that $B_n = -2n A_n + C_{n-1}$ since

$$-2n A_n + C_{n-1} = -2n(q_{3n} e - p_{3n}) + p_{3n-1} - q_{3n-1} e$$

$$= (p_{3n-1} + 2n p_{3n}) - (q_{3n-1} + 2n q_{3n}) e = p_{3n+1} - q_{3n-1} e$$

Finally, we need to then establish that $C_n = B_n - A_n$ since

$$B_n - A_n = (p_{3n+1} - q_{3n+1}\, e) - (q_{3n}\, e - p_{3n})$$

$$= (p_{3n} + p_{3n+1}) - (q_{3n} + q_{3n+1})\, e = p_{3n+2} - q_{3n+2}\, e$$

For part (a), $\frac{d}{dx}\left[\frac{x^n(x-1)^n}{n!}\, e^x\right] = \frac{x^n(x-1)^n}{n!}\, e^x + \frac{x^n(x-1)^{n-1}}{(n-1)!}\, e^x + \frac{x^{n-1}(x-1)^n}{(n-1)!}\, e^x$. Integrating both sides from 0 to 1 yields:

$$A_n + B_{n-1} + C_{n-1} = \int_0^1 \frac{d}{dx}\left[\frac{x^n(x-1)^n}{n!}\, e^x\right] dx = \frac{x^n(x-1)^n}{n!}\, e^x\Big|_0^1 = 0$$

Thus, $A_n = -B_{n-1} - C_{n-1}$.

For part (b), $\dfrac{d}{dx}\dfrac{x^n(x-1)^{n+1}}{n!}\, e^x = \left[\dfrac{x^n(x-1)^{n+1}}{n!} + \dfrac{nx^{n-1}(x-1)^{n+1}}{n!} + \dfrac{(n+1)x^n(x-1)^n}{n!}\right]e^x$

$$= \left[\frac{x^n(x-1)(x-1)^n}{n!} + \frac{nx^{n-1}(x-1)(x-1)^n}{n!} + \frac{(n+1)x^n(x-1)^n}{n!}\right]e^x$$

$$= \left[\frac{x^{n+1}(x-1)^n}{n!} - \frac{x^n(x-1)^n}{n!} + \frac{nx^n(x-1)^n}{n!} - \frac{nx^{n-1}(x-1)^n}{n!} + \frac{(n+1)x^n(x-1)^n}{n!}\right]e^x$$

$$= \left[\frac{x^{n+1}(x-1)^n}{n!} + 2n\frac{x^n(x-1)^n}{n!} - \frac{x^{n-1}(x-1)^n}{(n-1)!}\right]e^x$$

Integrating both sides from 0 to 1: $B_n + 2n\, A_n - C_{n-1} = 0$. That is, $B_n = -2A_n + C_{n-1}$.

But $\frac{x^n(x-1)^{n+1}}{n!}\, e^x = \left[\frac{x^{n+1}(x-1)^n}{n!} - \frac{x^n(x-1)^n}{n!}\right]e^x$.

Integrating both sides from 0 to 1 yields $C_n = B_n - A_n$. ∎

By the way, Euler's work contained some additional stunning results. Among them are the following:

$$\sqrt{e} = [1;\ 1,\ 1,\ 1,\ 5,\ 1,\ 1,\ 1,\ 9,\ 1,\ 1,\ 1,\ 13,\ 1,\ 1,\ 1,\ \ldots] \tag{6.30}$$

$$\frac{e^2 - 1}{2} = [3;\ 5,\ 7,\ 9,\ 11,\ 13,\ 15, \ldots]$$

$\frac{e+1}{e-1} = [2;\ 6,\ 10,\ 14,\ 18,\ 22,\ 26, \ldots]$. Quite extraordinary!

Exercise 6.6

1. Use Euler's continued fraction for e to approximate e to four decimal places.
2. (a) Find the tenth convergent to e from eq. (6.29). Then use it to approximate $\sqrt{e}$.
   (b) Compare the result in (a) with the tenth convergent given for $\sqrt{e}$ in eq. (6.30).

6.7 Algebraic and transcendental numbers

In this section, we give a basic introduction to the theory of algebraic and transcendental numbers, the study of which could fill several full lengthened textbooks. We begin with the definition of algebraic numbers and define transcendental numbers by default as being those real (or complex) numbers that are not algebraic. We then show that the algebraic numbers are countable (listable in some sense) while the set of reals is not. Hence there are uncountably many transcendental numbers. We then establish a famous theorem due to Joseph Liouville which leads to the explicit construction of some transcendental numbers.

Definition 6.10: An *algebraic number* is a number r that satisfies an algebraic equation $f(x) = a_n x^n + \cdots + a_1 x + a_0 = 0$ where $a_i \in \mathbf{Z}$ for all i (not all a_i being zero). If f is irreducible, then we say that r is of *degree n*. Real (and complex numbers) that are not algebraic are called *transcendental*.

The numbers $\Phi = \frac{1+\sqrt{5}}{2}$, $\sqrt[3]{17}$, and $\sqrt{2} + \sqrt{3}$ are all algebraic numbers being roots of the algebraic equations $x^2 - x - 1 = 0$, $x^3 - 17 = 0$, and $x^4 - 10x^2 + 1 = 0$ respectively.

Definition 6.11: A set S is *countable* (or *enumerable*) if it is either finite or can be put into one-to-one correspondence with the positive integers $\mathbf{Z}^+$.

In 1874, the German mathematician Georg Cantor (1845–1918) discovered that the algebraic numbers were countable while the reals were *uncountable*. From there, he made an extensive study of the nature of infinity and showed, for example, that the collection of subsets of any set (finite or infinite) is of a higher order of infinity than the set itself.

Proposition 6.25: The set of algebraic numbers is countable.

We begin with a useful definition.

Definition 6.12: The *rank* of the polynomial $f(x) = a_n x^n + \cdots + a_1 x + a_0$ is $n + |a_0| + |a_1| + \cdots + |a_n|$.

For example, the rank of $f(x) = x$ is 2, the rank of $f(x) = 3x^4 - 5x^3 + 2$ is 14, and the only polynomials of rank 3 are x^2, $-x^2$, $2x$, $-2x$, $x + 1$, $x - 1$, $-x + 1$, $-x - 1$. Furthermore, it follows from the Fundamental Theorem of Algebra that a polynomial of degree n has n roots (including complex roots and with possible repeated roots).

Proof of Proposition 6.25: There are only finitely many polynomials of any given rank. Hence, the set of polynomials with integer coefficients are countable by listing them by rank beginning with rank 2, 3, and so on. Next, each polynomial can be replaced by its set of roots (all necessarily finite by the Fundamental Theorem of Algebra). Despite the large number of repetitions included among our list, we have successfully enumerated all algebraic numbers. ∎

Since all rational numbers are algebraic, we have the following corollary:

Corollary 6.25.1: The set of rational numbers are countable.

Next, we show that not all sets of numbers are countable. Recall that every real number has an infinite decimal expansion. If the decimal expansion terminates, for the sake of definiteness and uniqueness, then we simply append an infinite tail of zeros. So the number $1/8 = 0.125$ will be written as $0.125000\bar{0}$ rather than the equivalent $0.124999\bar{9}$.

Proposition 6.26: The set of real numbers is uncountable.

Proof: It suffices to establish the result for the set of reals strictly between 0 and 1. Suppose, contrary to what we wish to show, that the set of such reals is countable. Then we could include all of them in some infinite list

$$x_1 = 0.a_{11}a_{12}a_{13} \ldots a_{1n} \ldots$$

$$x_2 = 0.a_{21}a_{22}a_{23} \ldots a_{2n} \ldots$$

$$\ldots$$

$$x_n = 0.a_{n1}a_{n2}a_{n3} \ldots a_{nn} \ldots$$

$$\ldots$$

Now construct the number $r = 0.r_1r_2r_3 \ldots r_{nn} \ldots$ where $r_i = a_{ii} + 1$ if $a_{ii} \neq 8$ or 9 and $r_i = 3$ if $a_i = 8$ or 9. The real number r lies between 0 and 1, but does not appear in our previous list. Hence, the set of reals is uncountable. ∎

Combining Propositions 6.25 and 6.26 establishes

Corollary 6.26.1: The set of transcendental numbers are uncountable.

Interestingly, explicitly exhibiting a transcendental number takes a bit more work. We begin with a definition.

Definition 6.13: Let r be a real number with $0 < r < 1$. Then r is *approximable by rationals to order n* if there exists a constant k (depending only on r) such that the inequality $|\frac{p}{q} - r| < \frac{k}{q^n}$ has an infinite number of solutions in relatively prime integers p and q.

If r is itself rational, then r is approximable to order 1 and no higher.

By some of our previous results, e.g., Corollory 6.14.1, we see that irrational numbers have order at least 2. In fact, quadratic surds are approximable to order 2 but no higher. The following is a key result due to J. Liouville (1851). The French mathematician Joseph Liouville (1809–1882) was a prolific mathematician who made significant advances in number theory, complex analysis, topology, and differential geometry. One of his many contributions was the development of the Riemann-Liouville integral which allows the extension of differentiation and integration to fractional orders. He also founded the prestigious *Journal de Mathématiques Pures et Appliquées.*

Theorem 6.27 (Liouville's Theorem): A real algebraic number of degree n is not approximable to any order greater than n.

Proof: Let r be an algebraic number of degree n. Hence, r satisfies an equation

$$f(x) = a_n x^n + a_{n-1} x^{n-1} + \cdots + a_0 = 0$$

where $a_0, a_1, \ldots, a_n \in \mathbf{Z}$. Its derivative, $f'(x)$ is continuous on the closed interval $[r - 1, r + 1]$ and differentiable on $(r - 1, r + 1)$. Hence, by the Extreme Value Theorem, there exists a constant M (dependent only on r) such that $|f'(x)| < M$ for any x with $r - 1 < x < r + 1$.

Now suppose that p/q is an approximation to r with $r - 1 < p/q < r + 1$ and such that p/q is closer to r than any other root of $f(x) = 0$. In particular, $f(p/q) \neq 0$. Then

$$|f(p/q)| = \frac{|a_n p^n + a_{n-1} p^{n-1} q + \cdots + a_0 q^n|}{q^n} \geq \frac{1}{q^n}$$

Furthermore, by the mean value theorem,

$$f(p/q) = f(p/q) - f(r) = \left(\tfrac{p}{q} - r\right) f'(x) \text{ where } x \text{ lies between } r \text{ and } p/q.$$

Hence, $|\tfrac{p}{q} - r| = \dfrac{\left|f\left(\tfrac{p}{q}\right)\right|}{|f'(x)|} > \dfrac{1}{Mq^n} = \dfrac{k}{q^n}.$

It follows that r is not approximable to any order greater than n. ∎

Liouville's Theorem affords us a method of constructing transcendental numbers by creating numbers exceeding any order.

Example: Let $x = \sum_{k=1}^{\infty} 10^{-k!} = \tfrac{1}{10} + \tfrac{1}{100} + \tfrac{1}{1,000,000} + \cdots = 0.11000100000000000000000010\ldots$

Given N, let $n > N$ and define $x_n = \sum_{k=1}^{n} 10^{-k!}$. Then $x_n = \tfrac{p}{q}$ for p some integer and $q = 10^{n!}$. $\neq$ Also, $0 < x - \tfrac{p}{q} = x - x_n = \sum_{k=n+1}^{\infty} 10^{-k!} < \tfrac{2}{10^{(n+1)!}} < \tfrac{2}{q^n}$. Hence, x is not an algebraic number of any degree less than N by Liouville's Theorem. But N is arbitrary. Thus, x is transcendental.

Although it is a bit beyond the methods provided here, it should be mentioned that the transcendence of e was established by C. Hermite in 1873 and the transcendence of π by F. Lindemann in 1882. Much progress has been made since then. To cite just one important theorem, Aleksandr Gelfond and Theodor Schneider (1934) independently proved that if α and β are algebraic, $\alpha \neq 0$ or 1 and β irrational, then α^β is transcendental. For example, $2^{\sqrt{2}}$ is transcendental as is say $r = \frac{\log 5}{\log 2}$ since $2^r = 5$ and r is irrational.

Exercise 6.7

1. Find the minimal polynomial (irreducible polynomial with positive leading coefficient having r as a root) for the algebraic numbers $\sqrt[5]{7}$, $\sqrt{2} + \sqrt{5}$, and $\sqrt{2} + \sqrt{3} + \sqrt{5}$.
2. Show directly that the set of positive rational numbers are countable by ordering them based on their sum of numerator and denominator
3. Use the Gelfond-Schneider Theorem to establish that $\frac{\log p}{\log q}$ is transcendental where p and q are distinct primes
4. Determine whether the set of rational points in the plane (points with both coordinates rational) is countable. What about rational points in n-space?
5. Hilbert Hotel Problem: Explain how the manager of a hotel with a countably infinite number of rooms can accommodate an infinite number of new customers from an infinite number of buses each with an infinite number of passengers even though the hotel is already fully booked.

Chapter 7
Factoring and primality testing

7.1 Primality and compositeness

In Chapter 2 we introduced some important notions dealing with primality testing and factoring. Two techniques for verifying the primality of a given integer n are the sieve of Eratosthenes and the converse of Wilson's theorem. In the first case, if n is not divisible by any prime $p \leq \sqrt{n}$, then n is prime. In the second case, if $(n-1)! \equiv -1 \pmod{n}$, then n is prime. Unfortunately, neither of these methods is at all practical for very large numbers because of the enormous amount of computations involved.

Theorem 5.1 states that a prime $p \equiv 1 \pmod 4$ has a unique representation as a sum of two squares. Hence if an integer $n \equiv 1 \pmod 4$ has more than one representation as a sum of two squares, then it must be composite. For example, $21037 = 141^2 + 34^2 = 106^2 + 99^2$. Thus 21037 must be composite (in fact, $21037 = 109 \times 193$). Of course, discovering the representations may involve a substantial amount of work. However, modifications of this observation have proven useful in both primality testing and factoring (see Exercise 7.6.14).

In this chapter we will describe a much more efficient method, which we call the Miller-Jaeschke primality test, for determining the primality of numbers less than 10^{14} say. We will also describe algorithms for determining the primality of huge numbers of very particular forms. In particular, we will discuss the Lucas-Lehmer test for the primality of integers $n = 2^p - 1$ for prime p and Pepin's primality test for the primality of $n = 2^m + 1$ where $m = 2^r$.

Determining that a given composite integer is in fact composite is often easier than proving the primality of a given prime. For example, if an odd number n were prime, then by Fermat's little theorem $2^{n-1} \equiv 1 \pmod{n}$. Consequently, if $2^{n-1} \not\equiv 1 \pmod{n}$, then n must be composite. Similarly, if $3 \nmid n$ and $3^{n-1} \not\equiv 1 \pmod{n}$, then n must be composite (what if $3 \mid n$?) Unfortunately, this method gives no information about the factors of n. In Section 7.2, we discuss the related concept of pseudoprimes (composites which masquerade as primes) and the related idea of Carmichael numbers.

Example 7.1: Verify that $n = 209$ is composite by checking $2^{n-1} \pmod n$.
Solution: Express 208 as a sum of powers of 2: $208 = 128 + 64 + 16$. So $(208)_{10} = (11010000)_2$. Now apply the binary exponentiation algorithm from Section 2.5 to compute $2^{208} \pmod{209}$:

$$1 \xrightarrow{1} 2 \times (1)^2 = 2 \xrightarrow{1} 2 \times (2)^2 = 8 \xrightarrow{0} 8^2 = 64 \xrightarrow{1} 2 \times (64)^2 = 8192 \equiv 41 \xrightarrow{0} (41)^2 = 1681 \equiv 9$$

$$\xrightarrow{0} (9)^2 = 81 \xrightarrow{0} (81)^2 = 6561 \equiv 82 \xrightarrow{0} (82)^2 = 6724 \equiv 36 \pmod{209}$$

https://doi.org/10.1515/9783111579283-007

Since $36 \neq 1 \pmod{209}$, it follows that 209 is composite.

Finding all the factors of a given composite integer n seems to be substantially more difficult than the problems discussed above. One method is to test for all prime factors $p \leq \sqrt{n}$ as in the sieve of Eratosthenes. Another method is Fermat's factorization method described in Section 2.4. Fermat's method is useful if n is known to have two prime factors approximately the same size. If not, then Fermat's method is considerably less useful (but see Exercise 7.1.9). In Section 7.6, we develop several additional factorization methods. The first two are the Pollard rho and Pollard p–1 factorization methods, both widely used by sophisticated mathematical programs such as Mathematica®. We also discuss another highly practical technique based on the continued fraction of $\sqrt{n}$.

The fact that factorization is difficult can be advantageous in some applications, an important one is described in Chapter 10. We now turn to an initial example that utilizes the intractability of factoring large numbers.

Example 7.2: (Coin tossing over the phone) Andrew and Gabby want to determine the outcome of a coin toss over the phone. If Andrew flips the coin and Gabby calls the flip, how can Gabby be sure beyond a reasonable doubt that Andrew reports the outcome accurately? Here's how:

1. Gabby chooses two large primes, say p and q, but tells Andrew only their product $G = pq$. The number G is so large that factoring it without any further information is considered essentially impossible.
2. Andrew chooses a number x that is relatively prime to G (verified by the Euclidean algorithm) and sends A to Gabby where $A \equiv x^2 \pmod{G}$ and $1 \leq A < G$. He does not tell Gabby the value x. (This is a mathematician's idea of a coin toss.)
3. Gabby calculates m_1 and m_2 for which $(\pm m_1)^2 \equiv A \pmod{p}$ and $(\pm m_2)^2 \equiv A \pmod{q}$. The fact that $A \equiv x^2 \pmod{G}$ guarantees the existence of m_1 and m_2. Next, she calculates four distinct solutions to $y^2 \equiv A \pmod{G}$ via the Chinese remainder theorem with $1 \leq y < G$. Certainly x and $G - x$ are among the four values. Gabby sends one of the four values to Andrew, call it y. (Gabby calls the flip.)
4. If y is x or $G - x$, then Andrew acknowledges that Gabby guessed correctly and she wins the flip, else she loses. Since there are two correct choices out of four equally likely alternatives, Gabby's chance of guessing correctly is $1/2$. If Gabby's value $y \not\equiv \pm x \pmod{G}$, then Andrew sends Gabby x to confirm she guessed incorrectly.

Notice that the probability of Andrew cheating is negligible since knowing x with $x \neq \pm y$ is equivalent to knowing p and q, the factors of G.

Knowing x, Andrew could simply calculate $\gcd(x + y, G)$. Since $x^2 \equiv y^2 \pmod{G}$, $G \mid (x + y)(x - y)$. But then $x + y$ and $x - y$ are congruent to p and $q \pmod{G}$.

In addition, if Gabby chooses p and q to be congruent to 3 (mod 4), then her work in step 3 is greatly reduced. If $p = 4k + 3$, then she lets $m_1 = A^{2k+2} \equiv x^{2(2k+2)} = x \cdot x^p \equiv x \cdot x = x^2 \pmod{p}$ by Fermat's little theorem. Of course, q is calculated analogously.

For example,

(1) Gabby secretly chooses $p = 11$ and $q = 19$ and sends Andrew the number $G = pq = 209$ (not exactly a huge number, but it will do for our demonstration).

(2) Andrew secretly chooses $x = 20$, confirms that 20 and 209 are relatively prime, and sends Gabby $A = 191 \equiv (20)^2 \pmod{209}$.

(3) Gabby notes that $191 \equiv 4 \pmod{11}$ and determines $m_1 = 2$ (or 9). Similarly, $191 \equiv 1 \pmod{19}$ and so $m_2 = 1$ (or 18). There are now four simultaneous solutions to $y \equiv \pm 2 \pmod{11}$ and $y \equiv \pm 1 \pmod{19}$. They are $y \equiv 20, 75, 134$, and $189 \pmod{209}$. (Note that Gabby need only calculate the first two values since the others are just their additive inverses mod 209.)

(4) Suppose Gabby guesses $y = 75$. Andrew says that Gabby is incorrect and sends $x = 20$, thus winning the toss. Equivalently, he calculates $\gcd(x + y, n) = \gcd(95, 209) = 19$ and sends 19 and 209/19 = 11.

Notice that this technique could be used repeatedly with several values x_i and $g_i = p_i \cdot q_i$ (say $1 \le i \le 20$). If Gabby had some secret information represented by knowing all the p_i's and q_i's and Andrew wanted to verify that Gabby is who she claims to be, then Andrew and Gabby could repeat the "protocol" described above twenty times. The probability that someone else lacking the secret information could accurately fake Gabby's responses to Andrew is just $1/2^{20}$ (less than one in a million). This method for accurate personal identification is used in the design of universal credit cards or "smart cards." Similar methods have been created for "zero knowledge" proofs where someone simply wants to prove they have some knowledge without divulging what it is.

Exercise 7.1

1. Determine whether the following integers n are prime by checking all prime factors at most $\sqrt{n}$:
   (a) $n = 133$ (b) $n = 1003$ (c) $n = 1331$ (d) $n = 1957$

2. Determine that the following integers are composite by calculating $2^{n-1} \pmod{n}$:
   (a) $n = 77$ (b) $n = 187$ (c) $n = 529$ (d) $n = 2479$

3. Determine that the following integers are composite by calculating $3^{n-1} \pmod{n}$:
   (a) $n = 55$ (b) $n = 253$ (c) $n = 343$ (d) $n = 703$

4. Work through the details of a coin toss over the telephone for $p = 3, q = 7$, and $x = 20$.

5. Work through the details of a coin toss over the telephone for $p = 7, q = 19$, and $x = 100$.

6. Work through the details of a coin toss over the telephone for $p = 11$, $q = 23$, and $x = 72$.

7. (a) Show that $n^4 + 4$ is composite for all $n \geq 2$ by verifying that
$n^4 + 4 = (n^2 + 2n + 2)(n^2 - 2n + 2)$. Use this to factor 50629.

   (b) Find a similar factorization of $n^4 + 4m^4$ (Euler – 1742). Use it to factor 949 and $60229 = 15^4 + 4 \times 7^4$.

   (c) Show that $2^{4n-2} + 1$ is composite for all $n \geq 2$ by verifying that $2^{4n-2} + 1 = (2^{2n-1} + 2^n + 1)(2^{2n-1} - 2^n + 1)$. Use this to completely factor 4194305.

8. (Ancient Egyptian multiplication algorithm)

   (a) Explain why the following algorithm works to calculate $m \cdot n$:

   (i) Form a column of numbers with m at the top and 1 at the bottom with each entry equal to the greatest integer less than or equal to half the entry directly above.

   (ii) Form a second column of equal length next to the first one with n at the top and each entry equal to double the entry directly above.

   (iii) Add all entries in the second column that are next to odd numbers in the first column. The result is $m \cdot n$.

   For example, to calculate 56×29:

56	29
28	58
14	116
7 →	23
3 →	464
1 →	928

   So $56 \times 29 = 232 + 464 + 928 = 1624$.

   (b) Use this algorithm to calculate 17×23, 112×89, and 156×1001.

9. (Generalization of Fermat's factorization method)

   (a) Let n be an odd composite and $k \geq 1$. Let $f(c) = c^2 - kn$ for $c^2 \geq kn$. Show that if $f(c) = r^2$, then $\gcd(n, c - r)$ is a nontrivial factor of n.

   (b) Compare the number of steps needed to factor $n = 3959$ using

   (i) trial division beginning with $p = 3$,

   (ii) Fermat's factorization method ($k = 1$),

   (iii) the generalized Fermat method with $k = 3$.

10. If n is an odd prime, how many steps does it take to discover that fact using Fermat's factorization method? Investigate the situation with several small primes using the generalized Fermat method.

7.2 Pseudoprimes and Carmichael numbers

Recall Fermat's little theorem which states that if p is prime and $p \nmid a$, then $a^{p-1} \equiv 1 \pmod{p}$. In Section 2.5 we gave an example of a composite integer $n = 341$ such that $2^{n-1} \equiv 1 \pmod{n}$. This demonstrates that the converse of Fermat's little theorem is not true. However, composite numbers n which behave like primes in that the congruence

$$2^{n-1} \equiv 1 \pmod{n} \tag{7.1}$$

holds are relatively rare compared to the preponderance of primes. For example, below 1,000,000 there are 78,498 primes but only 245 composites satisfying relation (7.1). Hence the chance that an arbitrary integer less than a million satisfying (7.1) is indeed prime is greater than 99.6%. (Take that, Ivory soap!) In this section we study congruences like eq. (7.1) and develop some limited converses to Fermat's little theorem.

Definition 7.1: If n is a composite number relatively prime to b and

$$b^{n-1} \equiv 1 \pmod{n} \tag{7.2}$$

then n is a *base b pseudoprime*, denoted psp(b). If $b = 2$, then n is called simply a *pseudoprime*.

From our discussion above, 341 is a pseudoprime. Below is another example. First recall Corollary 2.1.1(c) which states that if a and b are relatively prime and $a \mid c$ and $b \mid c$, then $ab \mid c$. This can easily be generalized so that if $a_1, \ldots, a_r$ are pairwise relatively prime and $a_i \mid c$ for all $i = 1, \ldots, r$, then $a_1 \cdots a_r \mid c$ (as in Exercise 2.1.17).

Example 7.3: Show that 1105 is a psp(3).
Solution: $1105 = 5 \times 13 \times 17$. We calculate 3^{1104} modulo 5, 13, and 17 in turn.
$3^{1104} = (3^4)^{276} \equiv 1^{276} = 1 \pmod{5}$. Similarly, $3^{1104} = (3^3)^{368} \equiv 1^{368} = 1 \pmod{13}$ and $3^{1104} = (3^8)^{138} \equiv -1^{138} = 1 \pmod{17}$. Since 5, 13, and 17 are all prime (and hence relatively prime), by our comments above $3^{1104} \equiv 1 \pmod{1105}$. Since 1105 is also composite, 1105 is a psp(3). In fact, 1105 is the smallest integer which is both a psp(2) and psp(3).

If there were a base b for which there were only finitely many base b pseudoprimes, then there would be some hope of listing all of the base b pseudoprimes. In this case, if n is not on the list and eq. (7.2) holds, then n would be prime. The next theorem shows, unfortunately, that there is no possibility of finding such a base b.

Theorem 7.1: There are infinitely many pseudoprimes to any given base.

Proof. Let $b > 1$ be a given base and let p be an odd prime with $p \nmid b(b^2 - 1)$. Define n by

$$n = \frac{b^2p - 1}{b^2 - 1} = b^{2p-2} + b^{2p-4} + \cdots + b^2 + 1 \tag{7.3}$$

Notice that b and n are relatively prime. Furthermore, n is composite since $n = \frac{bp-1}{b-1} \cdot \frac{bp+1}{b+1}$ and each factor is an integer greater than 1 since p is odd and $p \geq 3$ (Exercise 7.2.1). We will show that n is a psp(b). The result then follows by noting that there are infinitely many p satisfying the condition above (and hence infinitely many n).

$$b^{2p-2} = (bp-1)^2 \equiv 1^2 = 1 \pmod{p}$$

by Fermat's Little Theorem. So $p \mid b^{2p-2} - 1$. But eq. (7.3) implies that

$$(n-1)(b^2-1) = n(b^2-1) - (b^2-1) = (b^{2p}-1) - (b^2-1) = b^2(b^{2p}-2-1)$$

It follows that $p \mid (n-1)(b^2-1)$. Since $p \nmid (b^2-1)$, we have $p \mid (n-1)$.

By eq. (7.3), $n-1$ is expressible as the sum of $p-1$ terms of the same parity (that of b). Hence $n-1$ is even and $2p \mid (n-1)$. Write $n-1 = 2pk$. Again from eq. (7.3), $b^{2p} \equiv 1 \pmod{n}$ and so $b^{n-1} = (b^{2p})^k \equiv 1^k = 1 \pmod{n}$. Therefore, n is a psp(b). ∎

Although there are infinitely pseudoprimes to any given base, one might wish for some short list of bases to which no composite integer is a pseudoprime to all of them. For example. if there were no integers which were simultaneously psp(2), psp(3), psp(5), and psp(7), then there would be a good primality test based on testing relation (7.2) for $b = 2, 3, 5,$ and 7.

Unfortunately, such hope is in vain.

Definition 7.2: If n is an odd composite and n is a base b pseudoprime for all b relatively prime to n, then n is called a *Carmichael number* (or absolute pseudoprime).

Carmichael numbers are named after the American mathematician R.D. Carmichael (1879–1967) who characterized them in 1912 and gave the first 15 examples (some of them announced in 1910). A correct statement of their characterization without proof was actually given by A. Korselt in 1899 but included no examples and was not widely acknowledged. Over the course of a long and illustrious career, Carmichael made original contributions in number theory, group theory, differential equations, and physics.

Next, we show that our definition does not describe the empty set.

Example 7.4: Show that 561 is a Carmichael number.
Solution: $561 = 3 \cdot 11 \cdot 17$. Suppose that $\gcd(b, 561) = 1$. Then b is relatively prime to 3, 11, and 17. By Fermat's little theorem, $b^{560} = (b^2)^{280} \equiv 1^{280} = 1 \pmod{3}$. Similarly, $b^{560} = (b^{10})^{56} \equiv 1^{56} = 1 \pmod{11}$ and $b^{560} = (b^{16})^{35} \equiv 1^{35} = 1 \pmod{17}$. It follows that $b^{560} \equiv 1 \pmod{561}$. Therefore, 561 is a Carmichael number. In fact, 561 is the smallest Carmichael number.

Of course, there are fewer Carmichael numbers than pseudoprimes or base b pseudoprimes for a given base b. For example, there are only 43 Carmichael numbers below 10^6, as opposed to 245 such pseudoprimes. However, in 1992 Red Alford, Andrew Granville, and Carl Pomerance proved that in fact there are infinitely many Carmichael numbers by showing that there exists an N_0 such that for any $N > N_0$ there are more than $N^{2/7}$ Carmichael numbers n with $n < N$. Hence Carmichael numbers themselves aren't even so scarce!

Below we give a short table comparing the number of primes, pseudoprimes, and Carmichael numbers below certain limits. We use the standard notation $\pi(x)$ to denote the number of primes less than or equal to x. In addition, define $P\pi(x)$ to denote the number of pseudoprimes (base 2) less than or equal to x and let CN(x) be the number of Carmichael numbers at most x.

Table 7.1: Primes, pseudoprimes, and Carmichael numbers up to 10^{12}.

x	$\pi(x)$	$P\pi(x)$	CN(x)
10^1	4	0	0
10^2	25	0	0
10^3	168	3	1
10^4	1229	22	7
10^5	9592	78	16
10^6	78498	245	43
10^7	664579	750	105
10^8	5761455	2057	255
10^9	50847534	5597	646
10^{10}	455052512	14884	1547
10^{11}	4118054813	38975	3605
10^{12}	37607912018	101629	8241

Next, we characterize the set of Carmichael numbers. The following result is sometimes referred to as Korselt's criterion.

Theorem 7.2 (Carmichael's Theorem): The number n is a Carmichael number if and only if n is a product of three or more distinct odd primes where $(p-1) \mid (n-1)$ for all primes $p \mid n$.

For example, $1729 = 7 \times 13 \times 19$ and $6 \mid 1728, 12 \mid 1728$, and $18 \mid 1728$. Hence 1729 is a Carmichael number. We begin with a lemma.

Lemma 7.2.1: Let n be a product of distinct primes and let the prime $p \mid n$. For any positive integer k, p^k has a primitive root which is relatively prime to n. (If $p = 2$, then $k = 1$ or 2.)

Proof: Let $p \mid n$ and k be any natural number (with the above restrictions only if $p = 2$). By the primitive root theorem, p^k has a primitive root; call it g. Note that $\gcd(g, p) = 1$. Let $\gcd(g, n) = r$. If $r = 1$, then we are done. Otherwise, let $h = g + np^{k-1}/r$. Notice that h is a primitive root $(\bmod\, p^k)$ since $h \equiv g \pmod{p^k}$. In particular, $\gcd(h, p)$

$=1$. If q is a prime with $q \mid r$, then $q \mid g$ but q doesn't divide np^{k-1}/r. Hence, $\gcd(h, q) = 1$ for all primes q dividing n but not pr. Therefore, for all primes $q \mid n$, $\gcd(h, q) = 1$. It follows that h is relatively prime to n. ∎

Proof of Theorem 7.2:

($\Leftarrow$) Let $n = \prod_{i=1}^{r} p_i$ where the p_i are distinct primes arranged in ascending order. By hypothesis, n is composite. Let $p \mid n$. If $\gcd(b, n) = 1$, then $\gcd(b, p) = 1$. By Fermat's Little Theorem, $b^p - 1 \equiv 1 \pmod{p}$. Since $(p-1) \mid (n-1)$, there exists an s for which $n - 1 = s(p-1)$. But then

$$b^{n-1} = \left(b^{p-1}\right)^s \equiv 1^s = 1 \pmod{p}$$

Since n is a product of distinct primes, $b^{n-1} \equiv 1 \pmod{n}$. Thus n is a Carmichael number.

($\Rightarrow$) Let n be a Carmichael number and suppose $n = \prod_{i=1}^{r} p_i^{a_i}$ where the p_i are distinct primes arranged in ascending order. If n is not square-free, then there is a prime p with $p^a \mid n$ where $a \geq 2$. Write $n = p^2 m$. By Lemma 7.2.1, there is a primitive root $h \pmod{p^2}$ with $\gcd(n, h) = 1$. So $\text{ord}_{p2}\, h = \phi(p^2)$.

Since n is a Carmichael number,

$$h^{n-1} = h^{p^2 m-1} \equiv 1 \pmod{n}$$

But $p^2 \mid n$ and so

$$h^{p^2 m - 1} - 1 \equiv 1 \pmod{p^2}$$

By Proposition 4.2(b), $\phi(p^2) \mid (p^2\, m - 1)$. Hence, $p(p-1) \mid (p^2 m - 1)$ and, finally, $p \mid (p^2 m - 1)$. But this is an obvious fallacy and so n is square-free and $a_i = 1$ for all

$$i = 1, \ldots, r$$

Let p be any prime dividing n. By Lemma 7.2.1, there is a primitive root $g \pmod{p}$ that is relatively prime to n. So $\text{ord}_p g = p - 1$. Because n is a Carmichael number, $g^{n-1} \equiv 1 \pmod{n}$. As above, $g^{n-1} \equiv 1 \pmod{p}$ and by Proposition 4.2(b), $(p-1) \mid (n-1)$.

If $r = 1$, then n is prime, contrary to the definition of Carmichael number. But n is odd, since otherwise, $n - 1$ would be odd while $p_2 - 1$ would be even. This would contradict the fact $(p_2 - 1) \mid (n-1)$.

Finally, if $r = 2$, then

$$n - 1 = p_1 p_2 - 1 = p_1(p_2 - 1) + (p_1 - 1) \not\equiv 0 \pmod{p_2 - 1}$$

since $1 < p_1 - 1 < p_2 - 1$. Thus $(p_2 - 1) \nmid (n-1)$, a contradiction. ∎

It may appear from our discussion up until now that there is little hope of creating a primality test by utilizing some sort of a converse to Fermat's little theorem. However, there is a very nice limited converse due to the French number theorist Edouard

Lucas (1842–1891) which was published in 1891. Lucas was a high school teacher who did much to popularize mathematics through engaging public lectures and the creation of recreational problems. He is the creator of the Tower of Hanoi puzzle and won a gold medal at the 1889 world's fair for a collection of scientific puzzles.

Theorem 7.3 (Lucas' Primality Test): If there is an integer a for which $a^{n-1} \equiv 1 \pmod n$ but, for each prime $p \mid (n-1)$, $a^{\frac{n-1}{p}} \not\equiv 1 \pmod n$, then n is prime.

Proof: Let $s = \text{ord}_n a$. Since $a^{n-1} \equiv 1 \pmod n$, $s \mid (n-1)$ by Proposition 4.2(b). Write $n-1 = ks$. If $k > 1$, then there is a prime $p \mid k$. Hence $p \mid (n-1)$. It follows that $a^{(n-1)/p} = a^{ks/p} = (a^s)^{k/p} \equiv 1^{\frac{k}{p}} = 1 \pmod n$. This contradicts the hypothesis of the theorem and hence $k = 1$ and $s = n-1$. But by Proposition 4.2(a), $\text{ord}_n a \mid \phi(n)$ and $\phi(n) \leq n-1$. Therefore, $s = \phi(n) = n-1$. It follows that n must be prime. ∎

To use Lucas' primality test on a prime n it suffices to find a primitive root $a \pmod n$. Unfortunately for very large n there is no known algorithm which locates primitive roots in a reasonable amount of time. However, if n is not large, then Lucas' primality test can be quite effective.

Example 7.5: Use Lucas' primality test to verify that the following are prime:
(a) $n = 1301$, (b) $n = 7919$.
Solution:
(a) In this case, $n - 1 = 1300 = 2^2 \times 5^2 \times 13$. Let $a = 2$ and calculate $a^{\frac{n-1}{p}} \pmod n$ for $p = 2, 5$, and 13. (The binary exponentiation algorithm is useful.) We obtain $2^{650} \equiv -1 \pmod{1301}$, $2^{260} \equiv 163 \pmod{1301}$, and $2^{100} \equiv 78 \pmod{1301}$.
Hence $n = 1301$ is indeed prime.
(b) In this case, $n - 1 = 7918 = 2 \times 37 \times 107$. Unfortunately, $a^{(n-1)/2} = a^{3959} \equiv 1 \pmod n$ for $a = 2, 3$, and 5. However, $7^{3959} \equiv -1 \pmod n$. In addition, $7^{(n-1)/37} = 7^{214} \equiv 755 \pmod n$ and $7^{(n-1)/107} = 7^{74} \equiv 5549 \pmod n$.
Thus, $n = 7919$ is prime. (It is in fact the 1000th prime number.)

Lucas' primality test assumes that we know the complete factorization of $n-1$. The next related theorem weakens that assumption somewhat in that it only requires a partial factorization of $n-1$. It is due to the mathematician H.C. Pocklington (1870–1942), published in 1914.

Theorem 7.4 (Pocklington's Primality Test): Let $n - 1 = st$ where $\gcd(s, t) = 1$ and $s > t$. If for every prime $p \mid s$ there exists an integer b such that $\gcd(b^{(n-1)/p} - 1, n) = 1$ and $b^{n-1} \equiv 1 \pmod n$, then n is prime.

Notice that the integer b depends on p in Pocklington's Primality Test. Hence, we have some flexibility not available in Lucas' Primality Test where the integer a was fixed for all $p \mid (n-1)$.

Proof: Let q be any prime dividing n. As in the statement of the theorem, let $p \mid s$ and suppose there is a b such that $\gcd(b^{(n-1)/p} - 1, n) = 1$ and $b^{n-1} \equiv 1 \pmod{n}$. Let $a = \text{ord}_q b$. By Proposition 4.2(a), $a \mid (q-1)$. Since $b^{n-1} \equiv 1 \pmod{n}$, we know by Proposition 4.2(b) that $a \mid (n-1)$ too. But $b^{(n-1)/p} \not\equiv 1 \pmod{q}$. Hence, a does not divide $\frac{n-1}{p}$. So if $p^e \parallel (n-1)$, then $p^e \parallel a$. Since p is an arbitrary prime dividing s, it follows that $s \mid a$. But then $s \mid (q-1)$ and so $q-1 \geq s > \sqrt{n}$ because $s > t$. Since q is an arbitrary prime dividing n and $q > \sqrt{n}$, n must be prime by Proposition 2.14. ∎

Example 7.6: Use Pocklington's primality test to verify that $n = 1489$ is prime.
Solution: In this case $n - 1 = 1488 = 2^4 \cdot 3 \cdot 31$. Let $s = 2^4 \cdot 3 = 48$ and $t = 31$ (we need not know the factorization of t). Let $p_1 = 2$ and $b_1 = 7$; then $\gcd(b_1^{744} - 1, n) = 1$ ($b_1 = 2, 3$, and 5 do not work) and $b_1^{1488} \equiv 1 \pmod{n}$. Let $p_2 = 3$ and $b_2 = 2$; then $\gcd(b_2^{496} - 1, n) = 1$ and $b_2^{1488} \equiv 1 \pmod{n}$. Hence $n = 1489$ is prime.

Note that Lucas' Primality Test and Pocklington's primality test require some information about the factorization of $n - 1$. Beyond that, they require of us little more than the ability to exponentiate to large powers $(\bmod\, n)$ and the ability to calculate the greatest common divisor of potentially large integers. The binary exponentiation algorithm coupled with Fermat's Little Theorem is an exceedingly fast algorithm for exponentiation. In fact, it is roughly as fast as the Euclidean algorithm for determining the greatest common divisor of two integers.

There are other primality tests that utilize the factorization of $n^2 - 1$, $n^2 + 1$, $n^2 - n + 1$, $n^2 + n + 1$, and other quadratic polynomials. There are several relevant references in the bibliography for the interested student.

Exercise 7.2

1. Verify for all integers n that $(b-1) \mid (b^n - 1)$ and, if n is odd, then $(b+1) \mid (b^n + 1)$ as in the proof of Theorem 7.1.
2. (a) Verify that 1105 is a psp(2). Explain how 1105 could be the smallest psp(2) and psp(3) while 561 is the smallest Carmichael number.
    (b) Verify that 1541 is a psp(3) and psp(5). It is the smallest integer with that property.
    (c) Find the smallest psp(3). (It is less than 100.)
3. Show that 161038 is a pseudoprime in the sense that $2^{161038} \equiv 2 \pmod{161038}$. It is in fact the smallest even pseudoprime (discovered by D.H. Lehmer in 1950).
4. (a) Show that if n is both a psp(b_1) and a psp(b_2), then n is a psp($b_1 b_2$).
    (b) Let b^* be the arithmetic inverse of $b \pmod{n}$. Show that if n is a psp(b), then n is $app(b^*)$.
    (c) Show that if n is a psp(b), then n is a psp($n - b$).

5. (J. Chernick – 1939): Show that if $6k+1, 12k+1$, and $18k+1$ are all prime and $n = (6k+1)(12k+1)(18k+1)$, then n is a Carmichael number. Use this result to find a Carmichael number larger than 1729.

6. Verify that 1387 and 2701 are pseudoprimes.

7. Show that the a in Lucas' primality test must be a primitive root$(\bmod n)$.

8. Use Lucas' primality test to verify that the following are prime:
   (a) 31
   (b) 1307 (let $a=2$)
   (c) 2777 (let $a=3$)
   (d) 17389 (the 2000th prime) (let $a=2$)
   (e) 61129 (let $a=7$)

9. Use Pocklington's primality test to verify that the following are prime:
   (a) 31
   (b) 691
   (c) 1777

10. Explain why we can conclude that n is composite with factor c if in using Pocklington's Primality Test we obtain $\gcd(b^{(n-1)/p} - 1, n) = c$ where c does not equal 1 or n.

11. (a) Verify that 41041 is a Carmichael number. (It is the smallest Carmichael number divisible by four primes.)
    (b) Verify that $825265 = 5 \times 7 \times 17 \times 19 \times 73$ is a Carmichael number. (It is the smallest Carmichael number divisible by five primes.)

12. Verify that the following are Carmichael numbers:
    (a) 1105
    (b) 2465
    (c) 2821
    (d) 6601
    (e) 8911

13. It is an open question of D.H. Lehmer's (1931) whether there is a composite n for which $\phi(n) \mid (n-1)$.
    (a) Show that such an n is odd and square-free.
    (b) Show that such an n must be a Carmichael number.

14. A conjecture of John Selfridge (1927–2010) states that if $p \equiv \pm 2 \pmod 5$, then p is prime if and only if (i) $2^{p-1} \equiv 1 \pmod p$ and (ii) $p \mid F_{p+1}$ where F_n is the nth Fibonacci number. Verify the conjecture for all appropriate primes < 100. The composite number 6601 shows that the conjecture doesn't hold for $p \equiv 1 \pmod 5$.

7.3 Miller–Rabin–Jaeschke primality test

By Fermat's Little Theorem, if p is an odd prime and $p \nmid b$, then $bp^{-1} \equiv 1 \pmod p$. Since $p-1$ is even, $p \mid \left(b^{\frac{p-1}{2}} - 1\right)\left(b^{\frac{p-1}{2}} + 1\right)$. By Euclid's Lemma, p divides one of the two factors and so $b^{(p-1)/2} \equiv \pm 1 \pmod{p}$. If $\frac{p-1}{2}$ is even and $b^{(p-1)/2} \equiv 1 \pmod{p}$, then we can infer that $b^{(p-1)/4} \equiv \pm 1 \pmod{p}$. This process could be continued until either $(p-1)/2^k$ were odd or $b^{(p-1)/2^k} \equiv -1 \pmod{p}$ for some k. Of course, if b is a primitive root$\pmod{p}$, then this process terminates after the first step.

For example, $p = 97$ is prime and so $2^{96} \equiv 1 \pmod{97}$. In addition, $2^{48} \equiv 1 \pmod{97}$ while $2^{24} \equiv -1 \pmod{97}$. Another prime is $p = 1951$. Here $2^{1950} \equiv 1 \pmod{1951}$ and $2^{975} \equiv 1 \pmod{1951}$. The process terminates at this point.

Similarly, if $\gcd(b, n) = 1$, $b^{(n-1)/2^k} \not\equiv \pm 1 \pmod{n}$ for some k with $2^k \mid (n-1)$, and $b^{(n-1)/2^j} \equiv 1 \pmod{n}$ for all $j < k$, then n must be composite. For example, the Carmichael number $n = 561$ satisfies the condition that $2^{560} \equiv 1 \pmod{561}$ and $2^{280} \equiv 1 \pmod{561}$, but $2^{140} \equiv 67 \pmod{561}$. Hence 561 is composite.

The foregoing comments lead to a sufficient test for compositeness known as Miller's Test as propounded by G.L. Miller in 1976 and later modified by M.O. Rabin in 1980.

Miller's test: Let $n - 1 = 2^r m$ where $r \geq 1$ and m is odd. Let $\gcd(b, n) = 1$. If either $b^m \equiv \pm 1 \pmod{n}$ or $b^{2^k m} \equiv -1 \pmod{n}$ for some $k \leq r$, then we say that n passes *Miller's test* to base b.

Note that given n, it is very simple to determine r and m. We do not need much information at all regarding the factorization of $n - 1$. Also, notice that if $b^m \equiv \pm 1 \pmod{n}$, then $b^{2^k m} \equiv 1 \pmod{n}$ for all k with $1 \leq k \leq r$. Furthermore, if n is an odd composite relatively prime to b and n passes Miller's test to base b, then n is a psp(b). However, it does not follow that if n is a psp(b), then n must pass Miller's test to base b. As we saw above, 561 is a psp(2) but does not pass Miller's test to base 2.

The essential point is that all primes n pass Miller's test for $1 < b < n$. Furthermore, if n fails Miller's test for some b with $1 < b < n$, then n must be composite. We give a simple example below. Even so, there are composites that pass Miller's test for some values of b.

Example 7.7: Use Miller's test base 2 to prove that $n = 391$ is composite.
Solution: $n - 1 = 390 = (110000110)_2$. We use the binary exponentiation algorithm to calculate $2^{390} \pmod{391}$. In this case the sequence is

$$1 \xrightarrow{1} 2 \xrightarrow{1} 8 \xrightarrow{0} 64 \xrightarrow{0} 186 \xrightarrow{0} 188 \xrightarrow{0} 154 \xrightarrow{1} 121 \xrightarrow{1} 348 \xrightarrow{0} 285 \pmod{391}.$$

Since $2^{390} \equiv 285 \not\equiv \pm 1 \pmod{391}$, it follows that 391 is composite.

Definition 7.3: If n is an odd composite that passes Miller's test to base b, then n is a *base b strong pseudoprime*, denoted spsp(b). If $b = 2$ then n is called a *strong pseudoprime*.

Example 7.8: Let $n = 2047 = 2^{11} - 1 = 23 \times 89$. Then $n - 1 = 2046 = 2 \times 1023$. But $2^{1023} \equiv 1 \pmod{2047}$. Hence, $n = 2047$ is a spsp (2). It is in fact the smallest base 2 strong pseudoprime.

Let $\mathrm{SP}\pi(x)$ denote the *number of strong pseudoprimes (base 2)* less than or equal to x. Table 7.2 serves as an extension of Table 7.1.

Table 7.2: Primes, pseudoprimes, and strong pseudoprimes up to 10^{12}.

x	$\pi(x)$	$P\pi(x)$	$\mathrm{SP}\pi(x)$
10^1	4	0	0
10^2	25	0	0
10^3	168	3	0
10^4	1229	22	5
10^5	9592	78	16
10^6	78498	245	46
10^7	664579	750	162
10^8	5761455	2057	488
10^9	50847534	5597	1282
10^{10}	455052512	14884	3291
10^{11}	4118054813	38975	8607
10^{12}	37607912018	101629	22407

Hence the probability that a given odd integer less than 10^{10} which passes Miller's test to base 2 is indeed prime is $\frac{37607912017}{37607912017 + 22407} > 0.999999$ (we eliminate the prime 2). Could it be that there are only finitely many strong pseudoprimes base b for some b? Unfortunately, the answer is no as proved by Pomerance, Selfridge, and Wagstaff in 1980. We prove the result for $b = 2$.

Propostion 7.5: There are infinitely many strong pseudoprimes.

Proof: By Theorem 7.1, there are infinitely many pseudoprimes (base 2). Thus it suffices to prove that if n is a psp(2), then $s = 2^n - 1$ is a strong pseudoprime. Since n is an odd composite, it follows that $s = 2^n - 1$ is an odd composite. Since n is a psp(2), we have that $2^{n-1} \equiv 1 \pmod{n}$ and hence $2^{n-1} - 1 = kn$ for some odd integer k. But then $2^n - 2 = 2(2^{n-1} - 1) = 2kn$. Since $s = 2^n - 1$, it follows immediately that $2^n \equiv 1 \pmod{s}$. But then $2^{(s-1)/2} = 2^{kn} = (2^n)^k \equiv 1^k = 1 \pmod{s}$. Since $(s-1)/2$ is odd, s is a strong pseudoprime and the result follows. ∎

Despite the fact there are infinitely many strong pseudoprimes to any given base, there are no odd composite integers n which are strong pseudoprimes to all bases relatively prime to n. Hence there is no analog of Carmichael numbers for strong pseudoprimes. In particular, it can be shown that if n is an odd composite, then n passes

Miller's test for at most $(n-1)/4$ bases b with $1 < b < n-1$. This observation has led to the following probabilistic primality test due to Michael O. Rabin (1979).

Rabin's probabilistic primality test: Let n be an odd positive integer and let b_i for $i = 1, \ldots, k$ be such that $1 < b_i < n-1$ and $\gcd(b_i, n) = 1$. If n passes Miller's test for all bases b_i, then the probability that n is prime is at least $1 - 1/4^k$.

For example, if n passes Miller's test for 10 different bases, then the probability that n is composite is less than one-millionth.

Extensive computer searches have been made for strong pseudoprimes to various bases. Pomerance et al. [104] verified the following facts:
(1) The least integer that is a spsp(b) for $b = 2$ is 2047.
(2) The least integer that is a spsp(b) for $b = 2$ and 3 is 1,373,653.
(3) The least integer that is a spsp(b) for $b = 2, 3$, and 5 is 25,326,001.
(4) The least integer that is a spsp(b) for $b = 2, 3, 5$, and 7 is 3,215,031,751.
(5) The number in (4) is the only odd composite below $2.5 \cdot 10^{10}$ which is a spsp(b) for $b = 2, 3, 5$, and 7.

Combined this gives a primality test for all odd integers less than $2.5 \cdot 10^{10}$. If n passes Miller's test for all bases 2, 3, 5, and 7 and n is not 3,215,031,751, then n is prime. Gerhard Jaeschke [65] has extended these findings to add the following:
(6) The smallest integer that is a spsp(b) for $b = 2, 3, 5, 7$, and 11 is 2,152,302,898,747.
(7) The smallest integer that is a spsp(b) for $b = 2, 3, 5, 7, 11$, and 13 is 3,474,749,660,383.
(8) The smallest integer that is a spsp(b) for $b = 2, 3, 5, 7, 11, 13$, and 17 is 341,550,071, 728,321. It is also a spsp(19) but not a spsp(23).

Hence, we have a nice primality test for all odd integers less than $3 \cdot 10^{14}$.

Miller–Rabin–Jaeschke Primality Test: If n is an odd integer less than $3 \cdot 10^{14}$ and n passes Miller's test for the bases 2, 3, 5, 7, 11, 13, and 17; then n is prime.

Example 7.9: Prove that $n = 10^{11} + 3$ is prime.
Solution: $n - 1$ is divisible by 2, but not by 4. (In fact, $n - 1 = 2 \times 3 \times 7 \times 1543 \times 1543067$.) The number n passes Miller's test to the base b as long as $b^{(n-1)/2} \pmod{n}$ is either 1 or -1. We find that $2^{(n-1)/2} \equiv -1 \pmod{n}, 3^{\frac{n-1}{2}} \equiv -1 \pmod{n}, 5^{\frac{n-1}{2}} \equiv -1 \pmod{n}, 7^{(n-1)/2} \equiv -1 \pmod{n}, 11^{\frac{n-1}{2}} \equiv 1 \pmod{n}, 13^{\frac{n-1}{2}} \equiv -1 \pmod{n}$, and $17^{(n-1)/2} \equiv -1 \pmod{n}$. By the Miller–Jaeschke primality test, $n = 10^{11} + 3$ is prime.

In the past decade, several teams of programmers have extended our list to even greater bounds. It is now known that

(9) The least integer that is a pspb(b) for $p = 2, 3, 5, 7, 11, 13, 17, 19$, and 23 is 3,825,123, 056,546,413,051.

(10) The least integer that is a psps(b) for $p = 2, 3, 5, 7, 11, 13, 17, 19, 23, 29, 31$, and 37 is 18,446,744,073,709,551,616.

Let be an odd composite number. An integer b with $n \nmid b$ is called a *witness* (to the fact that n is composite) if n is not a spsp(b); that is, n fails Miller's test to base b. In 1956, P. Erdös showed that the number 2 is a witness for most odd composites. In fact, most odd composites are not even base 2 pseudoprimes. In the other direction, W.R. Alford, A. Granville, and C. Pomerance have shown that there are odd composites n having least witness arbitrarily large. Hence there will always be odd composites that pass Miller's test for any finite list of bases.

Finally, in a rather surprising development in 2002, M. Agrawal, N. Kayal, and N. Saxena developed the first provably unconditional deterministic polynomial time primality test (now called the AKS test). Although the running time has been shown to eventually be as fast as other current primality tests, its implementation has not surpassed others that we have discussed. Even so, the result is a stunning achievement.

Exercise 7.3

1. Show that the following numbers are composite by verifying that they fail Miller's test to base 2:
   (a) 529 (b) 961 (c) 1027 (d) 1271 (e) 1729
2. Show that the following numbers are composite by verifying that they fail Miller's test to base 3:
   (a) 91 (b) 841 (c) 1027 (d) 1105 (e) 5699
3. Verify that if $n = 1373653$, then n is a spsp(b) for $b = 2$ and $b = 3$.
4. Use the Miller-Jaeschke primality test to verify that the following are prime (for each prime, determine in advance which bases b are sufficient to check):
   (a) 811 (b) 7883 (c) 14929 (d) 137341 (e) $10^8 + 7$
5. Verify that if $n = 15841$, then n is both a Carmichael number and a strong pseudoprime.
6. (a) Let $n - 1 = 2^r m$ where $r \geq 1$ and m is odd. Let b be relatively prime to n. Suppose there exists a k with $k < r$ such that $b^{2^k m} \not\equiv \pm 1 \pmod{n}$, but $b^{2^{k+1} m} \equiv 1 \pmod{n}$. Show that $\gcd(b^{2^{k+1} m} - 1, n)$ is a nontrivial factor of n.
   (b) Use the result in part (a) to factor the pseudoprime $n = 1387$.
   (c) Use the result in part (a) to factor the base 3 pseudoprime $n = 1541$.
7. Define a base b *Euler pseudoprime* to be an odd composite integer n that satisfies the congruence $\left(\frac{b}{n}\right) \equiv b^{\frac{n-1}{2}} \pmod{n}$ where $\left(\frac{b}{n}\right)$ is a Jacobi symbol.
   (a) Show if n is prime and $n \nmid b$, then the congruence above is satisfied (see Proposition 4.13(d)).

(b) Verify that $n = 561$ is a base 2 Euler pseudoprime.

(c) Verify that $n = 1105$ is a base 2 Euler pseudoprime.

(d) Show that if n is a base b Euler pseudoprime, then n is a base b pseudoprime.

(e) Show that the converse to part (d) is false by considering $n = 341(b = 2)$.

8. Find the least witness for the following composites: 561, 1105, 2047, 7031.

7.4 Mersenne primes

In Section 3.4 we proved the fact known to Euclid that if $2^p - 1$ is prime, then $n = 2^{p-1}(2^p - 1)$ is a perfect number. Conversely, Euler proved that if n is an even perfect number, then n must be of the above form. Hence there is some interest in determining whether or not $2^p - 1$ is prime. Recall that we also noted that a necessary, but not sufficient, condition for $2^p - 1$ to be prime is that p be prime.

Marin Mersenne (1588–1648), a Minimite friar and voluminous scientific correspondent with some of the greatest European mathematicians of his day, made some far-reaching assertions about numbers of the form $2^p - 1$ in the preface of his book, *Cogitata Physica Mathematica* (1644). In particular, he claimed that $2^p - 1$ was prime for $p = 2, 3, 5, 7, 13, 17, 19, 31, 67, 127$, and 257 and for no other values of $p \leq 257$. The first seven values had already been discovered, but the rest of the assertion was new and Mersenne gave no evidence to support his claim. In any event, it is in his honor that we make the following definition.

Definition 7.4: The number $M_r = 2^r - 1$ is called the rth *Mersenne number*. If M_p is prime, then M_p is called a *Mersenne prime*.

Here is a table of all known Mersenne primes (as of 2025). Presently, the list contains 52 entries.

Table 7.3: All known Mersenne primes (as of 2025).

Prime p	Number of digits in M_p	Discoverer	Date of discovery
2	1	Unknown	Ancient
3	1	Unknown	Ancient
5	2	Unknown	Ancient
7	3	Unknown	Ancient
13	4	Regiomontanus	1456
17	6	Pieter A. Cataldi	1588
19	6	Pieter A. Cataldi	1588
31	10	Leonhard Euler	1772
61	19	I.M. Pervushin	1883
89	27	R. E. Powers	1911

Table 7.3 (continued)

Prime p	Number of digits in M_p	Discoverer	Date of discovery
107	33	E. Fauquembergue	1913
127	39	Edouard Lucas	1876
521	157	Raphael M. Robinson	1952
607	183	Raphael M. Robinson	1952
1279	386	Raphael M. Robinson	1952
2203	664	Raphael M. Robinson	1952
2281	687	Raphael M. Robinson	1952
3217	969	Hans Riesel	1957
4253	1281	Alexander Hurwitz	1961
4423	1332	Alexander Hurwitz	1961
9689	2917	Donald B. Gillies	1963
9941	2993	Donald B. Gillies	1963
11213	3376	Donald B. Gillies	1963
19937	6002	Bryant Tuckerman	1971
21701	6533	Curt Noll and Laura Nickel	1978
23209	6987	Curt Noll	1979
44497	13395	H. Nelson and D. Slowinski	1979
86243	25962	David Slowinski	1982
110503	33265	W.N. Colquitt and L. Welsch	1988
132049	39751	David Slowinski	1983
216091	65050	David Slowinski	1985
756839	227832	D. Slowinski and P. Gage	1992
859433	258716	D. Slowinski and P. Gage	1994
1257787	378632	D. Slowinski and P. Gage	1996
1398269	420921	Joel Armengaud	1996
2976221	895932	Gordon Spence	1997
3021377	909526	Roland Carlson	1998
6972593	2098960	Nayan Hajratwala	1999
13466917	4053946	Michael Cameron	2001
20996011	6320430	Michael Shafer	2003
24036583	7235733	Josh Findley	2004
25964951	7816230	Martin Nowak	2005
30402457	9152052	C. Cooper and S. Boone	2005
32582657	9808358	C. Cooper	2006
37156667	11185272	Hans-Michael Elvenich	2008
42643801	12837064	Odd M. Strindmo	2009
43112609	12978189	Edson Smith	2008
57885161	17425170	Curtis Coooper	2013
74207281	22338618	Curtis Cooper	2016
77232917	23429425	Jon Pace	2017
82589933	24862048	Patrick Laroche	2018
136279841	41024320	Luke Durant	2024

Euler's verification of the primality of $2^{31} - 1 = 2,147,483,647$, published along with complete factorizations of all M_n for $n \leq 37$, lent some credence to Mersenne's claim. (We will discuss Euler's technique following Proposition 7.6.) At the time M_{31} seemed enormous. In fact, in Peter Barlow's theory of numbers (1811) it is stated, "[The prime M_{31}] is the greatest that will ever be discovered, for as they are merely curious without being useful, it is not likely that any person will attempt to find one beyond it." Barlow's text also contains an erroneous proof of Fermat's last theorem.

In 1876 Lucas claimed that M_{67} was composite but M_{127} was indeed prime. To prove this required a theoretical breakthrough plus an incredible talent for computation. No larger prime was discovered prior to the development of electronic calculators and the modern computer. Lucas claimed that a machine could be designed that would nearly instantly determine whether M_p was prime. Such a single-purpose computing machine had to await the labors of D.H. Lehmer in the next century. In all Mersenne's claim contains five mistakes.

Though Lucas claimed that M_{67} was composite, he offered no factorization of it. Such a factorization was given by Frank N. Cole, long time secretary of the AMS and editor of the Bulletin, at the American Mathematical Society meeting in New York in October of 1903. Cole's "talk" simply consisted of silently calculating the decimal expansion of M_{67} on a chalkboard and then obtaining the same number by multiplying 193707721 by 761838257287! Reportedly, Cole received the only standing ovation at that meeting. He later admitted to having spent "three years of Sundays" working on its factorization.

A great number of people have been inspired to search for large primes. It may be of interest to you that Curt Noll and Laura Nickel were high school students when they discovered the primality of M_{21701}.

In subsequent decades, the verification of the primality of the largest Mersenne primes all relied on Theorem 7.9 and utilized the resources of the fastest supercomputers. For example, over 10,000 supercomputing hours were spent searching for the prime candidate M_{756839} and over 19 h on a Cray-2 were used in verifying its primality. Since 1996, all newly discovered Mersenne primes relied on a vast distributed network of computers referred to as GIMPS (Great Internet Mersenne Prime Search). In Table 7.3, I have given the name of the programmer whose computer was fortunate enough to have had a positive primality result.

It is unknown whether there are infinitely many Mersenne primes (though there are heuristic arguments to support this claim). In fact, there is yet no proof that infinitely many M_p are composite for prime p. It is known that $M_{57885161}$ is the 48th Mersenne prime, but there may be others smaller than the last three entries in Table 7.3. In 1876 E. Catalan conjectured that if M_p is a Mersenne prime, then $2^{M_p} - 1$ is prime. This assertion was disproved in 1953 by Wheeler who verified that $2^{M_{13}} - 1$ was composite. The computation took over 100 h on the Illiac computer.

Here is a primality test stated by Fermat (1640) in a letter to Mersenne. The first published proof is due to Euler (1747).

Proposition 7.6: All factors of $M_p = 2^p - 1$ are of the form $2np + 1$ for some n.

Lemma 7.6.1: Given positive integers a and b let $g = \gcd(a, b)$. Then $\gcd(M_a, M_b) = M_g$.

Proof of Lemma: Apply the Euclidean algorithm to the integers a and b:

$a = q_1 b + r_1$ where $0 \le r_1 < b$

$b = q_2 r_1 + r_2$ where $0 \le r_2 < r_1$

$r_1 = q_3 r_2 + r_3$ where $0 \le r_3 < r_2$

$r_{n-3} = q_{n-1} r_{n-2} + r_{n-1}$ where $0 \le r_{n-1} < r_{n-2}$

$r_{n-2} = q_n r_{n-1}$ and $\gcd(a, b) = r_{n-1}$.

By the division algorithm, for any integers s and t there exist q and r with $s = qt + r$ where $0 \le r < t$. Hence observe that $2^s - 1 = (2^t - 1)\left(2^{(q-1)t+r} + 2^{(q-2)t+r} + \ldots + 2^r\right) + (2^r - 1)$ where $0 \le 2^r - 1 < 2^t - 1$.

So the smallest positive residue of $M_t \pmod{M_S}$ is M_r.

Now apply the Euclidean algorithm to the integers M_a and M_b:

$$M_a = Q_1 M_b + M_{r_1}$$

$$M_b = Q_2 M_{r_1} + M_{r_2}$$

$$M_{r_{n-3}} = Q_{n-1} M_{r_{n-2}} + M_{r_{n-1}}$$

$$M_{r_{n-2}} = Q_n M_{r_{n-1}} \text{ and } \gcd(M_a, M_b) = M_{r_{n-1}} = M_g \qquad \blacksquare$$

Proof of Proposition 7.6: Let q be a prime with $q \mid M_p$. By Fermat's Little Theorem, $M_{q-1} \equiv 0 \pmod q$. Let $g = \gcd(p, q - 1)$. By Lemma 7.6.1, $\gcd(M_p, M_{q-1}) = M_g$.

But $q \mid M_p$ and $q \mid M_{q-1}$ implies that $q \mid M_g$. In particular, $g > 1$. But p is prime. So $g = p$. It follows that $p \mid (q - 1)$ and so there is an integer s for which $ps = q - 1$. Since q is odd, s is even and hence $s = 2n$ for some n, i.e., $q = 2np + 1$. Finally, note that all products of numbers of the form $2np + 1$ are of the same form. $\qquad \blacksquare$

In order to verify that M_{31} was prime, Euler had to rule out all possible prime divisors of the form $62n + 1$ that are less than $\sqrt{2^{31} - 1} = 46340.95\ldots$. In fact, based on some elementary congruence arguments, Euler was able to narrow the search to primes of the form $248n + 1$ and $248n + 63$. Even so, there remain a lot of hand (or in Euler's case-head) calculations.

Example 7.10: Completely factor $2^{29} - 1$ (à la Euler).
Solution: By Proposition 7.6, all factors of $M_{29} = 2^{29} - 1 = 536870911$ are of the form $58n + 1$. In fact, we only need to check primes of the form $58n + 1$ that are less than $\sqrt{2^{29} - 1} = 23170.47\ldots$ as possible divisors. The first such prime is 59 and $59 \nmid M_{29}$. The next prime of this form is $233 = 58(4) + 1$. Luckily, $233 \mid M_{29}$.$M_{29}/233 = 2304167$. To completely factor M_{29} we need to continue to check appropriate primes between 233

and $\sqrt{2304167} = 1517.948\ldots$. The primes 349, 523, and 929 do not divide M_{29}, but 1103 (corresponding to $n = 19$) does divide M_{29}. $M_{29}/(233)(1103) = 2089$. If 2089 were not prime, it would be divisible by an appropriate prime at most $\sqrt{2089} = 45.67\ldots$. But there are no such primes of the form $58n + 1$ and so 2089 is prime. Hence the complete factorization of $M_{29} = 233 \times 1103 \times 2089$.

A related theorem is the following due to none other than Euler (1732). It is a nice application of our results on quadratic residues.

Proposition 7.7: If $p \equiv 3 \pmod 4$ is prime with $p \geq 7$ and $2p + 1$ is prime, then $(2p + 1) \mid M_p$ (and hence M_p is composite).

Proof: Since $p \equiv 3 \pmod 4$, $q = 2p + 1 \equiv 7 \pmod 8$ and by Corollary 4.14.1, $\left(\frac{2}{q}\right) = (-1)$ $(q^2 - 1)/8 = 1$. By Euler's criterion (Proposition 4.13(d)), $2^{(q-1)/2} \equiv 1 \pmod q$. Hence $2^p \equiv 1 \pmod q$ and $q \mid M_p$. The condition $p \geq 7$ rules out the possibility that $q = M_p$. ∎

For example, M_{23} is composite since $2(23) + 1 = 47$ is prime. In addition, Proposition 7.7 applies to the primes $p = 83, 131, 179, 191, 239,$ and 251 (Exercise 7.4.1). This is consistent with Mersenne's conjecture and makes one wonder whether Mersenne himself had some knowledge of Proposition 7.7 (or Proposition 7.6 which is equally useful in such instances).

Odd primes p for which $2p + 1$ are also prime are called *Germain primes*, named after Sophie Germain (1776–1831). She showed that $x^p + y^p = z^p$ has no solutions in integers with $p \nmid xyz$ (the so-called first case of Fermat's Last Theorem) whenever p is a Germain prime. As a woman, Germain was banned from attending the Ecole Polytechnique, but she did secure lecture notes passed on for her benefit from Lagrange. In addition, she corresponded with Gauss under the nom de plume M. Leblanc. He was impressed with [her] results. Once he learned of her true identity, he recommended that she receive an honorary doctorate at the University of Göttingen. Unfortunately, she died before the honor could be bestowed. The first few Germain primes are 3, 5, 11, and 23. It is an unproven conjecture that there are infinitely many Germain primes. Many huge Germain primes have been found including $2{,}618{,}163{,}402{,}417 \cdot 2^{1{,}290{,}000} - 1$ identifiied by J.S. Brown and the PrimeGrid distributed server in 2016. The next proposition gives a sufficient condition for Germain primes.

Proposition 7.8: If p is prime and $2^p \equiv 1 \pmod{2p + 1}$, then $2p + 1$ is prime.

Proof: Let $q = 2p + 1$ and suppose r is a prime divisor of q. Since q is odd, so is r. By hypothesis, $2^p \equiv 1 \pmod q$ and so $2^p \equiv 1 \pmod r$. Thus $\text{ord}_r 2 \mid p$. But $\text{ord}_r 2 > 1$ implies $\text{ord}_r 2 = p$. Thus $p \mid (r - 1)$ by Proposition 4.2(a). But then $r \geq p + 1 > q/2 > \sqrt{q}$ since $q \geq 5$. But r is arbitrary and so every prime factor of q exceeds $\sqrt{q}$. So $q = r$ and q is prime. ∎

The next theorem is the main result of this section. A similar theorem was stated and proved by E. Lucas in 1878. However, several simplifications and improvements were made by D.H. Lehmer (1905–1991) in his doctoral dissertation (1930).

Theorem 7.9 (Lucas-Lehmer Test): Let $u_n \equiv u_{n-1}^2 - 2 \pmod{M_p}$ with $u_1 = 4$. Then M_p is prime if and only if $u_{p-1} \equiv 0 \pmod{M_p}$.

We will prove the useful half of this theorem, namely if $u_{p-1} \equiv 0 \pmod{M_p}$, then M_p is prime. Our proof is based directly on one given by J.W. Bruce listed in the bibliography. However, we need some basic algebraic preliminaries.

Definition 7.5: A group G is a nonempty set of elements together with a binary operation that satisfies the following:
(a) (Associativity) If $a, b, c \in G$, then $(a \times b) \times c = a \times (b \times c)$.
(b) (Identity) There is an element $e \in G$ such that $a \cdot e = e \cdot a = a$ for all $a \in G$.
(c) (Inverse) For every element $a \in G$ there is an element $b \in G$ such that $a \times b = b \times a = e$.

It is easy to show that the inverse of an element $a \in G$ is unique. Hence the inverse of a is usually written a^{-1}. A nonempty set together with a binary operation satisfying conditions (a) and (b) is called a *monoid*. There may be some elements in a monoid lacking inverses. Those elements of a monoid that do have inverses are called *units*. Please note that the binary operation itself is often suppressed to simplify expressions.

For example, let $G = \{1, 2, 3, 4, 5, 6\}$ with the operation of multiplication (mod 7). Then 1 is the identity element and the inverse of 2 is 4 (and vice versa), the inverse of 3 is 5 (and vice versa), and the inverse of 6 is itself. Much of Chapters 2 and 4 could be rewritten in terms of group theory, but it has been unnecessary until now to require the term "group" itself. If we let $M = \{0, 1, 2, 3, 4, 5, 6\}$ with multiplication (mod 7), then M is a *monoid* (0 lacks an inverse).

If G is a group with binary operation and $a \in G$, we define exponentiation as follows: $a^0 = e$, $a^1 = a$, and $a^n = a \cdot a^{n-1}$ for $n \geq 1$. In addition, we define $a^{-n} = (a^{-1})^n$.

Definition 7.6: If G is a finite group, then the *order of G*, denoted $|G|$, is the number of elements in G. If $g \in G$, then the order of g is the smallest positive integer s for which $g^s = e$.

The following two lemmas collect some simple facts we will need to prove Theorem 7.9.

Lemma 7.9.1: Let M be a monoid. Then the set of all units of M form a group with respect to the binary operation on M.

Proof: Let M^* denotes the set of all units of **M**. Clearly $e \in M^*$.

Furthermore, M^* inherits associativity from M. It remains to prove that M^* is closed with respect to the binary operation on M. However, if $a, b \in M^*$, then there exist a^{-1}

and $b^{-1} \in M^*$ with $aa^{-1} = e = bb^{-1}$. But then $(ab)(b^{-1}a^{-1}) = a(bb^{-1})a^{-1} = aea^{-1} = aa^{-1} = e$. Similarly, $(b^{-1}a^{-1})(ab) = e$. Hence $ab \in M^*$. ∎

Lemma 7.9.2: (a) If G is a finite group and $g \in G$, then the order of g is at most $|G|$.
(b) If $g \in G$ and $g^r = e$, then the order of g divides r.
Compare Lemma 7.9.2(b) with Proposition 4.2(b).

Proof:
(a) Let $g \in G$ and consider $g^0 = e, g^1 = g, g^2, \ldots, g^n$ where $|G| = n$. By the Pigeonhole Principle, there exist r and s with $0 \le r < s \le n$ with $g^r = g^s$. But then $g^{s-r} = g^s \cdot (g^r)^{-1} = g^s(g^s)^{-1} = e$. So g has order at most n.
(b) Let the order of g be m. Write $r = am + b$ where $0 \le b < m$. So $e = g^r = g^{am+b} = (g^m)^a g^b = e^a g^b = g^b$. If $0 < b < m$, then there is a contradiction to the assumption that m is the order of g. Hence $b = 0$ and $m \mid r$. ∎

Proof of Theorem 7.9: ($\Leftarrow$) Consider the quadratic surd $x = 2 + \sqrt{3}$ and its conjugate $x' = 2 - \sqrt{3}$. Note that $xx' = 1$. Let $a_k = x^{2^{k-1}}$ and $a'_k = x'^{2^{k-1}}$. Notice that $a_1 + a'_1 = 4 = u_1$ and $a_2 + a'_2 = x^2 + x'^2 = 2^2 + 2(1) + 3^2 = 14 = u_2$. In general, $a_k + a'_k = u_k$ for $k \ge 1$ (Exercise 7.4.7). Assume $u_{p-1} \equiv 0 \pmod{M_p}$; so $M_p \mid u_{p-1}$. Hence there is an r such that

$$M_p r = u_{p-1} = a_{p-1} + a'_{p-1}. \text{ So } x^{2^{p-2}} + x'^{2^{p-2}} = M_p r$$

Multiplying through by a_{p-1} and subtracting 1 gives $x^{2^{p-1}} = (M_p r)a'_{p-1} - 1$ and so

$$x^{2^{p-1}} \equiv -1 \pmod{M_p} \tag{7.4}$$

Squaring both sides of eq. (7.4) gives

$$x^{2^p} \equiv 1 \pmod{M_p} \tag{7.5}$$

Suppose now that M_p is composite and let q be an odd prime divisor with $q^2 \le M_p$. Let $S_q = \{0, 1, \ldots, q-1\}$ with the operation of multiplication $(\mathrm{mod}\ q)$. Let $T = \{a + b\sqrt{3}: a, b \in S_q\}$ with real multiplication (reduced mod q). Then T is a commutative monoid with identity $1 = 1 + 0\sqrt{3}$.

Let T^* be the set of all units in T. By Lemma 7.9.1, T^* is a group. Since T has q^2 elements and $0 = 0 + 0\sqrt{3} \notin T^*$, T^* has at most $q^2 - 1$ elements. By Lemma 7.9.2(a), the order of any element of T^* is at most $q^2 - 1$.

Now let $x = 2 + \sqrt{3}$ as above. Clearly $x \in T^*$. Since $q \mid M_p$, the number $(rM_p)x^{2^{p-2}} = 0$ in T. By eq. (7.5), $x^{2^p} = 1$ in T. By Lemma 7.9.2(b), the order of x in T^* is a divisor of 2^p. In fact, the order of x in T^* is precisely 2^p since otherwise it contradicts formula (7.4). So $2^p \le q^2 - 1$ by Lemma 7.9.2(a). But $q^2 - 1 < M_p - 1 = 2^p - 2$, a contradiction. Hence M_p is prime. ∎

Example 7.11: Use the Lucas-Lehmer test to verify that $M_{13} = 8191$ is prime.

Solution: We just have to find $u_{12}(\bmod\ 8191)$. We begin with u_1 and iterate 11 times:

$u_1 = 4, u_2 = 14, u_3 = 194, u_4 = 37634 \equiv 4870 \pmod{8191}, u_5 \equiv 3953 \pmod{8191},$
$u_6 \equiv 5970 \pmod{8191}, u_7 \equiv 1857 \pmod{8191}, u_8 \equiv 36 \pmod{8191},$
$u_9 \equiv 1294 \pmod{8191}, u_{10} \equiv 3470 \pmod{8191}, u_{11} \equiv 128 \pmod{8191},$ and
$u_{12} \equiv 0 \pmod{8191}.$ Therefore $M_{13} = 8191$ is prime.

Exercise 7.4

1. Show that M_p is composite for $p = 83, 131, 179, 191, 239,$ and 251 by using Proposition 7.7.

2. Use Proposition 7.8 to verify that 3 and 23 are Germain primes. What is the situation for the primes $5, 11, 29, 41,$ and 53?

3. (a) Use Proposition 7.6 to find the smallest prime factor of M_{37} (completely factored by Fermat in 1640).
   (b) Use Proposition 7.6 to find the smallest prime factor of M_{43} (completely factored by F. Landry in 1869).

4. Show that the set of nonzero integers forms a monoid under multiplication.

5. (a) Show that $M_p \equiv 7 (\bmod\ 8)$ for all $p > 2$ and hence M_p is divisible by a prime $q \equiv 7 \pmod{8}$.
   (b) Combine part (a) with Proposition 7.6 to show that M_{31} is divisible by a prime of the form $248t + 63$. (In fact, M_{31} is itself prime.)
   (c) Use a similar argument to show that if M_{47} is composite, then it must be divisible by a prime of the form $376t + 95$. Use this to find a nontrivial factor of M_{47}.

6. Use Lemma 7.6.1 together with the proof of Proposition 7.5 to show that there exist strong pseudoprimes with arbitrarily many prime divisors.

7. Use induction to verify that $u_k = a_k + a_k$ for $k \geq 1$ in the proof of the Lucas-Lehmer test.

8. Show that $M_n \equiv 7 (\bmod\ 24)$ for all $n \geq 3$.

9. (a) Show that if p and q are distinct primes, then M_p and M_q are relatively prime.
   (b) Show that if a and b are even perfect numbers, then $\gcd(a, b)$ is a power of 2.

10. It is unknown whether or not M_p is square-free for every prime p. Show that there is an integer n for which M_n is not square-free.

11. Show that $\gcd(M_{15}, M_{125}) = 31$.

12. Find a prime factor of M_{73} (Euler found it first).

13. Use the Lucas-Lehmer test to verify that M_{17} is prime.

14. Use Proposition 7.7 to verify that M_{3023} is composite.

15. Use Proposition 7.6 to demonstrate that there are infinitely many primes.

7.5 Fermat numbers

Fermat claimed (1640) that all numbers of the form $2^{2^n} + 1$ were prime after verifying this for $n = 0, 1, 2, 3$, and 4. In this section we consider Fermat's conjecture and develop some of the related theory.

Notice that if $m = ab$ with $a \geq 3$ odd, then

$$2^m + 1 = 2^{ab} + 1 = \left(2^b + 1\right)\left(2^{b(a-1)} - 2^{b(a-2)} + 2^{b(a-3)} - \cdots + 1\right)$$

Hence if $2^m + 1$ is prime, then m has no odd factors larger than 1 and $m = 2^n$ for some n. This leads to the following definition:

Definition 7.5: Let $f_n = 2^{2^n} + 1$ for $n \geq 0$. The number f_n is called the nth *Fermat number*. If f_n is prime, then it is called a *Fermat prime*.

Despite Fermat's generally accurate claims, in 1732 Euler showed that f_5 was composite and gave its factorization. In fact, $f_5 = 4294967297 = 641 \cdot 6700417$. Since that time much effort has gone into determining the nature of larger Fermat numbers. E. Lucas showed that f_6 was composite but was unable to find any factors. In 1880, the octogenarian F. Landry found that $f_6 = 274177 \cdot 67280421310721$. He later claimed to have verified that both factors were prime. To date, no other Fermat primes have been found. In fact, J. Selfridge conjectured that the only Fermat primes are the initial ones listed by Fermat. No one has proven that there are either an infinite number of Fermat primes or an infinite number of composite Fermat numbers (though at least one of those statements must be true).

At the present time, f_7, f_8, f_9, f_{10}, and f_{11} have all been completely factored. The number f_9 consists of 155-digits and wasn't completely factored until 1990 utilizing the collective efforts of a thousand computers over several month's time. Its factorization is 2,424,833 times two other primes, one of 49 digits and one of 99 digits! The 309-digit number f_{10} was factored in 1995 with the aid of a distributed computer search. The number f_{11} was completely factored in 1988 as the product of three primes, one of them of length 564 digits. The possibility of factoring such huge integers raises some interesting security questions with the cryptosystems described in Section 8.1, systems previously considered invulnerable to attack.

The Fermat numbers f_n for $5 \leq n \leq 32$ are all known to be composite, although for some of them like f_{20} and f_{24}, no prime factor has been identified. The number f_{20} consists of 315,653 digits. Additionally, several immense Fermat numbers have been proven composite. For example, W. Keller (1985) has shown that $\left(5 \times 2^{23473} + 1\right) \mid f_{23471}$. Much of this work relies on the following result (a special case of a theorem due to Euler concerning factors of numbers of the form $a^{2^n} + b^{2^n}$).

Proposition 7.10: Let $n \geq 2$. If $d \mid f_n$, then $d = k \cdot 2^{n+2} + 1$ for some k.

Proof: Let p be prime with $p \mid f_n$. So $2^{2^n} \equiv -1 \pmod{p}$ and hence $2^{2^{n+1}} \equiv 1 \pmod{p}$. It follows that $\mathrm{ord}_2 p = 2^{n+1}$. By Proposition 4.2(a), $2^{n+1} \mid (p-1) = \phi(p)$. So $8 \mid (p-1)$ and $p \equiv 1 \pmod 8$. By Corollary 4.14.1, the Legendre symbol $\left(\frac{2}{p}\right) = 1$. But by Euler's Criterion (Corollary 4.9.1), $2^{(p-1)/2} \equiv 1 \pmod p$. By Proposition 4.2($b$), $2^{n+1} \mid \left(\frac{p-1}{2}\right)$ and $2^{n+2} \mid (p-1)$. Therefore, $p = k \cdot 2^{n+2} + 1$ for some k. But products of numbers of the form $k \cdot 2^{n+2} + 1$ are of the same form (check). The result follows. ∎

Example 7.12: To factor $f_5 = 2^{32} + 1$, Euler needed to consider only primes less than 2^{16} of the form $p = k \cdot 2^7 + 1 = 128k + 1$. For $k = 1$, $p = 129$ is not prime and need not be checked. For $k = 2$, $p = 257$ is prime but does not divide f_5. Neither $k = 3$ nor $k = 4$ give prime values. For $k = 5$, $p = 641$ is prime and 641 does divide f_5. Hence f_5 is composite. $f_5/641 = 6700417$. In order to determine the nature of 6700417, Euler needed to check possible prime divisors of the form $p = 128k + 1$ for $p \leq \sqrt{6700417} \approx 2588.51$. Indeed, 6700417 is prime as well.

Another useful theorem in this regard is one due to F. Proth (1878) who was a French farmer, largely self-taught. This theorem has been used in discovering very large twin primes such as $2{,}996{,}863{,}034{,}895 \cdot 2^{1{,}290{,}000} \pm 1$, discovered in 2016 and currently the largest known pair of twin primes. Each consist of 388,342 digits.

Theorem 7.11 (Proth's Theorem): Let $n = k \cdot 2^m + 1$ where $m \geq 2$ and k is odd with $k < \sqrt{n}$. If there is an integer b for which $b^{(n-1)/2} \equiv -1 \pmod n$, then n is prime.

Proof: Apply Pocklington's primality test with $s = 2^m$ and $t = k < s$. The only prime divisor of s is $p = 2$. By hypothesis, $b^{n-1} \equiv 1 \pmod n$. In addition, since n is odd and $n \mid \left(b^{(n-1)/2} + 1\right)$, it follows that $\gcd\left(b^{(n-1)/2} - 1, n\right) = 1$. The result follows immediately. ∎

For example, for $n = 97 = 3 \times 2^5 + 1$, it happens that $5^{(n-1)/2} \equiv -1 \pmod n$ and so 97 is prime. However, neither $b = 2$ nor $b = 3$ works.

The practical usefulness of Proposition 7.10 and Proth's Theorem with regard to Fermat numbers is hampered by the rapid growth of f_n as n increases. The next theorem, dating to 1877, is in some sense the analog of the Lucas-Lehmer test for Fermat numbers. It has been the key result used in the study of f_n for all the larger values of n. Its creator, J.F.T. Pepin (1826–1904) was a Jesuit priest and mathematician.

Theorem 7.12 (Pepin's Primality Test): The Fermat number f_n is prime if and only if $3^{(f_n-1)/2} \equiv -1 \pmod{f_n}$.

Proof: ($\Rightarrow$) For $n \ge 1, f_n \equiv 5 \pmod{12}$ (Exercise 7.5.4). Thus $f_n \equiv 1 \pmod 4$ and $f_n \equiv 2 \pmod 3$. If f_n is prime, then the Legendre symbol $\left(\frac{3}{f_n}\right) = \left(\frac{f_n}{3}\right) = -1$ by the quadratic reciprocity law. By Euler's criterion (Proposition 4.13(d)),

$$\left(\frac{f_n}{3}\right) \equiv 3^{\frac{(f_n-1)}{2}} \pmod{f_n}$$

and so

$$3^{(f_n-1)/2} \equiv -1 \pmod{f_n}$$

($\Leftarrow$) Assume $3^{(f_n-1)/2} \equiv -1 \pmod{f_n}$.

Let $N = f_n$ and apply Lucas' primality test. The only prime dividing $N - 1$ is $p = 2$. Let $a = 3$. By hypothesis, $a^{N-1} = 3^{f_n-1} = \left[3^{(f_n-1)/2}\right]^2 \equiv (-1)^2 = 1 \pmod{f_n}$. But $a^{(N-1)/p} = 3^{(f_n-1)/2} \equiv -1 \equiv 1 \pmod{f_n}$. So f_n is prime. ∎

The number 3 in Pepin's primality test may be replaced by some other integers (see Exercise 7.5.6).

Example 7.13: Use Pepin's primality test to verify that $f_4 = 2^{16} + 1 = 65537$ is prime.
Solution: We need to calculate $3^{2^{15}} \pmod{65537}$. This may be accomplished by beginning with 3 and successively squaring 15 times (reducing modulo 65537 at each step). Here are the details:

$$3^{2^1} = 9, \; 3^{2^2} = 81, \; 3^{2^3} = 6561, \; 3^{2^4} \equiv 54449 \pmod{65537}, \; 3^{2^5} \equiv 618699 \pmod{65537},$$

$$3^{2^6} \equiv 19139 \pmod{65537}, \; 3^{2^7} \equiv 15028 \pmod{65537}, \; 3^{2^8} \equiv 282 \pmod{65537},$$

$$3^{2^9} \equiv 13987 \pmod{65537}, \; 3^{2^{10}} \equiv 8224 \pmod{65537}, \; 3^{2^{11}} \equiv 65529 \equiv -2^3 \pmod{65537},$$

$$3^{2^{12}} \equiv \left(-2^3\right)^2 = 2^6 \pmod{65537}, \; 3^{2^{13}} \equiv 2^{12} \pmod{65537}, \; 3^{2^{14}} \equiv 2^{24} \pmod{65537}, \text{and}$$

$$3^{2^{15}} \equiv 2^{48} \pmod{65537}.$$

But $\left(2^{16} + 1\right)\left(2^{32} - 2^{16} + 1\right) = 2^{48} + 1$ and so $3^{2^{15}} \equiv -1 \pmod{65537}$.
Therefore, $f_4 = 65537$ is prime.

Here is a delightful proof that there are infinitely many primes (Theorem 2.11) that utilizes Fermat numbers. It is commonly attributed to the great analyst and mathematical educator George Pólya (1887–1985).

Proof of the infinitude of primes: It suffices to exhibit an infinite sequence of pairwise relatively prime integers, for each must have at least one distinct prime divisor. We will show that the Fermat numbers form such an infinite sequence. Let us begin by using induction to show

$$f_n - 2 = f_0 \cdot f_1 \cdot \ldots \cdot f_{n-1} \text{ for } n \geq 1 \tag{7.6}$$

For $n = 1, f_1 - 2 = 5 - 2 = 3 = f_0$. Assume eq. (7.6) holds for some n. Then

$$f_{n+1} = 2^{2^{n+1}} + 1 = \left(2^{2^n}\right)^2 + 1$$

$$= \left[\left(2^{2^n} + 1\right)^2 - 2 \cdot 2^{2n} - 1\right] + 1$$

$$= \left(2^{2^n} + 1\right)^2 - 2\left(2^{2^n} + 1\right) + 2$$

$$= f_n^2 - 2f_n + 2 = f_n(f_n - 2) + 2$$

$$= f_0 \cdot f_1 \cdot \ldots \cdot f_{n-1} \cdot f_n + 2 \text{ by the inductive hypothesis.}$$

So eq. (7.6) holds for all $n \geq 1$. Now if the prime $p \mid f_j$ and $p \mid f_m$ with $j < m$, then $p \mid (f_m - 2)$ by eq. (7.6), But $2 = f_m - (f_m - 2)$ and so $p \mid 2$ (hence $p = 2$). But f_n is odd for all n, a contradiction. Therefore, the Fermat numbers form an infinite sequence of pairwise relatively prime integers. ∎

The Fermat numbers have a curious application in Euclidean geometry. The ancient Greeks were able to construct regular polygons with $2^k n$ sides for $n = 3, 4$, and 5 and $k \geq 0$ using only straight-edge and compass (the so-called Euclidean tools). Gauss made a startling discovery (noted in his journal March 30, 1796), namely that a regular 17-sided polygon (heptadecagon) could be so constructed. In his *Disquisitiones* (articles 365 and 366) Gauss proved that a regular n-sided polygon is constructible with Euclidean tools for any $n = 2^k \cdot p_1 \cdot \ldots \cdot p_m$ where the p_i's are distinct Fermat primes. The necessity of n being of the above form was also known to Gauss but was first published by P.L. Wantzel (1837).

There are several conjectures concerning Fermat numbers. In 1828, it was conjectured that the Fermat numbers $2 + 1, 2^2 + 1, 2^{2^2} + 1, 2^{2^{2^n}} + 1$, etc. are all prime. J. Selfridge disproved this conjecture by showing that f_{16} is composite. Another open question is whether all Fermat numbers are square-free. However, it has been shown that if p is prime and $p^2 \mid f_n$, then $2^{p-1} \equiv 1 \pmod{p^2}$. This last congruence occurs quite rarely (Proth erroneously conjectured (1876) that it was never satisfied.) It is related to a result of A. Wieferich (1909) that if the first case $(p \nmid xyz)$ of Fermat's Last Theorem is false for exponent p, then $2^{p-1} \equiv 1 \pmod{p^2}$. The first two so-called *Wieferich primes* are $p = 1093$ and $p = 3511$ discovered by W. Meissner (1913) and N.G.W.H. Beeger (1922), respectively. In fact, these are the only Wieferich primes below 6×10^9.

There is a wealth of related results and an abundance of fruitful directions for further study (excellent sources are Ribenboim and Riesel). One final footnote: In 1986, A. Granville proved that if the first case of Fermat's last theorem is false for exponent p, then $a^{p-1} \equiv 1 \pmod{p^2}$ for all primes $a \leq 89$. It then followed from extensive calculations that the first case of Fermat's last theorem holds for all $p < 714591416091389$. Of course, all of this is moot considering Wiles' complete proof of Fermat's last theorem.

Exercise 7.5

1. Use Proposition 7.10 to find a factor of f_{36} (Seelhoff–1886).
2. Use Proposition 7.10 to verify that Euler's factorization of f_5 is complete.
3. Use Proth's Theorem to verify that the following are prime:
   (a) 97 $(b=5)$
   (b) 193 $(b=5)$
   (c) 353 $(b=3)$
   (d) 449 $(b=3)$
   (e) 641 $(b=3)$
   (f) 1217 $(b=3)$
4. (a) Show that if $n \geq 1$, then $f_n \equiv 5 \pmod{12}$.
   (b) Show that if $n \geq 2$, then $f_n \equiv 17 \pmod{240}$.
5. Use Pepin's primality test to verify that f_3 is prime.
6. (a) Prove the following generalization of Pepin's primality test:
   Let $n, k \geq 2$. The following are equivalent:
   $$(i)\ f_n \text{ is prime and } \left(\tfrac{k}{f_n}\right) = -1 \qquad (ii)\ k^{(f_n-1)/2} \equiv -1 \pmod{f_n}.$$

   (b) Use part (a) to show that either 5 or 10 can be used in place of 3 in Pepin's primality test.
   (c) Show that 3 cannot be replaced by 7 in Pepin's primality test.
7. For $m \neq n$, show that $\gcd(a^{2^n}+1, a^{2^m}+1) = \begin{cases} 1 \text{ if } a \text{ is even} \\ 2 \text{ if } a \text{ is odd.} \end{cases}$
8. Verify the following congruences (due to C. Jacobi):
   (a) $3^{10} \equiv 1 \pmod{11^2}$
   (b) $9^{10} \equiv 1 \pmod{11^2}$
   (c) $14^{28} \equiv 1 \pmod{29^2}$
   (d) $18^{36} \equiv 1 \pmod{37^2}$
9. Find an odd integer n for which $\phi(n) = 2^{31}$.
10. At present, how many regular odd-sided polygons can be constructed with Euclidean tools?
11. Show if f_n is prime and g is a primitive root $(\bmod f_n)$, then $f_n - g$ is another primitive root $(\bmod f_n)$.
12. Show that if p is a Germain prime, then all the non-squares mod $(2p+1)$ except for $2p$ are primitive roots mod $(2p+1)$.

7.6 Factorization methods

We have developed several general-purpose primality tests such as Lucas' primality test, Pocklington's primality test, and the Miller-Rabin-Jaeschke primality test. In each case, if the test is positive, then it can be concluded that the given integer is prime. In addition, we have developed some specialized primality tests for integers of particular forms. These include Proposition 7.6, Proposition 7.8, the Lucas-Lehmer test, Proth's theorem, and Pepin's primality test. In many cases (say Pepin's primality test) if the test failed, then the given integer was proven composite. However, we still had no knowledge of any of its factors.

Given a composite integer with a small prime divisor the sieve of Eratosthenes can be used to find it (i.e., trial division by small primes). If the given composite integer has just two prime divisors roughly the same size, then Fermat's factorization method may be effective. In this section we describe three other factorization techniques useful in finding prime divisors where neither of the above methods are at all practical. They are the Pollard rho factorization method, the Pollard $p-1$ factorization method, and a continued fraction factorization method. All three are nondeterministic (or probabilistic) algorithms in the sense that the exact running time cannot be accurately determined in advance. However, their average (or expected) running times are much faster than trial division or Fermat's factorization method. Although the methods above have also been superseded recently, they will serve as worthwhile illustrations of many of the ideas used in the theory of factorization.

Let n be a large composite integer and d a nontrivial divisor of n (not known in advance). We consider 1 and n to be trivial divisors. The *Pollard rho factorization method* (or Monte Carlo method) was developed by J.M. Pollard in 1974. The idea is to generate a list of apparently random integers $x_0, x_1, \ldots$ such that for some $i < j$, $x_i \equiv x_j \pmod{d}$, but $x_i \not\equiv x_j \pmod{n}$. Then $\gcd(x_j - x_i, n)$ will be a nontrivial divisor of n.

In particular, we let x_0 be any "seed", say $x_0 = 2$, and let $x_{i+1} \equiv f(x_i) \pmod{n}$ where $f(x)$ is an irreducible polynomial. Typically, $f(x) = x^2 + 1$, but there are other appropriate polynomials. Notice that if $x_i \equiv x_j \pmod{d}$, then $x_{i+r} \equiv x_{j+r} \pmod{d}$ for all r (so the x_i's aren't really random). If i and j are the smallest indices for $x_i \equiv x_j \pmod{d}$, then let m be the smallest multiple of $j - i$ which is at least as large as i. It must be the case that $x_{2m} \equiv x_m \pmod{d}$. Then $\gcd(x_{2m} - x_m, n)$ will be a nontrivial divisor of n as long as $x_{2m} \not\equiv x_m \pmod{n}$.

Note that with this method as well as the others discussed in this section, n must be composite. Otherwise n would have no nontrivial factors and the algorithm would run endlessly.

Example 7.14: Use the Pollard rho method to factor $n = 10573$.
Solution: Let $x_0 = 2, f(x) = x^2 + 1$, and $x_{i+1} \equiv f(x_i) \pmod{n}$. Here are the initial x_i values: $x_0 = 2, x_1 = 5, x_2 = 26, x_3 = 677$, $x_4 = 458330 \equiv 3691 \pmod{n}, x_5 \equiv 5458 \pmod{n}$, $x_6 \equiv 5624 \pmod{n}$, $x_7 \equiv 5534 \pmod{n}, x_8 \equiv 5749 \pmod{n}$, and so on. Using the Euclidean algorithm, we obtain $\gcd(x_2 - x_1, n) = 1, \gcd(x_4 - x_2, n) = 1$, *and* $\gcd(x_6 - x_3, n) = 97$. Hence 97 is a nontrivial factor of 10573. It is now easy to completely factor n as 97×109.

Graphically, the situation is as shown in Figure 7.1 (hence the name "rho" method):

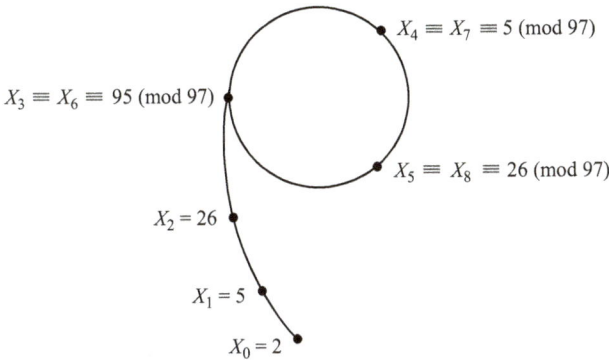

Figure 7.1: Pollard rho method on 10,573.

It is rather startling how quickly we found a factor of n. It can be shown that if p is the smallest prime divisor of n, then on average we expect to find a divisor of n after about $1.2\sqrt{p}$ steps. In our example we were somewhat luckier.

The next algorithm is the *Pollard p – 1 factorization method* (1975). Let n be a large odd composite integer and suppose p is a prime divisor of n. If $(p-1) \mid a$, then $2^a \equiv 1 \pmod{p}$ by Fermat's Little Theorem. If $d = \gcd(2^a - 1, n)$ and $n \nmid (2^a - 1)$, then d is a nontrivial factor of n (with divisor p). Specifically, let $L_k = lcm[2, 3, \ldots, k]$. (Sometimes $k!$ is used instead of L_k to reduce the number of computations.) If $(p-1) \mid L_k$ for k fairly small (i.e., $p-1$ has only small prime factors), then there is a good chance that L_k is a good choice for a.

In particular, this is how the algorithm works: Notice that if $k = p$ or $k = p^m$ for some prime p, then $L_k = p \cdot L_{k-1}$. Otherwise, $L_k = L_{k-1}$. To calculate 2^L, we proceed as follows:

Define
$$p_k = \begin{cases} p \text{ if } k = p \text{ or } k = p^m \\ 1 \text{ otherwise.} \end{cases}$$

Then $2^{L_k} = (2^{L_{k-1}})^{p_k}$. Let $q_k = 2^{L_k} \pmod{n}$ and $d_k = \gcd(q_k - 1, n)$. Notice that $q_k \equiv q_{k-1}^{p_k} \pmod{n}$. Fortunately, q_k can be reduced $\pmod{n}$ (Exercise 7.6.12). Calculate d_k until a nontrivial factor of has been found.

Example 7.15: Use the *Pollard p – 1 factorization method* to factor 20437. We present the process in tabular form.

If $p_k = 1$, then no new calculations need be done for q_k and we move on to $k + 1$. In the example, $d_{19} = 191$ and so 191 is a nontrivial factor of 20437. It can be readily shown that $20437 = 107 \times 191$. (Notice that $p-1$ had smaller prime factors for $p = 191$ than for $p = 107$.) The computational details are displayed in Table 7.4.

Table 7.4: Pollard $p - 1$ factorization method on 20,437.

k	p_k	$q_k \pmod{n}$	$q_k - 1 \pmod{n}$	d_k
2	2	4	3	1
3	3	64	63	1
4	2	4096	4095	1
5	5	11963	11962	1
6	1	– –	– –	–
7	7	5544	5543	1
8	2	19125	19124	1
9	3	7794	7793	1
10	1	– –	– –	–
11	11	18295	18294	1
12	1	– –	– –	–
13	13	4327	4326	1
14	1	– –	– –	–
15	1	– –	– –	–
16	2	2637	2636	1
17	17	11657	11656	1
18	1	– –	– –	–
19	19	7641	7640	191

The last factorization method we consider is the continued fraction factorization method (sometimes called CFRAC). Again let n be a large odd composite integer. Fermat's factorization method involved finding x and y such that $n = x^2 - y^2 = (x+y)(x-y)$. Using ideas going back to Legendre, D.H. Lehmer and R.E. Powers (1931) developed an algorithm using the continued fraction expansion of $\sqrt{n}$ to find x and y such that $x^2 \equiv y^2 \pmod{n}$. The next proposition describes the usefulness of finding such x and y.

Proposition 7.13: Let x and y be such that $x^2 \equiv y^2 \pmod{n}$ with $x \equiv \pm y \pmod{n}$. Then both $\gcd(x - y, n)$ and $\gcd(x + y, n)$ are nontrivial divisors of n.

Proof: Let $d = \gcd(x+y, n)$. Then $d \mid n$. If $x^2 \equiv y^2 \pmod{n}$, then $n \mid (x+y)(x-y)$. If $d = n$ then $n \mid (x+y)$, contradicting $x \not\equiv -y \pmod{n}$. If $d = 1$ then $n \mid (x-y)$, contradicting $x \equiv y \pmod{n}$. So $\gcd(x+y, n)$ is a nontrivial divisor of n. The proof that $\gcd(x-y, n)$ is a nontrivial divisor of n is analogous and is left as Exercise 7.6.10. ∎

Recall Theorem 6.21 (notation modified for our present purposes) concerning the continued fraction expansion for $\sqrt{n}$: $p_{k-1}^2 - nq_{k-1}^2 = (-1)^k t_k$. If t_k is a perfect square for even value of k, say $t_k = s^2$, then $p_{k-1}1^2 \equiv s^2 \pmod{n}$. If $p_{k-1} \not\equiv \pm s \pmod{n}$, then by Proposition 7.13, $\gcd(p_{k-1} - s, n)$ and $\gcd(p_{k-1} + s, n)$ are nontrivial divisors of n. The calculations of p_k, q_k, and t_k can be readily handled via Theorems 6.3 and 6.22. In addition, we reduce $p_{k-1} \pmod{n}$ at each step to control the size of our calculations.

Example 7.16: Use the continued fraction factorization method to completely factor $n = 9179$.

Solution: We begin by determining the continued fraction expansion for $\sqrt{n}$ using formula (6.11): $\sqrt{9179} = [a_0; a_1, a_2, \ldots]$ where $A_0 = \sqrt{9179}$, $a_0 = [A_0]$, and $A_k = \frac{1}{A_{k-1} - a_{k-1}}$, $a_k = [A_k]$ for $k \geq 1$. We readily obtain $\sqrt{9179} = [95; 1, 4, 5, 2, 3, 2, 1, 1, 1, 13, 32, 10, 1, 4, 5, \ldots]$. Next, we let $p_{-2} = 0$, $p_{-1} = 1$, $q_{-2} = 1$, $q_{-1} = 0$, and recursively calculate p_k, q_k, and t_k from the formulas

$$p_k = a_k p_{k-1} + p_{k-2}, \quad q_k = a_k q_{k-1} + q_{k-2},$$

and

$$(-1)^k t_k = p_{k-1}^2 - n q_{k-1}^2 \text{ for } k \geq 0$$

Recall that we reduce the p_k's, q_k's, and t_k's modulo n as we go until we find an even $k > 0$ for which t_k is a square. Details are provided in Table 7.5.

Table 7.5: Continued fraction factorization method on 9,179.

k	a_k	p_k (mod n)	q_k (mod n)	t_k (mod n)
−2	–	0	1	–
−1	–	1	0	–
0	95	95	1	1
1	1	96	1	154
2	4	479	5	37
3	5	2491	26	34
4	2	5461	57	77
5	3	516	197	50
6	2	6493	451	65
7	1	7009	648	98
8	1	4323	1099	73
9	1	2153	1747	115
10	13	4775	5452	14
11	32	8089	1810	11
12	10	3054	5194	4009
13	1	1964	7004	8127
14	–	–	–	$2116 = 46^2$

The dashes in the last row indicate that it is unnecessary to compute p_{14}, q_{14}, etc. once we noticed t_{14} was a perfect square. Of course, it was actually unnecessary to calculate t_k for odd k. Next, we determine that $\gcd(p_{13} - \sqrt{t_{14}}, n) = \gcd(1918, 9179) = 137$. In addition, $\gcd(p_{13} + \sqrt{t_{14}}, n) = \gcd(2010, 9179) = 67$. In fact, $9179 = 67 \times 137$.

It can be shown that the average number of expected steps is on the order of $\sqrt{p}$ where p is the smallest prime divisor of n. In our example we slightly exceeded that. If the algorithm does not find a square value of t_k for some even k, then one modification is to apply the continued fraction factorization method to $\sqrt{mn}$ where m is a small integer. A nontrivial factor of mn may indeed be a nontrivial divisor of n. Other significant improvements to the continued fraction method have been made by Daniel Shanks (1971) in his algorithm SQUFOF and the Brillhart–Morrison method (1975) which was initially used to completely factor f_7.

A more recent and commonly used factorization technique is Carl Pomerance's *quadratic sieve method* (1982) which utilizes parallel processing techniques to find x and y for which $x^2 \equiv y^2 \pmod{n}$. We give a brief introduction in the next section. Pollard has extended these ideas to a very efficient factorization algorithm known as the *number field sieve*, which has been further modified by D. Coppersmith and others. Another highly sophisticated technique is the elliptic curve method successfully implemented by A.K. Lenstra and H.W. Lenstra (1987). We save these interesting topics for more specialized works on this subject.

Exercise 7.6

1. Use the Pollard rho factorization method to factor $n = 8453$.
2. Use the Pollard rho factorization method to factor $n = 3799$.
3. Use the Pollard $p - 1$ factorization method to factor $n = 9943$.
4. Use the Pollard $p - 1$ factorization method to factor $n = 2279$.
5. Use the Pollard $p - 1$ factorization method to factor $n = 7811$. Compare with the Pollard rho factorization method for $n = 7811$.
6. Use the continued fraction factorization method to factor $n = 2881$.
7. Use the continued fraction factorization method to factor $n = 74104$ (first take out all factors of 2).
8. Completely factor the following numbers.
   (a) 20572 (b) 24566 (c) 473175 (d) 476182 (e) 24354330
9. Let $L_k = lcm[2, 3, \ldots, k]$. Show that if $k = p^m$ for some prime p and $m \geq 1$, then $L_k = p \cdot L_{k-1}$ and that $L_k = L_{k-1}$ otherwise.
10. Complete the proof of Proposition 7.13.
11. Find all values of $n \leq 11$ for which $n! + 1$ is prime. It is unknown if there are infinitely many such n (cf. Exercise 5.1.18).
12. Explain why we are justified in reducing modulo n in the algorithms in this section.
13. In the continued fraction factorization method it is only necessary to compute t_k for even k (and hence sufficient to compute p_k and q_k for odd k).
    (a) Show that the following formula holds: For $m \geq 1$

    $$p_{2m+1} = p_{2m-1} + a_{2m+1}(a_{2m}p_{2m-1} + a_{2m-2}p_{2m-3} + \cdots + a_2p_1 + a_0)$$

    and analogously for q_{2m+1}. (Note that the indices descend from $2m + 1$ to 0.)

    (b) Deduce that $p_{2m+1} = \left(1 + a_{2m+1}a_{2m} + \frac{a_{2m+1}}{a_{2m-1}}\right)p_{2m-1} - \frac{a_{2m+1}}{a_{2m-1}}p_{2m-3}$ and analogously for q_{2m+1}.

Hence it suffices to calculate p_k and q_k for odd k in implementing the continued fraction factorization method.

14. (*Euler's factorization method*)
    (a) Let $n = a^2 + b^2 = c^2 + d^2$ where $a > c$ are odd and $d > b$ are even. Let $r = \gcd(a - c, d - b), s = \gcd(a + c, b + d), t = \frac{a-c}{r}$, and $u = \frac{d-b}{r}$. Show that r and s are even and $\gcd(t, u) = 1$.
    (b) Show that $t(a + c) = u(b + d)$ and that $su = a + c$, st $= b + d$.
    (c) Verify that $n = \left[\left(\frac{r}{2} \right)^2 + \left(\frac{s}{2} \right)^2 \right] \left[t^2 + u^2 \right]$.

15. Use Euler's factorization method to factor the following integers:
    (a) $21037 = 141^2 + 34^2 = 99^2 + 106^2$
    (b) $22261 = 119^2 + 90^2 = 105^2 + 106^2$
    (c) $230701 = 349^2 + 330^2 = 99^2 + 470^2$
    (d) $1000009 = 235^2 + 972^2 = 3^2 + 1000^2$ (see Exercise 5.2.6).

7.7 Quadratic sieve factorization algorithm

Fermat's factorization method is fairly effective as a means to factor n if n is the product of two integers of relatively similar size. Over the years, there have been a host of improvements to provide for broader use. In this section we discuss two such modifications.

The Belgian mathematician M.B. Kraitchik (1882–1957) developed the following generalization of Fermat's factorization method in the early 1920s. Let n be an odd composite integer and let $k \geq 1$. Let $f(c) = c^2 - kn$ for $c^2 \geq kn$. (Fermat's method uses $k = 1$.) If there exists a c for which $f(c) = r^2$, then $(c + r)(c - r) = kn$ and $\gcd(n, c - r)$ is a nontrivial factor of n.

Example 7.17: Compare the number of steps required to factor $n = 3959$ using (i) trial division, (ii) Fermat's factorization method, and (iii) Kraitchik's generalization of Fermat's method with $k = 3$.

Solution:

(i) If we check each odd prime number beginning with 3 as a possible divisor of 3959, we have our first success on our eleventh attempt since $37|3959$. In fact, $3959 = 37 \times 107$.

(ii) Fermat's factorization method begins with determining that $62 < \sqrt{3959} < 63$. We calculate $f(c) = c^2 - 3959$ for $c \geq 63$ until $f(c)$ is a perfect square. In this case, $f(63) = 10$, $f(64) = 137$, $f(65) = 266$, $f(66) = 397$, $f(67) = 530$, $f(68) = 665$, $f(69) = 802$, $f(70) = 941$, $f(71) = 1082$, $f(72) = 1225 = 35^2$. Hence, $3959 = (72-35)(72 + 35) = 37 \times 107$. We saved but one step over trial division.

(iii) If $k = 3$, then $kn = 11{,}877$ and $108 < \sqrt{11{,}877} < 109$. We $f(c) = c^2 - 11{,}877$ for $c \geq 109$. In this case, we obtain a square immediately since $f(109) = 4$. Hence, $11{,}877 = (109 - 2)(109 + 2)$ from which we see that $11{,}877 = 107 \times 3 \times 37$ and hence $3959 = 37 \times 107$.

In this example, we were rather lucky to pick a value of k which produced a perfect square immediately. If a value of k doesn't readily lead to a perfect square, then additional values of k can be substituted for it. It has been shown that Kraitchik's method is a significant improvement beyond Fermat's factorization method. In fact, the average running time for Kraitchik's method is actually of the same magnitude as that for the Brillhart-Morrison method.

A further refinement to Fermat's factorization method was made by Carl Pomerance in 1981 and is known as the quadratic sieve. As a practical matter, it is still the fastest factorization method for integers of roughly 100 digits or less. Given a large composite integer n, the main idea is to factor $f(c) = c^2 - n$ for a range of values of c (say possibly c from $[\sqrt{n} - n^{1/4}]$ to $[\sqrt{n} + n^{1/4}]$ only saving those with small prime factors (so-called *smooth* numbers) and then see if a product of a carefully chosen subset of them produces a square. Though not necessary, including values of c less than $\sqrt{n}$ allows for twice as many candidates with small absolute values that can be more easily factored (though we must make sure to include an even number of them to obtain a positive square).

Here is an illustrative example.

Example 7.18: Let $n = 3193$. Here $56 < \sqrt{3193} < 57$. We calculate $f(c) = c^2 - 3193$ for $c = 50, 51,$ 52, and so on and keep track of those for which $f(c)$ has small prime factors only (say at most 11 in this case). We determine that $f(50) = -3^2 \times 7 \times 11$, $f(53) = -2^7 \times 3$, $f(55) = -2^3 \times 3 \times 7$, $f(57) = 2^3 \times 7$, and $f(59) = 2^5 \times 3^2$. This is the data collection phase which is easily implemented. Next we note that $f(53)f(55)f(57)f(59) = 2^{18} \times 3^4 \times 7^2 = (2^9 \times 3^2 \times 7)^2$, a perfect square. Finding such a product comprises the data processing phase which could require more computer time and memory. In our example, $(53^2 - 3193)(55^2 - 3193)(57^2 - 3193)(59^2 - 3193) = (32,256)^2$ and hence $x^2 \equiv y^2 \pmod{3193}$ where $x = (53)(55)(57)(59) = 9,803,145$ and $y = 2^9 \times 3^2 \times 7 = 32,256$.

We calculate $\gcd(x - y, 3193) = 103$ and $\gcd(x + y, 3193) = 31$ to obtain the prime factorization of $3193 = 31 \cdot 103$.

Generating the candidates $f(c)$, factorizing those that have small prime factors, and calculating the gcd of two numbers can all be accomplished efficiently. What is less clear is that we will find a suitable subset of candidates whose product is a perfect square. However, we can guarantee a solution in a reasonable amount of time. This is handled as follows in our previous example. We can denote the prime factorization of $f(c)$ as a binary vector to indicate whether it is positive or negative and whether our chosen primes occur to either an odd or even exponent, respectively. Recall that we only retain those numbers for which $f(c)$ is solely divisible by our short list of primes. The first entry of our vector is a 0 if the number is positive and 1 if negative. The other entries represent the exponents of 2, 3, 5, 7, and 11 with 0 representing an even power and 1 an odd power. We see that $f(50) = (1, 0, 0, 0, 1, 1)$, $f(53) = (1, 1, 1, 0, 0, 0)$, $f(55) = (1, 1, 1, 0, 1, 0)$, $f(57) = (0, 1, 0, 0, 1, 0)$, and $f(59) = (0, 1, 0, 0, 0, 0)$. We seek a sum (modulo 2) of vectors resulting in the additive identity $(0, 0, 0, 0, 0, 0)$ signifying that their product is a perfect square. Here $f(53) + f(55) + f(57) + f(59) = (0, 0, 0, 0, 0,$

0). Since all vectors are elements of the Galois field of order 2^6 (unique up to isomorphism) which has dimension six over its base field $\mathbb{Z}_2$, any seven such vectors must be linearly dependent and hence contain a subset having sum the additive identity.

In this case, we were fortunate to find such a combination from our first five vectors. In general, if we keep track of the sign of $f(c)$ and retain those having only prime factors from among $p_1, \ldots, p_n$ (where p_i is the ith prime), then our vectors will be members of $\mathbb{Z}_2^{n+1}$ consisting of $n+1$ entries all of which are either 0 or 1. Any set of $n+2$ such vectors must be linearly dependent which is all that is required.

Exercise 7.7

1. Use the quadratic sieve to factor the integer 13861. You will only need to check values of c from 110 to 120 retaining those values of c having prime factors of $f(c)$ $= c^2 - 13861$ at most 11.
2. Completely factor the integer 694449. Begin by removing small prime factors prior to applying the quadratic sieve method.
3. Factor the integer 17201.
4. Factor the integer 77468.

7.8 The AKS primality test

Let n be a large integer with m digits ($m = O(\log n)$). An algorithm runs in *polynomial time* if there is a $k \in \mathbb{Z}^+$ such that the number of basic operations needed to run the algorithm is $O(m^k)$ for any input n of m digits. Given any two integers, adding, subtracting, multiplying them, determining the residue of one modulo the other can all be done in polynomial time. Similarly, determining if a given integer is a perfect power can be determined in polynomial time.

A mathematical problem is said to be in class P if it can be solved by a polynomial time algorithm. Similarly, a problem is said to be in class NP if any solution can be verified in polynomial time. Clearly, P is a subset of NP. The notorious P versus NP problem is to determine whether P is a proper subset of NP. Its resolution is among the Millennium Prize Problems selected by the Clay Institute in 2000 along with a million-dollar prize for the first solution.

From our discussion of the Euclidean algorithm (Theorem 2.2), finding the greatest common divisor of two integers is in P. Furthermore, the Lucas-Lehmer test (Theorem 7.9) and Pepin's primality test (Theorem 7.12) are in P. Factorization is in NP, but at present it is unknown whether it is in P. Until fairly recently, the status of a generalized primality was also unknown. However, in 2002, three Indian computer scientists, M. Agrawal, N. Kayal, and N. Saxena, discovered an unconditional deterministic polynomial time algorithm to determine the primality of any input [2]. Thus, it was shown that primality testing is in P. (Showing that some previous select primality algorithms were in P relied on widely believed but unproven hypotheses such as the extended Riemann hypothesis.) We give a very brief overview of their algorithm, known as the AKS primality test.

A key observation is provided by the following lemma:

Lemma 7.14: Let $n \geq 2$ and a be an integer relatively prime to n. Then n is prime if and only if $(x + a)^n \equiv x^n + a \pmod{n}$.

Proof: By the binomial theorem (Theorem 1.8), $(x + a)^n = \sum_{k=0}^{n} \binom{n}{k} a^k x^{n-k}$.

If n is prime, then for $1 \leq k \leq n - 1$, $\binom{n}{k} = \frac{n(n-1)\cdots(n-k+1)}{k!}$ has a factor of n in the numerator which does not appear in the denominator. Hence, for all those terms, $\binom{n}{k} \equiv 0 \pmod{n}$ and so $(x + a)^n \equiv x^n + a^n \pmod{n}$. But $a^n \equiv a \pmod{n}$ by Corollary 2.16.1 to Fermat's little theorem.

If n is composite, let q be a prime dividing n. Then

$$\binom{n}{q} = \frac{n(n-1)\cdots(n-q+1)}{q!} = \frac{\frac{n}{q}(n-1)\cdots(n-q+1)}{(q-1)!}.$$

But if $q^r || n$, then $q^{r-1} || \binom{n}{q}$ since $q^{r-1} || \frac{n}{q}$ while q is not a divisor of any other factor in the numerator or denominator in our previous equation. Hence, $\binom{n}{q} \neq 0 \pmod{n}$.

Hence, $(x + a)^n \neq x^n + a \pmod{n}$. ∎

By Lemma 7.14, we can test for primality by checking if $(x + a)^n \equiv x^n + a \pmod{n}$. However, this is rather impractical since it could require computing on the order of n binomial coefficients. The breakthrough involved verifying congruences of the form $(x + a)^n \equiv x^n + a \pmod{x^r - 1, a}$ for r from a set of relatively small integers. If the congruence relation doesn't hold for some r, then by Lemma 7.14, n must be composite. The AKS algorithm shows that if the congruence relation does hold for a suitably chosen r and all values a below some specified bound, then n is indeed prime. Here is the outline of the algorithm which we merely state. The proof can be found in the original paper [].

Step 1: Given n, check to see if $n = a^k$ for some integers a and k. (i.e., check to see if n is a perfect power.) If so, n is composite.

Step 2: Find the smallest positive integer r such that $\text{ord}_r(n) > \log^2 n$.

Step 3: If $1 < \gcd(a, n) < n$ for some $a \le r$, then n is composite.

Step 4: If $n \le r$, then n is prime.

Step 5: If for some a with $1 \le a < \sqrt{\phi(r)} \log n$, we have $(x + a)^n \equiv x^n + a \pmod{x^r - 1, a}$, then n is composite.

Step 6: Otherwise, n is prime.

Exercise 7.8

1. Show that the natural number n has exactly $\left\lfloor \frac{\log n}{\log 10} \right\rfloor + 1$ digits.
2. Use the AKS primality test to verify that 13 is prime.
3. Use the AKS primality test to verify that 23 is prime.

Chapter 8
Some applications

8.1 Introduction to cryptology

The need for and methods used in sending secret messages have a long and interesting history dating back at least as far as Julius Caesar. Over the centuries a multitude of techniques have been used to efficiently alter a message into a form unintelligible except to its intended recipient. The study of techniques used to encode a message is called *cryptography*. The study of techniques used to decode a previously encoded message is called *cryptanalysis*. Collectively, the study of cryptography and cryptanalysis is called *cryptology*.

Until recently most cryptosystems involved two parties with a key which was used to encrypt messages and a similar (if not the same) key to decrypt messages. The security of the system hinged completely on the ability of the two parties to keep the key secret. Discovering the encoding key was equivalent to cracking the system altogether. Such cryptosystems are called symmetric cryptosystems.

One of the simplest systems is a *shift cipher* such as that used by Julius Caesar. We begin with the message or literal plaintext. Translating the literal plaintext to numerical form is called *formatting*. In this setting, one common method is to let A be 01, B be 02, $\ldots, Z$ be 26 and to ignore spaces, punctuation, as well as capitalization. If P_1 is the number representing a plaintext letter, then it's enciphered as $C_1 \equiv P_1 + 3 \pmod{26}$. In the Caesar cipher, each letter is replaced with the letter three positions beyond it in the alphabet (where we cycle from Z back to A). Hence the message MEET YOU AT NOON becomes PHHWBRYDWQRRQ. The intended recipient can simply recover the original message since $P_1 = C_1 - 3 \pmod{26}$. Of course, an eavesdropper can readily decipher the message as well if it's known that a shift cipher is being used.

A slight improvement can be made by using a *keyword cipher*. There are many variants of it. The sender and intended recipient share a keyword, say TRAPEZIUM. The encoding is done by writing out the alphabet and directly under it writing out the keyword (with any repeated letters ignored) followed by the rest of the alphabet in normal order. In this case we have

A B C D E F G H I J K L M N O P Q R S T U V W X Y Z
T R A P E Z I U M B C D F G H J K L N O Q S V W X Y

In this case, the message MEET YOU AT NOON becomes FEEOXHQTOGHHG.

Decoding works by simply using the chart backward from bottom to top rather than top to bottom. To expedite the process, the rows can be flipped and rearranged so that the letters on top are still in alphabetical order. In this case the decoding chart would look like this

https://doi.org/10.1515/9783111579283-008

A B C D E F G H I J K L M N O P Q R S T U V W X Y Z
C J K L E M N O F P Q R I S T D U B V A G W X Y Z F

Another system involves multiplying each letter by a constant or to combine addition and multiplication via a linear transformation. This is called an *affine cipher*. For example, if the plaintext letter is denoted by P_1, we might encipher it by $E_1 \equiv 3P_1 + 5$ (mod 26). The multiplier 3 is called the *decimation* constant since it spreads out the alphabet by that constant. The decimation constant must be relatively prime to 26 to ensure that each letter is enciphered in a unique way. For example, the letter *A* is replaced with the eighth letter *H* since $3 \cdot 1 + 5 = 8$. Our alphabet chart now takes this form

A B C D E F G H I J K L M N O P Q R S T U V W X Y Z
H K N Q T W Z C F I L O R U X A D G J M P S V Y B E

In this case, SEND ME THE NOTES TODAY is enciphered as JTJTUQRTMCTUXMTJMXQHB.

To decrypt the message, one must solve the linear congruence $E_1 \equiv 3P_1 + 5$ (mod 26) for P_1. In this case, $P_1 \equiv 9E_1 + 7$ (mod 26) by utilizing techniques from Section 2.2.

The examples given so far are *monoalphabetic ciphers* which are not at all secure. Each letter is enciphered the same way throughout the entire message with double letters remaining double letters and the like. Although there are 26! = 403,291,461,126,605,635,584,000,000 ways to permute the alphabet, as with all languages, the English language uses some letters more often than others. The most common letters in order of usage are E, T, O, A, N, I, R, S, H and some words like The, And, To, Of, That, With appear often. For example, the letters *E* and *T* appear about 12.7% and 9.1% of the time, respectively, while *X* and *Z* each appear only about 0.1% of the time. Hence, a frequency analysis can often crack the code. A somewhat more sophisticated system is to use a *polyalphabetic cipher* where the same letter in the plaintext message can be enciphered in several different ways and double letters are better disguised. Such an improvement was made to the Caesar cipher by Leon Battista Alberti in 1467 who made use of a movable cipher disk for such a purpose.

An improved polyalphabetic cipther is the *Vigenère cipher* due to the French diplomat and cryptanalyst Blaise de Vigenère (1523–1596). Historically, the letter *A* corresponded to 0, *B* to 1, *C* to 2, and so on. We work modulo 26, but with this slight adjustment of the alphabet. Again a keyword is used and is shared with the intended recipient. For example, suppose the keyword is IMAGINATION. The keyword is written on top disallowing any letter repetition. Beneath it the message is written using as many lines as necessary (ignoring spaces and punctuation). For example, the message PLEASE SEND HELP QUICKLY would be written out as follows:

```
I M A G N T O
P L E A S E S
E N D H E L P
Q U I C K L Y
```

To encipher the message, the first letter P (which lies below I in the keyword line) would be replaced by the letter 8 letters beyond P in the cyclic alphabet since the letter I corresponds to the number 8 in our shifted alphabet. So P would be replaced by X. Similarly, L lies below M and so is replaced by X as well since M corresponds to the number 12. The message would be encrypted as

X X E G F X G M Z D N R E D Y G I I X E M.

The fact that the same letter in the plaintext can get encypted to different letters in the ciphertext and that the same letter in the ciphertext can be generated from different plaintext letters certainly adds to the complexity of breaking this code. And of course, no doubt you can think of ways to scramble the message even further. Despite its complexities, the Vigenère cipher was essentially cracked in the 1800s independently by Friedrich Variski and Charles Babbage.

A further attempt to add greater complexity is to use *digraphs* as our basic units, two letters at a time, rather than individual letters. Since there are 26 letters in the English alphabet, there are $26^2 = 676$ digraphs (though some such as *VH* or *QZ* never actually appear). Even so, the number of possibilities is on the order of 676! which is approximately 1.8×10^{163}, an astronomical number. One general system is again based on linear transformations. We choose four secret numbers a, b, c, d as our key. For each pair of letters in our plaintext, we replace it with their corresponding numbers in the alphabet P_1, P_2 (using $A = 1$, $B = 2$, and so on once again) and then calculate the numbers C_1, C_2 via

$$C_1 \equiv aP_1 + bP_2 \pmod{26}$$

$$C_2 \equiv cP_1 + dP_2 \pmod{26}$$

The sender then transmits the letters in order corresponding to the C numbers.

This can be rewritten as a matrix product

$$\begin{bmatrix} C_1 \\ C_2 \end{bmatrix} = \begin{bmatrix} a & b \\ c & d \end{bmatrix} \begin{bmatrix} P_1 \\ P_2 \end{bmatrix} \pmod{26}$$

To decipher the message, the recipient must solve for P_1 and P_2. The solution is

$$\begin{bmatrix} P_1 \\ P_2 \end{bmatrix} = (ad - bc)^* \begin{bmatrix} d & -b \\ -c & a \end{bmatrix} \begin{bmatrix} C_1 \\ C_2 \end{bmatrix} \pmod{26}$$

where $(ad - bc)^*$ is the arithmetic inverse of $ad - bc \pmod{26}$.

For example, suppose $a = 2$, $b = 5$, $c = 3$, $d = 4$ and the message to be transmitted is NEED MONEY NOW. The message is written numerically as 14–5 5–4 13–15 14–5 25–14 15–23. (If there is an odd number of letters in the message, an arbitrary letter can be added at the end.)

Letting $P_1 = 14$ and $P_2 = 5$ leads to $\begin{bmatrix} C_1 \\ C_2 \end{bmatrix} = \begin{bmatrix} 2 & 5 \\ 3 & 4 \end{bmatrix} \begin{bmatrix} 14 \\ 5 \end{bmatrix} \pmod{26}$

Hence, $C_1 = 1$ and $C_2 = 10$. Thus, NE becomes AJ. All told, the encrypted message is AJ-DEWUAJPAOG. Though the message is well scrambled, a cryptanalyst might note the repeated AJ which could suggest the use of digraphs.

To recover the message, the recipient must solve

$$\begin{bmatrix} P_1 \\ P_2 \end{bmatrix} = (8 - 15)^* \begin{bmatrix} 4 & -5 \\ -3 & 2 \end{bmatrix} \begin{bmatrix} C_1 \\ C_2 \end{bmatrix} \pmod{26}$$

Since $-7 \equiv 19 \pmod{26}$, the Euclidean algorithm can be used to determine that 11 is its arithmetic inverse. Hence, $\begin{bmatrix} P_1 \\ P_2 \end{bmatrix} = \begin{bmatrix} 18 & 23 \\ 19 & 22 \end{bmatrix} \begin{bmatrix} C_1 \\ C_2 \end{bmatrix} \pmod{26}$ or perhaps more simply

$$\begin{bmatrix} P_1 \\ P_2 \end{bmatrix} = -\begin{bmatrix} 8 & 3 \\ 7 & 4 \end{bmatrix} \begin{bmatrix} C_1 \\ C_2 \end{bmatrix} \pmod{26}$$

Various improvements can be made to make the system more secure such as using trigraphs or combinations of some of the methods we have outlined. However, for ease of use, the system cannot be too cumbersome and most of the systems can be broken through frequency analysis, decimation searches, pattern recognition, and the like. In the next section, we discuss a vast improvement still used widely today.

Exercise 8.1

1. Use a shift cipher with shift 5 to write the following plaintext in ciphertext (due to Martin Luther King, Jr.): The arc of the moral universe is long, but it bends toward justice.
2. (a) Try to decipher the following text which was created with a shift cipher: DTZFWJAJWDLTTIFZIJHNUMJWNSL.
   (b) Decipher the following text which was created with a shift cipher: OZGRCGEYYKKSYOSVUYYOHRKATZOROZOYHSTK.
3. What message was sent using the keyword cipher TRAPEZIUM if the message is NDTPXHQAHQDPAHFE?
4. Use the Vigenère cipher with keyword TRANSFORMATION to write out the message: Need military transport immediately.
5. (a) Using digraphs and the system $C_1 \equiv aP_1 + bP_2 \pmod{26}$, $C_2 \equiv cP_1 + dP_2 \pmod{26}$ with $a = 2$, $b = 3$, $c = 1$, and $d = 4$ leads to the message WGUVWAGOPEODOG. Decipher the message!
   (b) Use the same system to decipher LZYYNDVMODOGWIQHDU.

8.2 RSA algorithm

In recent times the need for asymmetric cryptosystems has arisen due to the need for security over networks involving many users. The idea, first developed by W. Diffie and N.M.E. Hellman in 1976, was to create a public key cryptosystem where all the users' encoding keys were listed in an open directory (much like a phone book). However, each user had a secret individual decoding key. The public knowledge of an individual's encoding key was of little to no help in determining that individual's decoding key. Hence any user could send any other user a secret message only the intended recipient could possibly read. The concept that it is much easier to multiply two large primes together rather than factor their product is the essential ingredient in creating such a cryptosystem.

In 1977, R. Rivest, A. Shamir, and L. Adelman developed a public key cryptosystem, known as the *RSA algorithm* [114]. In many respects, this has become the standard worldwide due to its ease in implementation and generally high level of security (though see comments at end of this section.)

Suppose Mike and Bob want to communicate over a network without eavesdroppers reading their messages. Here are the rough details of the RSA algorithm:

1. Mike and Bob each choose two very large primes, say p_M, q_M and p_B, q_B, respectively. Let $m = p_M \cdot q_M$ and $b = p_B \cdot q_B$. Mike chooses a positive integer e_M relatively prime to $\phi(m)$ and Bob chooses a positive integer e_B relatively prime to $\phi(b)$. The ordered pairs (e_M, m) and (e_B, b) are the public encryption keys for Mike and Bob. The prime factors must be kept secret.

2. If Bob wants to send Mike a message (known as literal plaintext), he first translates it via some simple mutually agreed upon scheme to a numerical plaintext, call it P. Then he calculates P^{e_M} (mod m). This is the numerical ciphertext, call it C, which is sent to Mike over the network.

3. Mike calculates the arithmetic inverse of e_M(mod $\phi(m)$), call it d_M. So $d_M \cdot e_M \equiv 1$ (mod $\phi(m)$). He then computes C^{d_M} (mod m), giving him P. He then readily translates P back to readable form (literal plaintext).

Now let us describe the details and give fuller explanations. Recall that translating the literal plaintext to numerical form P is called formatting. We will let A be 01, B be 02, ...,Z be 26, and "space" be 27. For present purposes we treat capital and small letters equally and ignore all punctuation. For example, NUMBER THEORY IS FUN TO LEARN becomes 14211302051827200805151825270919270621142720152705011814. This is the plaintext P from which the original sentence can be recovered easily. Of course, there are many other schemes for representing a message in numerical form. Another one often used is the ASCII code which is consistent with a computer's built-in system.

Next the plaintext is broken into blocks consisting of a specified even number of digits. In our example the blocks would be 1421, 1302, 0518, ..., 1400 (where 00 is appended to the last digits so all blocks are the same size). The largest possible block number (in our scheme, 2727) must be smaller than m. Rather than calculating $P^{e_M} \pmod{m}$ all at once, Bob actually computes $B_i^{e_M} \pmod{m}$ for each block B_i and then juxtaposes all the results to create the numerical ciphertext C. For the sake of consistency, for each i, $B_i^{e_M}$ is reduced to the least positive residue $\pmod{m}$. (For simplicity's sake, in the rest of our discussion we will assume P is just one block in length.)

Mike's calculation of d_M can be readily accomplished by the Euclidean algorithm. Recall that if $gcd(e_M, \phi(m)) = 1$, then we can find integers a and b such that $a_M + b\phi(m) = 1$. Let $d_M \equiv a \pmod{\phi(m)}$ with $0 < d_M < \phi(m)$. Then d_M is an arithmetic inverse of $e_M \pmod{\phi(m)}$ and there is a k for which $e_M d_M = k\phi(m) + 1$. Hence

$$C^{d_M} \equiv (P^{e_M})^{d_M} = P^{k\phi(m)+1} = (P^{\phi(m)})^k P \equiv 1^k P = P \pmod{m}$$

by the Euler-Fermat theorem.

Note that $\phi(m)$ must be kept secret as well. In general, if $m = pq$ then

$$p + q = pq - (p-1)(q-1) + 1 = m - \phi(m) + 1$$

And

$$p - q = \sqrt{(p+q)^2 - 4pq} = \sqrt{[m - \phi(m) + 1]^2 - 4m}$$

We then obtain p and q via the relations

$$p = \frac{1}{2}[(p+q) + (p-q)] \text{ and } q = \frac{1}{2}[(p+q) - (p-q)]$$

Thus knowing both m and $\phi(m)$ is tantamount to knowing the factorization of m.

Example 8.1: Suppose Bob wants to send Mike the message LUNCH IS ON ME. Let Mike's secret primes be $p_M = 61$ and $q_M = 71$. So $m = 61 \cdot 71 = 4331$. He chooses $e_M = 143$ which is relatively prime to $\phi(m) = 4200$. Mike's public encyphering key is $(e_M, m) = (143, 4331)$.

The literal plaintext is translated to the string $P = 12211403082709192715142711305$ and then broken into blocks of length 4. The blocks are 1221, 1403, 0827, 0919, 2715, 1427, 1305. Bob next computes $B_i^{e_M} \pmod{m}$ for each block B_i. Using the binary Exponentiation algorithm, the results are as follows:

$1221^{143} \equiv 1892 \pmod{4331}, 1403143 \equiv 2684 \pmod{4331}, 827143 \equiv 2338 \pmod{4331},$
$919143 \equiv 3699 \pmod{4331}, 2715^{143} \equiv 4203 \pmod{4331}, 1427^{143} \equiv 343 \pmod{4331},$ and
$1305^{143} \equiv 3637 \pmod{4331}$. Hence $C = 1892268423383699420303433637$.

Mike breaks up C into blocks 1892, 2684, and so on and decodes by raising each block to the power d_M. The Euclidean algorithm implies that $1 = 27 \cdot 4200 - 793 \cdot 143$. Hence $d_M = 3407 \equiv -793 \pmod{4200}$. Mike obtains $1892^{3407} \equiv 1221 \pmod{4331}, 2684^{3407} \equiv$

1403 (mod 4331), etc. from which the message is recovered (and Mike gets the free lunch).

Another beautiful application of the RSA algorithm is the ability to sign a message. In Example 8.1, Bob can send a digital signature to Mike so that Mike knows the message is authentic. Bob writes his signature as a numerical block (or blocks) s. Next, he calculates $t \equiv s^{d_B} \pmod{b}$. (Bob is the only one who knows d_B.) Finally, he computes $S \equiv t^{e_M} \pmod{m}$ and sends S to Mike at the end of his message.

When Mike decodes the end of the message by computing $S^{d_M} \equiv t \pmod{m}$, he will get a seemingly meaningless string of digits.

However, he then uses Bob's encryption key (e_B, b) to compute $t^{e_B} \pmod{b}$ to recover Bob's signature s.

Suppose $p_B = 53$, $q_B = 97$, and $b = 53 \cdot 97 = 5141$ in Example 8.1. Let $e_B = 25$ (check that e_B is relatively prime to $\phi(b)$). Bob's public encryption key is $(e_B, b) = (25, 5141)$. Bob computes $d_B = 4393$. If Bob's signature is 02150200, then he separates it into two blocks $s_1 = 0215$ and $s_2 = 0200$. He calculates $t_1 = 3321 \equiv 215^{4393} \pmod{5141}$ and $t_2 = 0879 \equiv 200^{4393} \pmod{5141}$. Hence $S_1 = 0590 \equiv 3321^{143} \pmod{4331}$ and $S_2 = 0300 \equiv 879^{143} \pmod{4331}$. Bob appends 05900300 to his message. When Mike receives it and decodes the message he gets $t_1 = S_1^{d_M} \equiv 3321 \pmod{4331}$ and $t_2 = S_2^{d_M} \equiv 0879 \pmod{4331}$. The string 33210879 does not translate back to an alphabetic message and so Mike knows this might be Bob's signature. He computes $s_1 = 0215 = t_1^{e_B} \pmod{b}$ and $s_2 = 0200 = t_2^{e_B} \pmod{b}$, thus verifying Bob's signature.

The idea behind authenticating a digital signature has real world application to nuclear treaty verification. Suppose countries R and U have a treaty that limits the number, timing, and severity of underground nuclear tests. Each places a group of deep seismic sensing devices within the other's territory which gathers data, encrypts it (to avoid eavesdroppers), and then transmits the data (plain and encrypted) to its intended recipient back to its home country. However, each country wants to make sure (i) that the data collected in its country is correct and untampered and also (ii) wants to make sure that the data it receives from the other country has not been altered in transmission. The protocol closely follows that for the digital signature.

Country U chooses two large primes p_u and q_u and determines their product $n = p_u q_u$. Country U also chooses an encryption key e_u relatively prime to $\phi(n)$ and calculates its arithmetic inverse $d_u \pmod{\phi(n)}$. Country U shares n and e_u with country R but keeps n and d_u secret. When data x is collected in country R, it is then encrypted locally to $y \equiv x^{d_u} \pmod{\phi(n)}$, sort of a reverse RSA scheme. The information x and y is shown to country R who then verifies that recovers the data x. Now country R knows that what is being transmitted to country U is accurate. When country U receives x and y, it verifies that $x \equiv y^{e_u} \pmod{n}$. Hence, it knows that the data has not been corrupted. A similar process is simultaneously being conducted in the other direction as well.

In practice the primes p and q should be 200 digits or more to ensure security for a long time. In addition, it is wise to choose primes whose lengths vary to avoid an assault by Fermat's factorization method. Finding 200-digit primes is difficult, but not nearly as difficult as factoring a 400-digit number! Generally, probabilistic primality tests like Rabin's primality test are used to generate probable primes (called "industrial grade" primes by Henri Cohen). When Rivest, Shamir, and Adelman created the RSA algorithm in 1977, they presented a 129-digit composite integer (known as RSA 129) as proof of the security of their system. At that time they estimated that it would take others at least 40 quadrillion years to factor it. However, by using the quadratic sieve factoring algorithm and parceling out pieces of the calculation to thousands of users over Internet, RSA 129 was factored in 1994 – just 17 years rather than 40 quadrillion (see the exercises)!

Beyond RSA 129, in 1991, RSA Laboratories published a long list of large semiprimes (numbers that are the product of two primes) in order to demonstrate the resiliency of the RSA scheme and to challenge and encourage others to develop factorization methods. In 2001, the list was extended along with monetary prizes for many of the numbers. RSA 140 was factored in 1999 by the Dutch mathematician Herman te Riele et al. using the quadratic number sieve. Several others have been factored using the number field sieve. The largest payout was given to Thorsten Kleinjung et al for their factorization of RSA 768 in 2009. However, at the current time, many 200 and 300-digit RSA semiprimes have not been publicly factored.

As computers get faster the size of the primes needed must grow as well. Yet it appears that the discovery of large primes will always be easier than factoring large integers. As Len Adelman has said, "Improvements in computer technology always favor the cryptographer over the cryptanalyst."

Exercise 8.2

1. (a) Use our scheme to translate I CAME I SAW I CONQUERED into numerical plaintext.
   (b) Use our scheme to translate 04152772515212723011420270618090519272309200 82720080120 to literal plaintext.

2. Work through the details of encrypting and decrypting the message I GOT AN A to a recipient whose public encryption key is $(e, m) = (3, 55)$.

3. Work through the details of encrypting and decrypting the message I LOVE VERMONT to a recipient whose public encryption key is $(e, m) = (5, 91)$.

4. (a) If my numerical signature is 16052005, what is its encoded form if my encryption key is $(3, 55)$ and my recipient's key is $(5, 91)$.
   (b) Encode your signature if your encryption key is $(5, 51)$ and I am your intended recipient with key $(3, 55)$.

5. (a) If $n = 177581$ is the product of two primes and $\phi(n) = 176700$, then what are the prime factors of n?

(b) If $n = 126911$ is the product of two primes and $\phi(n) = 126024$, then what are the prime factors of n?

6. (a) If $n = pq$ and $\sigma(n)$ are known, how can we find p and q?

   (b) If $n = pq = 31007$ and $\sigma(n) = 31416$, then what are p and q?

7. Show that if $n = pqr$, then $p + q + r = \frac{1}{2}[\phi(n) + \sigma(n) - 2n]$.

8. Encrypt the message EVERYTHING IS AOK if the encryption key is $(e, m) = (11, 17513)$.

9. Encrypt the message ROSEBUD if the encryption key is $(e, m) = (11, 2881)$.

10. Decipher the following message sent to you if your encryption key is $(e, m) = (7, 3149)$:

    005318440837252900502760077205002236272827692755208214451750121720552796

11. Let A be 01, B be 02, ..., and Z be 26 (ignore spaces and punctuation). In a *Caesar cipher* with shift n, the numerical plaintext is encrypted by replacing each digit d by $(d + n)\pmod{26}$ and then the literal ciphertext is formed from the above substitution. Decipher the messages

    (a) WXYHCRYQFIVXLISVCERHCSYAMPPEPAECWFIMRCSYVTVMQI

    (b) DROZBYYPSCSXDROZENNSXQ.

12. RSA 129 is 114,381, 625, 757, 888, 867, 669, 235, 779, 976, 146, 612, 010, 218, 296, 721, 242, 362, 562, 561, 842, 935, 706, 935, 245, 733, 897, 830, 597, 123, 563, 958, 705, 058, 989, 075, 147, 599, 290, 026, 879, 543, 541. Verify that is the product of 3, 490, 529, 510, 847, 650, 949, 147, 849, 619, 903, 898, 133, 417, 764, 638, 493, 387, 843, 990, 820, 577 with 32, 769, 132, 993, 266, 709, 549, 961, 988, 190, 834, 461, 413, 177, 642, 967, 992, 942, 539, 798, 288, 533.

    (The secret message was: "The magic words are squeamish ossifrage.")

13. Decipher the following message sent to you if your encryption key is $(e, m) = (11, 5183)$: 25920509467721483726497843474581244629013827230703863631006670517.

14. Decipher the following message sent to you if your encryption key is $(e, m) = (11, 5183)$: 154420973632434728753227232239930.

8.3 Random number generation

Applications for generating long lists of seemingly random numbers extends to statistical sampling, Monte Carlo experimentation, computer simulation, cryptography, user account verification, gambling machine implementation, and any situation where predictability is eschewed. Entropy sources such as fluctuations in atmospheric noise or a Geiger counter's recording of radioactive decay are ways to use natural phenomena to create a truly random generator. Though seemingly paradoxical, it's usually desirable to have a deterministic random generator instead that is reliable and can be checked as used for computer program debugging for example. Hence, some sort of mathematical algorithm is usually employed which leads to what are called *pseudo-random number generators* (PRNGs). In this section, we give a brief overview of some of the main ideas.

A good PRNG must satisfy a fair number of the following to some degree (depending on its intended use): a long cycle length (to avoid recognizable repeated patterns), independence of the numbers generated (so that the knowledge of previous numbers seem to have no predictive value for current or future numbers), uniformity (so that the probability of any number within a given range is equally likely), and perhaps satisfy some additional statistical and correlation tests. Please note that any output of integers between 1 and N say can be mapped bijectively to a set of real numbers in the interval [0, 1] by simply placing a decimal point in front of each number.

One of the earliest such techniques is known as the *middle-square method* as used and described by John von Neumann in a talk in 1949 where he admitted that "anyone who considers arithmetical methods of producing random digits is, of course, in a state of sin." To create a seemingly random set of n digit numbers, pick an n-digit number X_1 as a seed for starters, then square the number and pick out the middle digits of the 2n-digit number n^2 as the succeeding number X_2 (with 0s padded at the beginning if necessary). Repeat the process as long as desired. For example, with $n = 4$, let our seed be the number 1729. Its square is 2989441 which we write as 02989441 and extract the middle number 9894. Our sequence begins 1729 $\rightarrow$ 9894 $\rightarrow$ 8912 $\rightarrow$ 4237 $\rightarrow$ 9521 $\rightarrow$ 6494 $\rightarrow$ 1720 $\rightarrow$ 9584 $\rightarrow$ 8530 $\rightarrow$ 7609 $\rightarrow$ 8968. The sequence appears random and this method has been used successfully where a low level of randomness and secrecy is required. For user verification numbers, letting $n = 6$ is appropriate since most people can remember a 6-digit number as two groups of three for a short period of time. However, there are some pitfalls. Depending on the seed, there may be a very short period or subsequent numbers might decrease to 0. For example, with $n = 4$, if the sequence hits any of 0100, 2500, 3792, or 7600, then that same number will be generated repeatedly. Additionally, if the first half of the number n^2 consists solely of 0s, then the sequence will continue to decrease until it gets stuck at 0000.

However, modifications of the middle-square sequence has been made by combining it with what is known as a Weyl sequence using uniformity ideas as described in Section 6.4. The basic idea is that a number from a separate list of provably uniform numbers is added to each number before squaring. A successful implementation has been described by Bernard Widynski in 2022.

Another technique still commonly used utilizes a *linear congruential generator* (LCG). Again, there is a seed value X_1 and a linear recurrence relation $X_{n+1} = (aX_n + b)$ (mod m). If $b = 0$, then this is called a multiplicative congruential generator as established by D.H. Lehmer in 1951 and if $c \neq 0$, then it's a mixed congruential generator which was subsequently developed. The values a, c, and m must be carefully chosen in order to satisfy as many desirable properties as possible.

For a simple example that can be checked by hand, let $a = 3$, $b = 1$, $m = 32$, and $X_1 = 5$. Our sequence begins in a seemingly random manner, but repeats after 16 iterations:

$5 \to 16 \to 17 \to 20 \to 29 \to 24 \to 9 \to 28 \to 21 \to 0 \to 1 \to 4 \to 13 \to 8 \to 25 \to 12 \to 5 \ldots$

There is a vast literature related to this, but most common is to choose a high power of 2 for m making use of a computer's hardware to limit storage space and run time. Even so, LCGs are not considered robust enough in very high security situations such as that needed for cryptographic purposes. Notice that some information about a, b, and m can be gleaned by knowing a few of the outputs in order.

Fibonacci numbers and their generalizations have found use in creating PRNG's as well. If we consider the sequence of Fibonacci numbers mod n, the sequence must cycle within n^2 steps since there are n^2 possible ordered pairs of integers mod n and each pair determines the next Fibonacci number. The sequence will repeat once a previous pair appears. However, depending on the application, a fairly long sequence of seemingly random integers can be created with relatively small modulus. For a given modulus m, the length of the cycle before repetition is known as the *Pisano period* of that modulus, which we will denote by π_m. For example, the sequence of Fibonacci numbers modulo 5 are $\{1, 1, 2, 3, 0, 3, 3, 1, 4, 0, 4, 4, 3, 2, 0, 2, 2, 4, 1, 0, 1, 1, 2, \ldots\}$ and so $\pi_5 = 20$. Since $\pi_{mn} = \mathrm{lcm}[\pi_m, \pi_n]$, we can easily create longer sequences with predictable cycle lengths. The study of Pisano periods is reduced to that for prime powers, $m = p^k$. It can be shown that π_{p^k} divides $p^{k-1}\pi_p$. In fact, in all known cases,

$$\pi_{p^k} = p^{k-1}\pi_p \qquad (8.1)$$

Essentially, this reduces the analysis of Pisano periods to that of π_p for prime p.

Other than $m = 2$ where $\pi_2 = 3$, π_m is an even number. This we establish as follows: let $M = \begin{bmatrix} F_2 & F_1 \\ F_1 & F_0 \end{bmatrix} = \begin{bmatrix} 1 & 1 \\ 1 & 0 \end{bmatrix}$ be the *Fibonacci matrix*. $M_2 = \begin{bmatrix} F_1+F_2 & F_2 \\ F_2 & F_1 \end{bmatrix} = \begin{bmatrix} F_3 & F_2 \\ F_2 & F_1 \end{bmatrix} = \begin{bmatrix} 2 & 1 \\ 1 & 1 \end{bmatrix}$ and by induction $M^n = \begin{bmatrix} F_{n+1} & F_n \\ F_n & F_{n-1} \end{bmatrix}$. Since π_m exists, $F_{r+\pi_m} \equiv F_r \pmod{m}$ for all $r \geq 0$ and hence $M^{\pi_m+1} = M$. But M is invertible and hence $M^{\pi_m} \equiv I = \begin{bmatrix} 1 & 0 \\ 0 & 1 \end{bmatrix} \pmod{m}$. But $\det(M) = -1$. Hence $\det(M^{\pi_m}) = (-1)^{\pi_m} = 1$ by the multiplicativity of the determinant. Thus, π_m is even.

Formula (8.1) is known to hold for $p = 2$ and $p = 5$. For example, $\pi_{100} = \mathrm{lcm}[\pi_4, \pi_{25}] = \mathrm{lcm}[2 \cdot 3, 5 \cdot 20] = 300$. Hence the Fibonacci numbers generate a somewhat random sequence of numbers between 0 and 99 (or from 1 to 100) of length 300.

Variations on the operation used to create Fibonacci numbers leads to another class of PRNG called *lagged Fibonacci generators* of LFG. The Fibonacci recurrence $F_n = F_{n-1} + F_{n-2}$ is replaced with the relation

$$S_n \equiv S_{n-k} \mathbin{\#} S_{n-l}(\bmod\, m) \tag{8.2}$$

where $1 \leq k < l$, m is a large number (often a power of 2 such as 2^{32} or 2^{64}) and $\#$ is a binary operation such as addition, subtraction, or multiplication. Perhaps surprisingly, the multiplication case generally gives shorter periods than the additive ones. Certain pairs of values for k and l have been established as being especially good and have been published widely.

For cryptographic purposes, more intricate PRNGs have been formulated. One such example is the Blum Blum Shub RNG which was developed by L. Blum, M. Blum, and M. Shub in 1986. The fact that factoring large numbers is difficult lies at the basis of this method. In this case, let $m = pq$ where p and q are two large primes each congruent to 3 (mod 4). Let $X_1 > 1$ be the seed value where $\gcd(X_1, m) = 1$. Then let

$$X_{n+1} \equiv X_n^2(\bmod\, m) \text{ for } n \geq 1$$

Each X_n in binary can then be used to create another output S_n which is then used in some fashion. For example, S_n might be the bit parity of X_n, namely the parity of the number of 1s that appear when X_n is expressed in binary. So S_n will give a long string of seemingly random bits (0s and 1s) from which we collect them in groups 2^k for some k to create integers in the range from 1 to 2^k. By way of explication, let's take a rather trivial example.

Let $m = 19 \cdot 47 = 893$, $X_1 = 3$. Our sequence $\{X_i\}$ begins 3, 9, 81, 310, 549, 460, 852, ..., 541, ... with $X_{32} = 541$. Next, we write our sequence $\{X_i\}$ in binary obtaining 11, 1001, 1010001, 100110110, 1000100101, 1000100101, 111001100, 1101010100, ..., 1000011101, ... Let S_n equal 1 if the number of 1s in the binary expansion of X_n is odd and equal 0 if it's even. (Think of S_n as a check digit.) The sequence $\{S_n\}$ is readily determined to be 0011011011001101110000000101011 ... We can interpret our string of bits as our "random" sequence or we can associate them in some way. For example, we can take a *nibble* at a time, namely collect four bits at a time (half a byte) and then switch back to base 10 to get a string of random integers between 0 and 15. In this case our sequence begins 3, 6, 12, 13, 12, 0, 6, 11, ... Unlike sequences based on something like the Fibonacci sequence, note that the same number or pair of numbers won't necessarily have the same subsequent number.

No doubt you can devise other ways to create a sequence from our original output. Depending on the application in mind, care must be taken in choosing the general method and the specific parameters that form the basis for the random sequence. Enjoy your own experiments!

Exercise 8.3

1. What sequence is produced from the middle square method from the following seeds?

    (a) 0199   (b) 1953   (c) 2027

2. Determine the sequence produced from the LCG $X_{n+1} = (aX_n + b) \pmod{m}$ with the following parameters:

    (a) $a = 2$, $b = 5$, $m = 32$, $X_1 = 1$

    (b) $a = 3$, $b = 11$, $m = 64$, $X_1 = 2$

3. (a) Determine the Pisano periods for $n = 2, \ldots, 20$.

    (b) Use formula (8.1) to determine π_n for $n = 625$, $n = 1000$, $n = 1050$, $n = 2700$.

4. Generate the sequence based on the lagged Fibonacci generator $S_n = F_{n-7} + F_{n-11}$ (mod 100) for $n \geq 1$ where F_k is the kth Fibonacci number defined for negative integers as well.

5. Verify the details in our example of the Blum Blum Shub algorithm.

Chapter 9
Introduction to analytic number theory

9.1 The infinitude of primes and the zeta function

The notion that analytical methods (calculus, real analysis, and complex analysis) could play a part in number-theoretic investigations goes back to Euler. It was he who realized that functions of a real variable could be instrumental in studying primes and other discrete objects. We begin with an important definition.

Definition 9.1: Let $s > 1$. Define the *zeta function* as

$$\zeta(s) = \sum_{n=1}^{\infty} \frac{1}{n^s}$$

From elementary calculus (*p*-test), $\zeta(s)$ converges for $s > 1$. Euler made an extensive study of the zeta function during the 1730s and 1740s. The precise value of $\zeta(2)$ had been sought after since the mid-1600s. Euler developed exact formulas for $\zeta(2n)$ for all positive integral n and computed the values of the zeta function for several odd arguments to 15 decimal places. In fact, the zeta function can be analytically extended beyond the range $s > 1$. By 1749 he had verified the functional equation for $\zeta(s)$ for several real values of s. In modern notation the functional equation takes the form

$$\pi^{-s/2}\Gamma(s/2)\zeta(s) = \pi^{-(1-s)/2}\Gamma((1-s)/2)\zeta(1-s)$$

where $\Gamma(s)$ is the gamma function. It is defined by

$$\Gamma(s) = \int_0^{\infty} e^{-t}t^{s-1}dt \ \text{ for } s > 0$$

and satisfies $\Gamma(n+1) = n!$ Euler created the gamma function in 1729.

If we allow s to take on complex values (so $s = \sigma + it$ with σ and t real), then $\zeta(s)$ becomes the famous *Riemann zeta function* which was beautifully developed by G. Bernard Riemann (1826–1866) in an eight-page memoir in 1859. In it the analytic continuation of $\zeta(s)$ to the entire complex plane was given along with a thorough proof of the functional equation for $\zeta(s)$.

Furthermore, several key conjectures were made which relate the zeros of the Riemann zeta function to the distribution of the primes. All have since been rigorously proven save for one, known as the Riemann hypothesis. By the functional equation, the Riemann zeta function has zeros at all negative even integers (the so-called trivial zeros). The rest of the zeros must occur on the critical strip $0 \le \sigma \le 1$. The *Rie-*

https://doi.org/10.1515/9783111579283-009

mann hypothesis states that all the nontrivial zeros of the zeta function lie on the critical line $\sigma = 1/2$ in the complex plane.

In 1914, G.H. Hardy proved that infinitely many zeros lie on $\sigma = 1/2$. In 1942, Atle Selberg proved that a positive proportion of zeros lie there. In 1974, N. Levinson proved that at least one-third of the zeros lie on the critical line. The lower bound was substantially improved. J. Brian Conrey who proved (1989) that over 2/5 of the nontrivial zeros lie on $\sigma = 1/2$. More recently, Pratt et al. [106] have pushed the proportion to 5/12. In particular, let $N(T)$ be the number of zeros of $\zeta(s)$ for $0 < t \leq T, 0 < \sigma < 1$, and let $N_0(T)$ be the number of zeros of $\zeta(s)$ for $0 < t \leq T, \sigma = 1/2$. Then

$$\liminf_{T \to \infty} \frac{N_0(T)}{N(T)} > \frac{5}{12}$$

In addition, Platt and Trudgian (2021) have confirmed that the first 3 trillion nontrivial zeros lie on the critical line. And key work on zero free regions on the critical strip continue to be improved, most notably by Guth and Maynard (2024). We save further discussion of these exciting developments for a future course.

In this chapter we will deal solely with functions of a real variable, infinite series of real numbers, and with real power series. It may be helpful to recall the basic notions of convergence and absolute convergence of infinite series along with some convergence tests like the ratio test, integral test, and limit comparison test. Furthermore, power series may be differentiated and integrated term by term in the interior of their interval of convergence. Finally, the terms of a convergent series can be associated in any way without altering the sum of the series and the terms of an absolutely convergent series can be rearranged in any way without disturbing the sum.

Euclid proved that there are infinitely many primes (Theorem 2.11). Euler established the same result somewhat indirectly. First, he established an analytic version of the fundamental theorem of arithmetic known as the Euler product. Second, he used the identity to show that the sum of the reciprocal of the primes diverges. It then follows as a corollary that there are infinitely many primes.

Before proceeding, let us introduce the following notation: for a positive integer n, let $P(n)$ denote the largest prime factor of n. For example, $P(100) = 5$ and $P(154) = 11$.

Proposition 9.1 (Euler Product): Let $s > 1$. Then

$$\zeta(s) = \prod_p (1 - p^{-s})^{-1} \tag{9.1}$$

where the product is over all primes p.

Proof: For fixed p, the geometric series

$$1 + \frac{1}{p^s} + \frac{1}{p^{2s}} + \cdots \text{ converges absolutely to } (1 - p^{-s})^{-1}. \text{ Hence}$$

$$\prod_{p \leq N} (1 - p^{-s})^{-1} = \prod_{p \leq N} \left(\sum_{r=0}^{\infty} \left(\frac{1}{p} \right)^{rs} \right)$$

$$= \sum_{p(n) \leq N} \frac{1}{n^s}$$

where the sum is over all integers n having only prime factors $p \leq N$. Now

$$\zeta(s) = \sum_{P(n) \leq N} \frac{1}{n^s} + R(N)$$

where

$$R(N) < \sum_{n > N} \frac{1}{n^s}$$

But $\lim_{N \to \infty} R(N) = 0$ since $s > 1$.

Hence

$$\prod_{\text{all } p} (1 - p^{-s})^{-1} = \lim_{N \to \infty} \prod_{p \leq N} (1 - p^{-s})^{-1}$$

$$= \lim_{N \to \infty} \sum_{P(n) \leq N} \frac{1}{n^s} = \zeta(s) \qquad \blacksquare$$

Theorem 9.2 (Euler – 1737): The sum of the reciprocals of the primes diverges.

Proof: If $|x| < 1$, then $\log(1 - x) = -\sum_{r=1}^{\infty} \frac{x^r}{r}$ (Exercise 9.1.2(b)). So

$$\log \zeta(s) = \log \prod_p (1 - p^{-s})^{-1} \text{ (by the Euler product)}$$

$$= -\sum_p \log \left(1 - \frac{1}{p^s} \right)$$

$$= \sum_p \sum_{r=1}^{\infty} \frac{1}{r p^{rs}}$$

by the above.

But $\zeta(s) = \sum_{n=1}^{\infty} \frac{1}{n^s}$ implies that $\lim_{s \to 1+} \zeta(s) = \infty$ since the harmonic series $\sum_{n=1}^{\infty} \frac{1}{n}$ diverges. It follows that $\lim_{s \to 1+} \log \zeta(s) = \infty$ as well. Now

$$\sum_{p}\sum_{r=1}^{\infty}\frac{1}{rp^{rs}} = \sum_{p}\frac{1}{p^s} + \sum_{p}\sum_{r=2}^{\infty}\frac{1}{rp^{rs}}$$

But

$$\sum_{p}\frac{1}{p^s} + \sum_{p}\sum_{r=2}^{\infty}\frac{1}{rp^{rs}} < \sum_{p}\frac{1}{p^s} + \sum_{p}\sum_{r=2}^{\infty}\frac{1}{p^r}$$

Furthermore,

$$\sum_{r=2}^{\infty}\frac{1}{p^r} = \frac{1}{p(p-1)} \quad (\text{Exercise 9.1.3(a)})$$

Hence,

$$\sum_{p}\sum_{r=2}^{\infty}\frac{1}{rp^{rs}} < \sum_{p}\frac{1}{p(p-1)} < \sum_{n}\frac{1}{n(n-1)}$$

which converges (limit comparison test). So

$$\lim_{s\to1^+}\sum_{p}\frac{1}{p^s} = \sum_{p}\frac{1}{p} \quad \text{diverges} \qquad ∎$$

For comparison's sake, in 1919 Viggo Brun proved that the sum of the reciprocals of the twin primes converges. Hence there are significantly fewer twin primes than primes in general. His proof involved intricate sieve methods but failed to determine whether there are infinitely many twin primes.

Now we establish a stronger link between the primes and the zeta function. Recall that the prime counting function, $\pi(N)$, represents the number of primes less than or equal to N.

Proposition 9.3: For $s > 1$,

$$\log \zeta(s) = s\int_2^{\infty}\frac{\pi(x)}{x(x^s-1)}dx \qquad (9.2)$$

Proof. From the Euler product for the zeta function,

$$\log \zeta(s) = \log\prod_p (1-p^{-s})^{-1}$$

$$= -\sum_p \log\left(1-\frac{1}{p^s}\right)$$

$$= -\sum_{n=2}^{\infty}\{\pi(n)-\pi(n-1)\}\log\left(1-\frac{1}{n^s}\right)$$

(since $\pi(n)$ is a step function with integer steps at each prime)

$$= -\sum_{n=2}^{\infty} \pi(n)\log\left(1-\frac{1}{n^s}\right) + \sum_{n=3}^{\infty} \pi(n-1)\log\left(1-\frac{1}{n^s}\right)$$

$$= -\sum_{n=2}^{\infty} \pi(n)\left[\log\left(1-\frac{1}{n^s}\right) - \log\left(1-\frac{1}{(n+1)^s}\right)\right]$$

by a change of index. But $\frac{d}{dx}\log\left(1-\frac{1}{x^s}\right) = \frac{sx^{-s-1}}{1-x^{-s}} = \frac{s}{x(x^s-1)}$.

By the fundamental theorem of calculus,

$$\int_n^{n+1} \frac{s}{x(x^s-1)}\,dx = \log\left(1-\frac{1}{(n+1)^s}\right) - \log\left(1-\frac{1}{n^s}\right)$$

Hence

$$\log\zeta(s) = \sum_{n=2}^{\infty} \pi(n) \int_n^{n+1} \frac{s}{x(x^s-1)}\,dx$$

$$= s\int_2^{\infty} \frac{\pi(x)}{x(x^s-1)}\,dx$$

as desired ∎

Before we can study the primes in greater depth, we need to develop a useful analytical tool which allows us to relate arithmetic sums to definite integrals.

Theorem 9.4 (Abel's Summation Formula): Let $t(n)$ be an arithmetic function and $T(x) = \sum_{n\leq x} t(n)$ be its summatory function with $T(x) = 0$ if $x < 1$.

Assume the function $f(x)$ has a continuous first derivative on the closed interval $[a, b]$ where $0 < a < b$. Then

$$\sum_{a<n\leq b} t(n)f(n) = T(b)f(b) - T(a)f(a) - \int_a^b T(x)f'(x)dx \tag{9.3}$$

Proof: Let $A = [a]$ and let $B = [b]$. Then

$$\sum_{a<n\leq b} t(n)f(n) = \sum_{n=A+1}^{B} t(n)f(n)$$

$$= \sum_{n=A+1}^{B} \{T(n) - T(n-1)\}f(n)$$

$$= \sum_{n=A+1}^{B} T(n)f(n) - \sum_{n=A}^{B-1} T(n)f(n+1)$$

by a change of index

$$= \sum_{n=A+1}^{B-1} T(n)\{f(n) - f(n+1)\} + T(B)f(B) - T(A)f(A+1)$$

$$= -\sum_{n=A+1}^{B-1} T(n) \int_{n}^{n+1} f'(x)dx + T(B)f(B) - T(A)f(A+1)$$

But $T(x) = T(n)$ for $x \in [n, n+1)$. So

$$\sum_{a<n\leq b} t(n)f(n) = -\sum_{n=A+1}^{B-1} \int_{B}^{n+1} T(x)f'(x)dx + T(B)f(B) - T(A)f(A+1)$$

$$= -\int_{A+1} T(x)f'(x)dx + T(B)f(B) - T(A)f(A+1)$$

But $T(B)f(B) = T(b)f(B) = T(b)[f(b) + f(B) - f(b)]$

$$= T(b)f(b) - \int_{B} T(x)f'(x)dx.$$

Similarly,

$$T(A)f(A+1) = T(a)f(A+1) = T(a)[f(a) + f(A+1) - f(a)]$$

$$= T(a)f(a) + \int_{a} T(x)f'(x)dx.$$

So

$$\sum_{a\leq n\leq b} t(n)f(n) = -\int_{A+1}^{B} T(x)f'(x)dx + T(b)f(b)$$

$$- \int_{B}^{b} T(x)f'(x)dx - T(a)f(a) - \int_{a}^{A+1} T(x)f'(x)dx$$

$$= T(b)f(b) - T(a)f(a) - \int_{a}^{b} T(x)f'(x)dx. \qquad \blacksquare$$

Corollary 9.4.1: $\sum_{1\leq n\leq b} t(n)f(n) = T(b)f(b) - \int_{1}^{b} T(x)f'(x)dx.$

Proof: Let $a = 1/2$ and recall $T(x) = 0$ for $x < 1$. $\qquad \blacksquare$

Corollary 9.4.2 (Euler's Summation Formula): If $f(x)$ has a continuous first derivative on the closed interval $[a, b]$ where $0 < a < b$, then

$$\sum_{a<n\leq b} f(n) = \int_a^b f(x)dx + \int_a^b (x-[x])f'(x)dx + (a-[a])f(a) - (b-[b])f(b) \qquad (9.4)$$

Proof: Let $t(n) = 1$ for all $n \geq 1$. Then $T(b) = [b] = B$ and $T(a) = [a] = A$. Abel's summation formula implies

$$\sum_{a<n\leq b} f(n) = Bf(b) - Af(a) - \int_a^b [x]f'(x)dx$$

Integrating by parts,

$$\int_a^b xf'(x)dx = bf(b) - af(a) - \int_a^b f(x)dx$$

But

$$\int_a^b [x]f'(x)dx = \int_a^b xf'(x)dx - \int_a^b (x-[x])f'(x)dx$$

$$= bf(b) - af(a) - \int_a^b (x-[x])f'(x)dx - \int_a^b f(x)dx$$

Hence

$$\sum_{a<n\leq b} f(n) = Bf(b) - Af(a) - bf(b) + af(a) + \int_a^b (x-[x])f'(x)dx + \int_a^b f(x)dx$$

$$= \int_a^b f(x)dx + \int_a^b (x-[x])f'(x)dx + (a-[a])f(a) - (b-[b])f(b) \qquad ∎$$

Before proceeding it is helpful to introduce some notation which serves to describe the rate of growth of functions. Let $g(x) > 0$ for $x > a$. The following definitions were introduced by the great German analytic number theorist Edmund Landau (1877–1938) in 1927.

Definition 9.2:

(a) *(Big Oh)* $f(x) = O(g(x))$ as $x \to \infty$ if there is a constant $K > 0$ such that $\frac{|f(x)|}{g(x)} < K$ for all $x > a$.

(b) *(Little oh)* $f(x) = o(g(x))$ as $x \to \infty$ if $\lim_{x\to\infty} \frac{|f(x)|}{g(x)} = 0$.

(c) *(Asymptotic)* $f(x) \sim g(x)$ as $x \to \infty$ if $\lim_{x\to\infty} \frac{f(x)}{g(x)} = 1$.

We say that $f(x)$ is of order at most $g(x)$ if $f(x) = O(g(x))$. If $f(x) = o(g(x))$, then f is of order less than $g(x)$. If there are positive constants K_1 and K_2 such that $K_1 < \frac{|f(x)|}{g(x)} < K_2$ for all sufficiently large x, then we say that $f(x)$ is of order $g(x)$.

For example, if $f(x) = 2x^2 + \frac{1}{x}, g(x) = x^2$, and $h(x) = x^3$ then $f(x) = O(g(x)), f(x) = o(h(x)), f(x) \sim 2g(x)$, and $f(x)$ is of order $g(x)$. Note that $F(x) = O(1)$ means that $|F|$ is a bounded function. For example, $F(x) = 2\sin x + 3\cos x = O(1)$. In fact, since $-5 < F(x) < 5$ for all x, $F(x)$ is of order 1.

Notice that the limit "as $x \to \infty$" is usually suppressed.

Proposition 9.5:

(a) If $f_1(x) = O(g(x))$ and $f_2(x) = O(g(x))$, then
  $(f_1 + f_2)(x) = O(g(x))$.
(b) If $f_1(x) = o(g(x))$ and $f_2(x) = o(g(x))$, then
  $(f_1 + f_2)(x) = o(g(x))$.
(c) If $f(x) \sim g(x)$, then $f(x) = O(g(x))$.
(d) If $f(x) = o(g(x))$, then $f(x) = O(g(x))$.
(e) If $f_1(x) = O(g_1(x))$ and $f_2(x) = O(g_2(x))$, then $(f_1 f_2)(x) = O(g_1 g_2(x))$.

Proof: The proof is left as Exercise 9.1.5. ∎

Next, we introduce Euler's constant. Notice that if $N \geq 2$, then

$$1 + \frac{1}{2} + \frac{1}{3} + \cdots + \frac{1}{N} - \log N = \left(1 + \cdots + \frac{1}{N-1} - \log N\right) + \frac{1}{N}$$

$$> \frac{1}{2}\left[\left(1 - \frac{1}{2}\right) + \left(\frac{1}{2} - \frac{1}{3}\right) + \cdots + \left(\frac{1}{N-1} - \frac{1}{N}\right)\right] + \frac{1}{N}$$

$$= \frac{1}{2}\left(1 - \frac{1}{N}\right) + \frac{1}{N} = \frac{1}{2} + \frac{1}{2N}$$

(see Figure 9.1).

But $f(N) = 1 + \frac{1}{2} + \frac{1}{3} + \cdots + \frac{1}{N} - \log N$ is a monotonically decreasing function (Exercise 9.1.7) bounded below by $\lim_{N \to \infty} \left(\frac{1}{2} + \frac{1}{2N}\right) = \frac{1}{2}$. By the monotone convergence theorem, $\lim_{N \to \infty} f(N)$ exists.

Definition 9.3: *Euler's constant* $\gamma = \lim_{N \to \infty} \left(1 + \frac{1}{2} + \frac{1}{3} + \cdots + \frac{1}{N} - \log N\right)$.

The arithmetic nature of γ is still an outstanding problem in mathematics. No one has even been able to prove that γ is irrational. Euler's constant γ is approximately 0.5772157 (which can be easily remembered by the phrase "Euler blinded greatly by a magic formula").

Next, we prove a pair of technical results which will facilitate our discussion in Section 9.2. The sum $\sum_{n \leq N} \frac{1}{n}$ is often abbreviated as H_n, the nth *harmonic number*.

Proposition 9.6: If $r \geq 0$ and $N \geq 1$, then
(a) $\sum_{n \leq N} \frac{1}{n} = \log N + \gamma + O\left(\frac{1}{N}\right)$
(b) $\sum_{n \leq N} n^r = \frac{N^{r+1}}{r+1} + O(N^r)$.

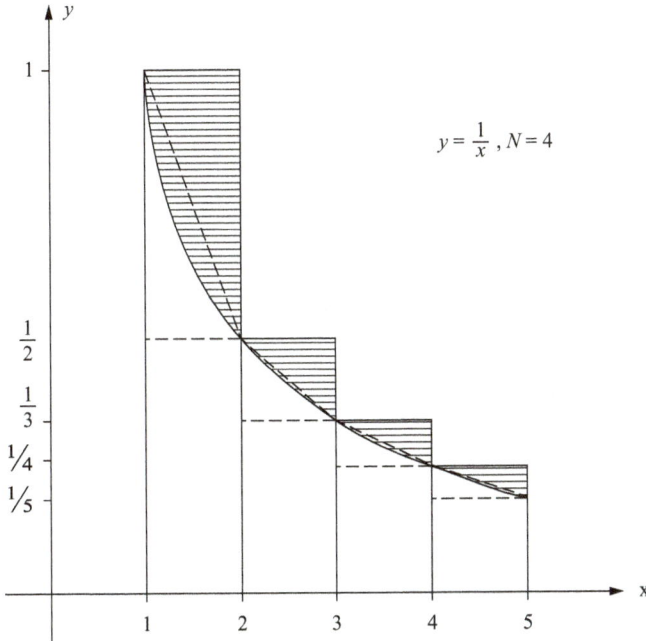

Figure 9.1: Graph of $y = 1/x$ showing bounds on $f(N)$.

Proof: (a) Let $f(x) = \frac{1}{x}$, $a = 1$, and $b = N$ in Euler's summation formula. Then

$$\sum_{n \leq N} \frac{1}{n} = 1 + \int_1^N \frac{1}{x} dx - \int_1^N (x - [x]) \frac{1}{x^2} dx - (N - [N]) \frac{1}{N}$$

$$= 1 + \int_1^N \frac{1}{x} dx - \int_1^N (x - [x]) \frac{1}{x^2} dx + O\left(\frac{1}{N}\right)$$

But

$$(x - [x]) \frac{1}{x^2} < \frac{1}{x^2} \text{ for all } x \text{ and } \int_1^N \frac{1}{x^2} dx = 1 - \frac{1}{N}$$

So the improper integral $\int_1^\infty \frac{1}{x^2} dx = 1$ and $\int_1^\infty (x - [x]) \frac{1}{x^2} dx$ converges to $c < 1$ say. Hence,

$$\sum_{n \leq N} \frac{1}{n} = \log N + 1 - \int_1^N (x - [x]) \frac{1}{x^2} dx + O\left(\frac{1}{N}\right)$$

But

$$y = \lim_{N \to \infty} \left(\sum_{n \le N} \frac{1}{n} - \log N \right)$$

Hence,

$$\sum_{n \le N} \frac{1}{n} = \log N + y + O\left(\frac{1}{N}\right)$$

(b) Let $f(x) = x^r, a = 1, b = N$ in Euler's summation formula. Then

$$\sum_{n \le N} n^r = 1 + \int_1^N x^r dx + \int_1^N (x - [x]) r x^{r-1} dx - (N - [N]) N^r$$

But $1 = O(N^r)$ and $(N - [N]) N^r = O(N^r)$. Furthermore,

$$(x - [x]) r x^{r-1} < r x x^{r-1} \text{ for all } x \text{ and } \int_1^{rx} r^{-1} dx = O(N^r)$$

So

$$\sum_{n \le N} n^r = \frac{N^{r+1}}{r+1} + O(N^r) \qquad \blacksquare$$

Last, we present a somewhat surprising application of harmonic numbers to a probabilistic problem.

Example 9.1 (Matching cards): If n distinguishable cards are flipped over from a deck one a time without replacement and we guess each card prior to its exposure, how many correct matches can we expect? (The number n need not be 52.) If we never see any of the cards, then the answer is exactly 1 no matter the size of the deck. The reason is that the chance of guessing any individual card is $1/n$ and so the expected number is $n \cdot (1/n) = 1$ since the expected value of a sum of identically distributed independent events is the sum of the expected values. For example, with a standard deck, we could guess queen of hearts every time to guarantee being correct exactly once. If we choose some other strategy, despite a greater variance of outcomes, our expectation doesn't change.

The problem we now consider is the expected number of matches if we see each card once it's been revealed. In this case, let E_n be the expected number of matches with a deck of size n. Trivially, $E_1 = 1$ since there is only one possibility. $E_2 = 2 \cdot (1/2) + 1 \cdot (1/2) = 3/2$ since we have an even chance of guessing the first card correctly and are guaranteed to guess the second card no matter what. You may wish to verify (via a tree diagram perhaps) that $E_3 = 11/6$. The pattern might not yet be recognizable. Here is the result:

Solution: The expectation $E_n = H_n$, the nth harmonic number.

Proof. Consider the situation with n cards. The probability of correctly guessing the first card is $1/n$. If guessed correctly, then the expected number of additional correct guesses is E_{n-1}. If the first card is guessed incorrectly, one card is revealed and hence the expected number of correct guesses is now E_{n-1} as well. So $E_n = \frac{1}{n}(1+E_{n-1}) + \frac{n-1}{n} E_{n-1} = E_{n-1} + 1/n$. Since $E_1 = 1$, $E_n = 1 + 1/2 + 1/3 + \cdots + 1/n = H_n$. Perhaps more simply, note that we always have probability of $1/k$ of guessing correctly when there are k cards remaining, and each reveal diminishes the remaining possibilities by one.

In particular, for a standard deck of cards, $E_{52} \approx 4.54$.

Exercise 9.1

1. Show that $\lim_{s \to 1^+} \prod_p (1-p^{-s})^{-1} = \infty$.
2. (a) Show that $\frac{1}{1-x} = 1 + \sum_{r=1}^{\infty} x^r$ for $|x| < 1$.
   (b) Show that $\log(1-x) = -\sum_{r=1}^{\infty} \frac{x^r}{r}$ for $|x| < 1$.
   (c) Show that $\log(1+x^2) = \sum_{r=1}^{\infty} (-1)^{r+1} \frac{x^{2r}}{r}$ for $|x| < 1$.
3. (a) Show that $\sum_{r=2}^{\infty} \frac{1}{p^r} = \frac{1}{p(p-1)}$ for $p > 1$.
   (b) Show that $\sum_{r=k}^{\infty} \frac{1}{p^r} = \frac{1}{p^{k-1}(p-1)}$ for $p > 1$.
4. (a) Show that $\tan^{-1} x = x - \frac{x^3}{3} + \frac{x^5}{5} - \cdots$ for $|x| < 1$.
   (b) Show that $\pi/4 = 1 - 1/3 + 1/5 - 1/7 + \cdots$ (James Gregory – 1671).
5. Prove Proposition 8.5.
6. Show that $f(x) \sim g(x)$ if and only if $f(x) = g(x) + o(g(x))$.
7. Show that $f(N) = \sum_{n=1}^{N} \frac{1}{n} - \log N$ is a monotonically decreasing function.
8. Give an example where $f(x) \sim g(x)$, but $f'(x)$ is not asymptotic to $g'(x)$.
9. Show that for $s > 1$, $1/\zeta(s) = \sum_{n=1}^{\infty} \frac{\mu(n)}{n^s}$ where $\mu(n)$ is the Mobius μ function. (Hint: Use the Euler Product.)
10. Show that for $s > 1$, $\zeta^2(s) = \sum_{n=1}^{\infty} \frac{\tau(n)}{n^s}$ where $\tau(n)$ is the divisor function.
11. Define the von Mangoldt function by $\Lambda(n) = \log p$ if $n = p^t$ and $\Lambda(n) = 0$ otherwise (see Exercise 3.3.12). Show that $\sum_{n=1}^{\infty} \frac{\Lambda(n)}{n^s} = -\zeta'(s)/\zeta(s)$.
12. Show that Euler's constant γ satisfies $\gamma < 1$.
13. (a) Show that the gamma function satisfies $s\Gamma(s) = \Gamma(s+1)$ for $s > 0$.
    (b) Show that $\Gamma(n+1) = n!$.
14. Verify directly (via a tree diagram) that $E_3 = 11/6$ and $E_4 = 25/12$ in the card matching problem 9.1.

9.2 Average order of the lattice and divisor functions

For any integer n, the number $n^2 + 1$ is expressible as the sum of two squares (we just did it). However, any integer congruent to 3 (mod 4) is not expressible as the sum of two squares. Thus, there are integers arbitrarily large which have no representation as the sum of two squares and integers arbitrarily large which are the sum of two squares. The following definition allows us to speak more concretely.

Definition 9.4: The *lattice point function* $r(n)$ is the total number of representations of n as a sum of two squares.

In Definition 9.4, we include all solutions (x, y) with $x^2 + y^2 = n$, not just essentially distinct solutions. For example, $r(8) = 4$ since $2^2 + 2^2 = 8$, $2^2 + (-2)^2 = 8$, $(-2)^2 + 2^2 = 8$, and $(-2)^2 + (-2)^2 = 8$. The arithmetic function $r(n)$ is not multiplicative (Exercise 9.2.1(b)), but it has a simple geometric interpretation. The value $r(n)$ is the number of lattice points in the plane that are a distance $\sqrt{n}$ from the origin. A lattice point is a point having integral coordinates. Equivalently, $r(n)$ is the number of squares of area n with corners at lattice points, one of them being the origin. The geometric visualization of $r(n)$ is what allows us to study its behavior further.

Our comments prior to Definition 9.4 may be summarized by the statement lim $inf_{n\to\infty} r(n) = 0$. By Exercise 4.2.3 there are infinitely primes congruent to 1 (mod 4). By Exercise 5.2.9 there is no constant upper bound for $r(n)$ appropriate for all n. Hence it is true that lim $sup_{n\to\infty} r(n) = \infty$. Thus, the behavior of the function $r(n)$ is extremely erratic (much like many of the arithmetic functions we have studied).

To smooth out the jumpy behavior of an arithmetic function $t(n)$, it may be more appropriate to study its average behavior. Define

$$T(N) = \sum_{n=1}^{N} t(n)$$

as the summatory function of $t(n)$. Then we make the following definition:

Definition 9.5: If $\frac{T(N)}{N} \sim F(N)$ as $N \to \infty$, then the *average order* of $t(n)$ (over the interval $[1, n]$) is $F(n)$.

Consequently, we define $R(N) = \sum_{n=1}^{N} r(n)$ and study its rate of growth in order to get a handle on the average order of $r(n)$. Table 9.1 gives the initial growth rate of R(n) (any conjectures?).

The next theorem is due to Gauss.

Theorem 9.7:

$$R(N) = \pi N + O\left(\sqrt{N}\right) \tag{9.5}$$

Proof: The lattice points in the plane are in one-to-one correspondence with the unit squares for which they form the top right vertex. But each lattice point (x, y) corresponds to a sum of two squares, namely $x^2 + y^2 = n$ (where $\sqrt{n}$ is the distance between

n	1	2	3	4	5	6	7	8	9	10
r(n)	4	4	0	4	8	0	0	4	4	8
R(n)	4	8	8	12	20	20	20	24	28	36
R(n)/n	4	4	2.67	3	4	3.33	2.86	3	3.11	3.6

n	11	12	13	14	15	16	17	18	19	20
r(n)	0	0	8	0	0	4	8	4	0	8
R(n)	36	36	44	44	44	48	56	60	60	68
R(n)/n	3.27	3	3.38	3.14	2.93	3	3.29	3.33	3.16	3.2

the lattice point and the origin). It follows that $R(N)$ is equal to the sum of the areas of the squares having top right vertex on or inside the circle $x^2 + y^2 = N$ (see Figure 9.2).

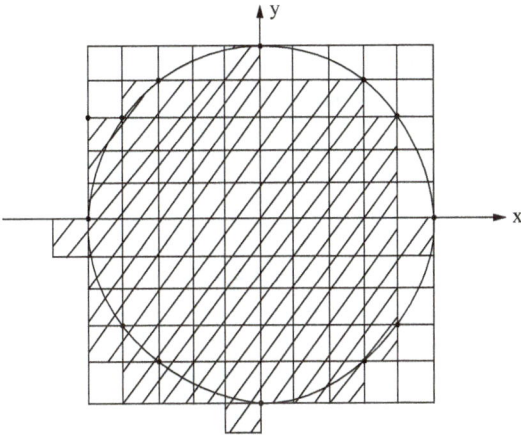

Figure 9.2: Two-dimensional visualization of R(N) for N = 25.

$$x^2 + y^2 = 25$$

The sum of the areas of appropriate squares is not exactly the same as the area of the circle $x^2 + y^2 = N$. However, the length of the diagonal of each square is $\sqrt{2}$. Hence the larger circle $x^2 + y^2 = (\sqrt{N} + \sqrt{2})^2$ contains all appropriate squares. In addition, for $N \geq 2$, the smaller circle $x^2 + y^2 = (\sqrt{N} - \sqrt{2})^2$ is contained in the union of squares. It follows that

$$\pi(\sqrt{N} - \sqrt{2})^2 < R(N) < \pi(\sqrt{N} + \sqrt{2})^2$$

Thus

$$\pi\left(N - 2\sqrt{2N} + 2\right) < R(N) < \pi\left(N + 2\sqrt{2N} + 2\right)$$

Therefore,

$$R(N) = \pi N + O(\sqrt{N}) \qquad \blacksquare$$

Corollary 9.7.1: The average order of the lattice function $r(n)$ is π.

Proof: By Theorem 9.7, $\frac{R(N)}{N} = \pi + O(N^{-1/2})$. So the average order of $r(n)$ is

$$\lim_{N \to \infty} \frac{R(N)}{N} = \pi \qquad \blacksquare$$

Significant improvements to the error term in Gauss' lattice point problem (Theorem 9.7) have been made by many mathematicians. The best result to date is $O(N^{23/73} + \varepsilon)$ due to M.N. Huxley (2003).

Next, we turn to a similar discussion of the divisor function, $\tau(n)$. We are interested in an estimate for $\sum_{n \leq N} \tau(n)$. It is helpful to realize that $\tau(n)$ is the number of lattice points on the hyperbola $rd = n$ where $r, d \geq 1$. In 1849 Dirichlet proved the following result:

Theorem 9.8: For all $N \geq 1$,

$$\sum_{n \leq N} \tau(n) = N\log N + (2\gamma - 1)N + O\left(N^{1/2}\right) \qquad (9.6)$$

Proof:

$$\sum_{n \leq N} \tau(n) = \sum_{n \leq N} \sum_{d \mid n} 1 = \sum_{rd \leq N} 1$$

where the last sum is over all lattice points (r, d) in the first quadrant with $rd \leq N$.

The appropriate lattice points are pictured below in Figure 9.3. Notice that the lattice points are symmetrically placed about the line $d = r$. For fixed k with $1 \leq k \leq [\sqrt{N}]$, there are the same number of lattice points on the vertical line $r = k$ above (k, k) as there are lattice points on the horizontal line $d = k$ to the right of (k, k). This is due to the fact that if d is a divisor of n with $d < \sqrt{n}$, then n/d is another divisor of n. So, the total number of lattice points (r, d) with $rd = n \leq N$ is twice the shaded number of lattice points plus the number on the line $d = r$. But the number of shaded lattice points on the line $d = k$ is $[N/k] - k$. Further, there are $[\sqrt{N}]$ lattice points on the line $d = r$. Hence,

$$\sum_{n \leq N} \tau(n) = 2 \sum_{d \leq \sqrt{N}} \left\{ \left[\frac{N}{d}\right] - d \right\} + [\sqrt{N}] \qquad (9.7)$$

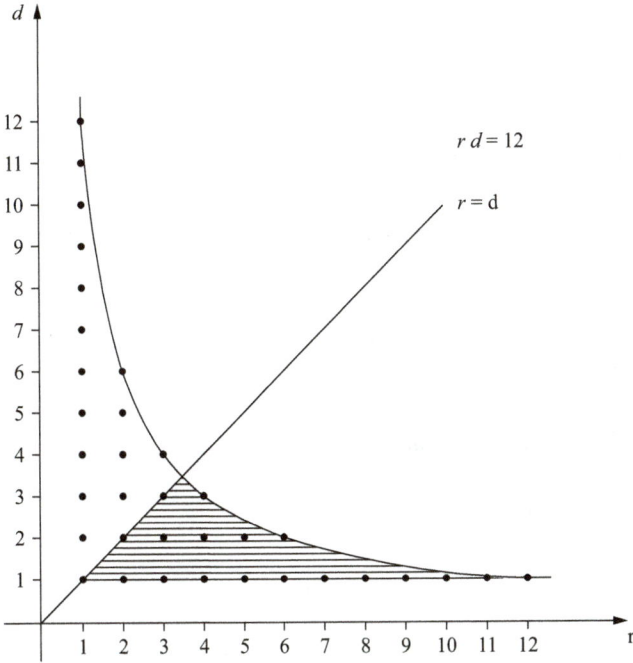

Figure 9.3: Lattice point visualization of $\tau(n)$.

But $\left[\sqrt{N}\right] = O\left(N^{1/2}\right)$ and, in general, $[x] = x + O(1)$. So eq. (9.7) becomes

$$\sum_{n \leq N} \tau(n) = 2 \sum_{d \leq \sqrt{N}} \left\{ N/d - d + O(1) \right\} + O\left(N^{1/2}\right)$$

$$= 2N \sum_{d \leq \sqrt{N}} \frac{1}{d} - 2 \sum_{d \leq \sqrt{N}} d + 2 \sum_{d \leq \sqrt{N}} O(1) + O\left(N^{1/2}\right)$$

By Proposition 9.6(a),

$$\sum_{d \leq \sqrt{N}} \frac{1}{d} = \log\left(N^{1/2}\right) + \gamma + O\left(N^{-1/2}\right)$$

By Proposition 9.6(b) with $r = 1$,

$$\sum_{d \leq \sqrt{N}} d = \frac{N}{2} + O\left(N^{\frac{1}{2}}\right)$$

Hence

$$\sum_{n \le N} \tau(n) = 2N\left(\log N^{1/2} + \gamma + O\left(N^{-1/2}\right)\right) - 2\left(\frac{N}{2} + O\left(N^{1/2}\right)\right) + 2\sum_{d \le \sqrt{N}} O(1)$$

$$= N\log N + (2\gamma - 1)N + O\left(N^{\frac{1}{2}}\right) \qquad \blacksquare$$

Improvements to the error term $O\left(N^{1/2}\right)$ have been made by several mathematicians. In 1904, the Russian G. Voronoi (1868–1908) proved that the error term could be replaced by $O\left(N^{1/3}\log N\right)$. A more recent result of Iwaniec and Mozzochi (1988) gives $N^{\frac{7}{22}+\varepsilon}$ for any $\varepsilon > 0$. In 1916, G.H. Hardy proved that the error is not $O\left(N^{1/4}\right)$. It is conjectured that the "truth" may be $O\left(N^{1/4+\varepsilon}\right)$.

Corollary 9.8.1: The divisor function, $\tau(n)$, is of average order $\log n$.

Proof: The proof is left as Exercise 9.2.3. $\qquad \blacksquare$

Exercise 9.2

1. (a) Explain why if p is a prime $\equiv 3 \pmod 4$, then $r(p) = 0$ while if p is a prime $\equiv 1$ $\pmod 4$, then $r(p) = 4$.
    (b) Show that $r(n)$ is not a multiplicative function.
2. The volume of a unit sphere in m-space is given by $V_m = \frac{\pi^{m/2}}{\Gamma(m/2+1)}$. In particular, $V_4 = \pi^2/2$. Let $r_4(n)$ be the total number of representations of n as a sum of four squares.
    Let $R_4(N) = \sum_{n=1}^{N} r_4(n)$.
    (a) Mimic the proof of Theorem 8.7 to show that

    $$\frac{\pi^2}{2}\left(\sqrt{N} - 2\right)^4 < R_4(N) < \frac{\pi^2}{2}\left(\sqrt{N} + 2\right)^4$$

    (b) Conclude that $R_4(N) = \frac{\pi^2}{2}N^2 + O(N^{3/2})$. Hence the average order of $r_4(n)$ is $\pi^2 n/2 = (4.9348\ldots)n$.
3. Prove Corollary 9.8.1.
4. A key result due to Jacobi (1828) states that if $d_1(n)$ and $d_3(n)$ are the number of divisors of n congruent to 1 and 3 $\pmod 4$ respectively, then $r(n) = 4(d_1(n) - d_3(n))$. Verify this result for $n = 5$, 7, 10, 12, 15, and 21.
5. Another result of Jacobi's (1828) states that if S is the sum of the odd divisors of n, then $r_4(n) = 24S$ if n is even and $r_4(n) = 8S$ if n is odd where $r_4(n)$ is as defined in exercise 9.2.2. Verify this result for $n = 5$, 7, 10, 12, 15, and 21.
6. (Dirichlet) Show that $[N] + [N/2] + [N/3] + \cdots = N\log N + (2\gamma - 1)N + O(N^{1/2})$.

9.3 Average order of $\phi(n)$ and applications

Attempts to find the exact value of $\zeta(2)$ have a long and interesting history. Jakob Bernoulli (1654–1705) knew that $\sum_{n=1}^{\infty} \frac{1}{n^2}$ converges since it is bounded by $\frac{1}{1\cdot1} + \sum_{n=1}^{\infty} \frac{1}{n(n+1)} = 2$. The Oxford don, John Wallis (previously mentioned in connection with Pell's equation) calculated $\zeta(2)$ to three decimal places, 1.645. Christian Goldbach claimed correctly that $1 + \frac{16223}{25200} < \zeta(2) < 1 + \frac{30197}{46800}$. Attempts by Johann Bernoulli (1667–1748) to evaluate $\zeta(2)$ led him to study integrals of the form $\int x^m (1+x)^n dx$, but his attempts were inconclusive.

In 1731 Euler published a paper which related $\zeta(2)$ to the sum of two integrals involving the logarithm function. In 1732 he developed the formula

$$\zeta(2) = (\log 2)^2 + 2 \sum_{n=1}^{\infty} \frac{1}{2^n n^2}$$

whose rapid convergence allowed Euler to show that $\zeta(2)$ is 1.644934 to six decimal places. This calculation was extended to 1.64493406684822643647 in 1733. By 1734 Euler discovered that $\zeta(2) = \frac{\pi^2}{6}$. His reasoning is described below.

An algebraic theorem of I. Newton's states that if a polynomial has constant term 1, then the sum of the reciprocals of its roots is the negative of the coefficient of the linear term (Exercise 9.3.1). The power series representation for the sine function is

$$\sin x = x - \frac{x^3}{3!} + \frac{x^5}{5!} - \frac{x^7}{7!} + \cdots \text{ for all real } x$$

Let $x \neq 0$ be such that $\sin x = 0$. Then

$$0 = x - \frac{x^3}{3!} + \frac{x^5}{5!} - \frac{x^7}{7!} + \cdots$$

$$0 = 1 - \frac{x^2}{3!} + \frac{x^4}{5!} - \frac{x^6}{7!} + \cdots$$

Let $z = x^2$,

$$0 = 1 - \frac{z}{3!} + \frac{z^2}{5!} - \frac{z^3}{7!} + \cdots =: p(z)$$

Although the right-hand side is not a polynomial, if we apply Newton's theorem we get

$$\frac{1}{6} = \sum_{n=1}^{\infty} \frac{1}{(n\pi)^2}$$

since the roots of $p(z)$ are $\pi^2, (2\pi)^2, (3\pi)^2$, etc. The result now follows by multiplying through by π^2.

Like other great mathematicians, Euler not only had the creative spark necessary to discover deep results but he also insisted on placing them in a firm, logical framework. Results were often refined and proved several times over – each time paying greater attention to basic assumptions and to the range of applicability of his methods. By 1736, Euler had refined his argument above to a degree considered rigorous by modern standards. In fact, he generalized the above to include all even arguments of the zeta function. The main result is that if n is a positive integer, then

$$\zeta(2n) = (-1)^{n+1} \frac{(2\pi)^{2n} B_{2n}}{2(2n)!} \tag{9.8}$$

where B_k is the kth Bernoulli number defined by

$$B_0 = 1 \text{ and } (k+1)B_k = -\sum_{m=0}^{k-1} \binom{k+1}{m} B_m \text{ for } k \geq 1 \tag{9.9}$$

We evaluate $\zeta(2)$ in a neat, but clever, way presented by Don Zagier as part of the introduction to his Hedrick lectures in 1989.

Proposition 9.9: $\zeta(2) = \pi^2/6$.

Proof: Since $\zeta(2) = \sum_{n=1}^{\infty} \frac{1}{n^2}$, it follows that $\frac{1}{4}\zeta(2) = \sum_{n=1}^{\infty} \frac{1}{(2n)^2}$. Subtracting

$$\left(1 - \frac{1}{4}\right)\zeta(2) = \sum_{n=0}^{\infty} \frac{1}{(2n+1)^2}$$

So

$$\zeta(2) = \frac{4}{3} \sum_{n=0}^{\infty} \frac{1}{(2n+1)^2} \tag{9.10}$$

The double integral $\int_0^1 \int_0^1 (xy)^{2n} dx dy = \frac{1}{(2n+1)^2}$ (Exercise 9.3.11).
Hence,

$$\zeta(2) = \frac{4}{3} \sum_{n=0}^{\infty} \int_0^1 \int_0^1 (xy)^{2n} dx dy$$

$$= \frac{4}{3} \int_0^1 \int_0^1 \sum_{n=0}^{\infty} (x^2 y^2)^n dx dy$$

$$= \frac{4}{3} \int_0^1 \int_0^1 \frac{1}{1 - x^2 y^2} dx dy$$

by summing the geometric series for $0 < xy < 1$. (This is a convergent improper integral.)

Now let $x = \sin u / \cos v$ and $y = \sin v / \cos u$ for $0 \le u \le \pi/2$ and $0 \le v \le \pi/2$.

(Equivalently, $u = \cos^{-1} \sqrt{\frac{1-x^2}{1-x^2y^2}}$ and $v = \cos^{-1} \sqrt{\frac{1-y^2}{1-x^2y^2}}$.) The Jacobian

$$\frac{\partial(x,y)}{\partial(u,v)} = 1 - \frac{\sin^2 u}{\cos^2 v} \cdot \frac{\sin^2 v}{\cos^2 u} = 1 - x^2 y^2$$

Hence we get,

$$\zeta(2) = \frac{4}{3} \iint\limits_{\Delta} du\, dv$$

where Δ is the triangular region with vertices at $(0,0), (0, \pi/2)$, and $(\pi/2, 0)$.

It follows that $\zeta(2) = \frac{4}{3} \cdot \frac{\pi^2}{8} = \frac{\pi^2}{6}$. ∎

Next, we calculate the average order of $\phi(n)$ by considering its summatory function

$$\Phi(N) = \sum_{n=1}^{N} \phi(n)$$

The main result is due to Franz Mertens (1874).

Theorem 9.10: For all $N \ge 1$,

$$\Phi(N) = \frac{3N^2}{\pi^2} + O(N \log N) \tag{9.11}$$

Proof: $\Phi(N) = \sum_{n=1}^{N} \phi(n) = \sum_{n=1}^{N} n \sum_{d|n} \frac{\mu(d)}{d} = \sum_{n=1}^{N} \sum_{d|n} \frac{n}{d} \mu(d)$ by formula (3.8)

$= \sum_{rd \le N} r\mu(d)$ summing over all lattice points (r,d) with $rd \le N$.

$$= \sum_{d=1}^{N} \mu(d) \cdot \sum_{r=1}^{[N/d]} r$$

Now $\sum_{r=1}^{\{\frac{N}{d}\}} r = [N/d]([N/d] + 1)/2 = N^2/2d^2 + O(N/d)$. Thus

$$\Phi(N) = \sum_{d=1}^{N} \mu(d) \left\{ \frac{N^2}{2d^2} + O\left(\frac{N}{d}\right) \right\}$$

$$= \frac{N^2}{2} \sum_{d=1}^{N} \frac{\mu(d)}{d^2} + O\left(N \sum_{d=1}^{N} \frac{1}{d} \right) \tag{9.12}$$

But $\sum_{d=1}^{N} \frac{1}{d} = \log N + \gamma + O\left(\frac{1}{N}\right)$ by Proposition 9.6(a). Furthermore,

$$\sum_{d=1}^{N} \frac{\mu(d)}{d^2} = \sum_{d=1}^{\infty} \frac{\mu(d)}{d^2} - \sum_{d>N} \frac{\mu(d)}{d^2} = 1/\zeta(2) - \sum_{d>N} \frac{\mu(d)}{d^2}$$

by Exercise 9.1.9. But

$$\sum_{d>N} \frac{\mu(d)}{d^2} = O\left(\sum_{d>N} \frac{1}{d^2}\right) \text{ and } \sum_{d>N} \frac{1}{d^2} = O\left(\int_N^\infty x^{-2}dx\right) = O(1/N)$$

Substituting into eq. (9.12) yields

$$\Phi(N) = \frac{N^2}{2\zeta(2)} + O(N/2) + O(N \log N)$$

$$= \frac{3N^2}{\pi^2} + O(N \log N)$$

by Proposition 9.9. ∎

Corollary 9.10.1: The average order of $\phi(n)$ is $\frac{3n}{\pi^2}$.

Proof: The result follows immediately from the theorem. ∎

Corollary 9.10.2: The number of elements in the Farey sequence $\mathcal{F}_n$ is $\frac{3N^2}{\pi^2} + O(N \log N)$.

Proof: Recall that the number of elements in $\mathcal{F}_n$ is $1 + \Phi(n)$. ∎

We now turn to some additional applications of Theorem 9.10. Let us define a lattice point $(a, b) \neq (0, 0)$ as being *visible from the origin* (or simply visible) if the line segment joining the origin to (a, b) contains no other lattice points. Certainly, the lattice point (a, b) is visible from the origin if and only if $\gcd(a, b) = 1$ (Exercise 9.3.6).

Consider the square S_N centered at the origin with sides of length $2N$. So $S_N = \{(x, y): |x| \leq N, |y| \leq N\}$. Let $L(N)$ be the total number of lattice points on or in S_N (excluding the origin) and let $V(N)$ be the number of visible lattice points on or in S_N. For example, $L(3) = 48$ and $V(3) = 32$ (see Figure 9.4.)

The ratio $V(N)/L(N)$ gives the proportion of visible lattice points in the square S_N from among all the lattice points in the square. The value $\lim_{N \to \infty} V(N)/L(N)$ gives the proportion (or density) of visible lattice points in the plane.

Corollary 9.10.3: The density of visible lattice points in the plane is $6/\pi^2$ (approximately 0.608).

Proof: The eight lattice points on the boundary of S_1 are all visible from the origin. There are no other visible lattice points on the lines $y = 0, x = 0, y = x$, or $y = -x$.

Consider the wedge $W_N = \{(x, y): 0 < y < x \leq N\}$ (see Figure 9.4).

It follows that $V(N) = 8 + 8 \cdot$ # visible lattice points in W_N. But # visible lattice points in W_N is $\sum_{a=2}^{N} \sum_{b=1}^{\prime a} 1$ where $\sum^{\prime}$ means that the sum is over all b relatively prime to a. Hence,

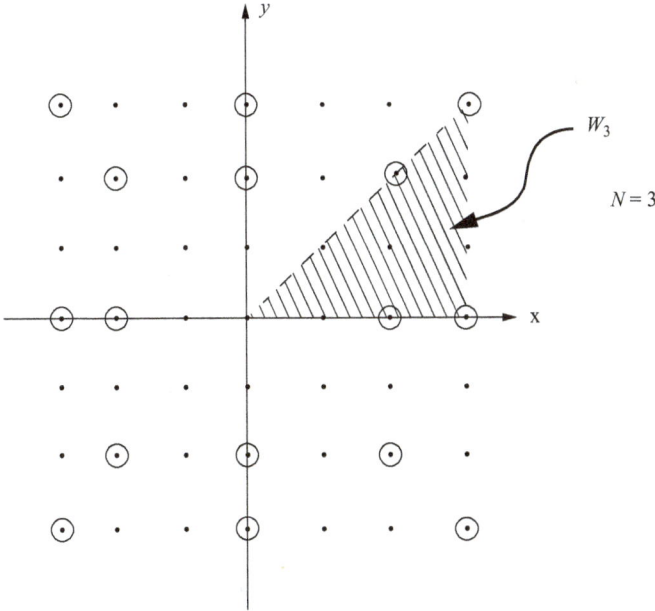

$$V(N) = 8 + 8 \cdot \sum_{a=2}^{N} \sum_{b=1}^{\prime\,a} 1 = 8\left(1 + \sum_{a=2}^{N} \phi(a)\right) = 8 \sum_{a=1}^{N} \phi(a)$$

By Theorem 9.10,

$$V(N) = \frac{24N^2}{\pi^2} + O(N \log N)$$

In addition,

$$L(N) = (2N+1)^2 - 1 = 4\,N^2 + 8N = 4\,N^2 + O(N)$$

Hence,

$$\frac{V(N)}{L(N)} = \frac{24N^2/\pi^2 + O(N\log N)}{4N^2 + O(N)} = \frac{6/\pi^2 + O(\log N/N)}{1 + O(1/N)}$$

Taking limits as $N \to \infty$ yields the desired result. ∎

Corollary 9.10.4: The likelihood that two random independently selected integers are relatively prime is $6/\pi^2$.

Proof: The proof is immediate. ∎

If S is a subset of $\mathbf{N}$, let $|S(n)|$ be the *number of elements* r of S with $1 \le r \le n$.

Definition 9.6: If $\lim_{n \to \infty} \frac{|S(n)|}{n} = p$, then we say that S has *natural density* p. We denote the natural density of S by $\delta_n(S)$.

Notice that if S has a natural density p, then $0 \le p \le 1$. If S has natural density p, then we say that the probability that a randomly selected natural number is in S is p. Conversely, if the probability is p that a randomly selected natural number is in S, then S has natural density p. For example, the set of integers congruent to $3 \pmod 4$ has natural density $1/4$ since every fourth positive integer is congruent to $3 \pmod 4$. Even though the set of natural numbers do not admit a probability density function, by a minor abuse of language, we often say that a set of positive integers has probability p if its natural density is p.

The next result is actually a corollary to Proposition 9.9, but we record it here.

Corollary 9.10.5: The probability that a randomly selected integer is square-free is $6/\pi^2$.

Proof: An integer n is square-free if and only if n is not divisible by the square of any prime. For a given prime p, n is not divisible by p^2 if and only if n is incongruent to $0 \pmod{p^2}$. The probability that n is congruent to $0 \pmod{p^2}$ is $1/p^2$. Hence the probability that n is incongruent to $0 \pmod{p^2}$ is $1 - 1/p^2$. It follows that the probability that n is square-free is

$$\prod_{\text{all } p} \left(1 - \tfrac{1}{p^2}\right) = 1/\zeta(2) = 6/\pi^2 \qquad \blacksquare$$

It should be noted that we can readily generalize Corollary 9.10.5. The probability that a randomly selected integer is nth power-free is $1/\zeta(n)$. Similarly, the probability that n random independently chosen integers are relatively prime is $1/\zeta(n)$.

Exercise 9.3

1. Prove Newton's theorem which states that if a polynomial has constant term 1, then the sum of the reciprocals of its roots is the negative of the coefficient of the linear term.
2. Use formula (9.9) to calculate the Bernoulli numbers $B_2, \dots, B_8$.
3. Use the exercise above and formula (8.8) to calculate $\zeta(n)$ for $n = 2, 4, 6$, and 8.
4. Use formula (9.9) to show that the Bernoulli numbers $B_3, B_5, B_7, \dots = 0$.
5. Calculate $\sum_{n=1}^{\infty} \frac{1}{(2n)^2}$.
6. (a) Show that the lattice point (a, b) is visible from the origin if and only if $\gcd(a, b) = 1$.
   (b) Under what condition is the lattice point (a, b) visible from the point $(1, 2)$?

(c) Under what condition is the lattice point (a, b, c) invisible from the origin in 3-space?

7. Show that the probability that a randomly selected integer is nth power-free is $1/\zeta(n)$.

8. Show that the probability that n random independently chosen integers are relatively prime is $1/\zeta(n)$.

9. For m, n positive integers, let (Sum of Powers Formula – Jakob Bernoulli)

$$S_m(n) = 1^m + 2^m + \cdots + n^m$$

(a) Use the power series for e^{kt} to show that

$$\sum_{k=0}^{t} e^{kt} = \sum_{m=0}^{\infty} S_m(n) \frac{t^m}{m!}$$

(b) Use geometric series to show that the left-hand side above is $\frac{e^{(n+1)t}-1}{e^t-1}$.

(c) Show that $\frac{e^{(n+1)t}-1}{e^t-1} = \sum_{k=1}^{\infty} \frac{(n+1)^k t^{k-1}}{k!} \cdot \sum_{j=0}^{\infty} B_j \frac{t^j}{j}$.

(d) Equate coefficients of t^m to obtain $\frac{S_m(n)}{m!} = \sum_{k=0}^{m} \frac{B_k}{k!} \frac{(n+1)^{m+1-k}}{(m+1-k)!}$.

(e) Multiply through by $(m+1)!$ to obtain the sum of powers formula:

$$(m+1)S_m(n) = \sum_{k=0}^{m} \binom{m+1}{k} B_k (n+1)^{m+1-k}$$

10. Use the sum of powers formula to derive formulas for $S_m(n)$ for $m = 2$, 3, and 4 with n arbitrary.

11. Evaluate $\int_0^1 \int_0^1 (xy)^{2n} dx\, dy$.

12. Let $S_k(n) = \sum_{j=1}^{n} j^k$ be the sum of the first n kth powers. Let $\mu_k(n) = \frac{S_k(n)}{n}$ be the average of the first n kth powers. Let $B_k = \frac{N_k}{D_k}$ be the kth Bernoulli number with $\gcd(N_k, D_k) = 1$ and $D_k > 0$.

By a theorem of P. Damianou and Schumer (2003), if k is odd then $\mu_k(n)$ is an integer $\Leftrightarrow n \not\equiv 2 \pmod 4$ and if k is even, then $\mu_k(n)$ is an integer $\Leftrightarrow n$ is not divisible by any prime p where $p \mid D_k$. Investigate this result for $k = 1, 2, 3, 4, 6$.

9.4 Chebyshev's theorems and the distribution of primes

Much of the deepest work in number theory deals with the distribution of the primes. Recall that there are infinitely many primes (Theorem 2.11) but yet arbitrarily large gaps between them (Proposition 2.12). It also appears that there are infinitely many twin pairs and good heuristic arguments to support it (though no proof to date.) We have shown that there is no non-constant polynomial that produces only primes (Proposition 2.15) and Fermat's attempt at a prime producing function, $f(n) = 2^{2^n} + 1$, proved to be woefully inadequate. The erratic behavior of the primes suggests that it may be more fruitful to investigate on average how large we expect the nth prime p_n to be. Equivalently, what is the order of magnitude of $\pi(x)$, the prime counting function?

In the years 1792 and 1793, Gauss made a careful analysis of the distribution of primes. He determined that an excellent approximation to $\pi(x)$ was the *logarithmic integral*

$$\text{Li}(x) = \int_2^x \frac{dt}{\log t},$$

conjecturing that

$$\pi(x) \sim \text{Li}(x)$$

Gauss made ample use of his prodigious computational facility. He wrote that "[he] frequently . . . spent an idle quarter of an hour to count another chilliad here and there." Apparently a chilliad is a block of integers of length 1,000. Eventually, Gauss checked his hypothesis up to $x = 3000000$ and found excellent corroboration for his conjecture. In 1798 Legendre made a similar (but less accurate) conjecture concerning the distribution of primes. With modern methods and technology, these calculations have been carried out to far greater limits. For example, $\pi(10^{23}) = 1{,}925{,}320{,}391{,}606{,}803{,}968{,}923$ and Li $(10^{23}) \approx 1{,}925{,}320{,}391{,}614{,}054{,}155{,}138$ showing impressive agreement.

If we integrate by parts, we obtain

$$\text{Li}(x) = \frac{x}{\log x} - \frac{2}{\log 2} + \int_2^x \frac{dt}{\log^2 t}$$

For $x \ge 4$, we can break up the last integral as

$$\int_2^x \frac{dt}{\log^2 t} = \int_2^{\sqrt{x}} \frac{dt}{\log^2 t} + \int_{\sqrt{x}}^x \frac{dt}{\log^2 t}$$

Since the function $y = \log^2 x$ is monotone increasing for $x \ge 1$, we find that

$$\int_2^{\sqrt{x}} \frac{dt}{\log^2 t} \le \frac{\sqrt{x}}{\log^2 2} \quad \text{and} \quad \int_{\sqrt{x}}^x \frac{dt}{\log^2 t} \le \frac{x}{\log^2(\sqrt{x})}$$

But

$$\frac{\sqrt{x}}{\log^2 2} = o\left(\frac{x}{\log x}\right) \quad \text{and} \quad \frac{x}{\log^2(\sqrt{x})} = \frac{4x}{\log^2 x} = o\left(\frac{x}{\log x}\right)$$

Hence

$$\text{Li}(x) \sim \frac{x}{\log x} \tag{9.13}$$

Although Li(x) gives a superior numerical estimate to $\pi(x)$, we write Gauss's conjecture in the following form:

Prime Number Theorem (PNT): $\pi(x) \sim \frac{x}{\log x}$ as $x \to \infty$.

Neither Gauss nor Legendre were able to prove the Prime Number Theorem. The first substantial contributions toward a proof were made by the Russian mathematician P. L. Chebyshev (1821–1894) in 1851 and 1852. Chebyshev proved that for x sufficiently large,

$$0.92129 \leq \frac{\pi(x)}{x/\log x} \leq 1.10555$$

In this section, we will prove a related result with somewhat weaker constants. In addition, Chebyshev proved that if $\lim_{x \to \infty} \frac{\pi(x)}{x/\log x}$ exists, then it must equal 1. By the way, Chebyshev was a brilliant and influential mathematician even outside the realm of number theory. He wrote on roots of algebraic equations, multiple integrals, best approximations to functions, elliptic functions, probability, the theory of finite differences, several branches of physics, cartography, and mechanical engineering. Chebyshev even built a workable calculating machine in the late 1870s.

The next major breakthrough was due to G.B. Riemann, who made fundamental advances in our understanding of the primes in his memoir, "On the Number of Primes Less than a Given Magnitude" completed in 1859. In his work he showed how the distribution of the primes is directly related to the zeros of the Riemann zeta function, $\zeta(s)$. Riemann also indicated how the establishment of a zero free region including the real line Re $s = 1$ would imply the PNT. In fact, Riemann made several astonishing assertions concerning the distribution of the zeros of $\zeta(s)$ in the critical strip. None were proven for over 30 years and some mathematicians concluded that Riemann had made a great number of lucky guesses. However, in the 1930s C.L. Siegel carefully scrutinized Riemann's notes deposited in the library at Göttingen University. From the notes it was abundantly clear that Riemann had a much deeper knowledge of $\zeta(s)$ than is apparent from his sole publication in number theory. A work by Siegel in 1932 included an asymptotic formula known as the Riemann-Siegel formula for $Z(t)$, a function closely related to $\zeta(\frac{1}{2} + it)$. The key ideas were all contained in Riemann's personal notes.

In 1896, the French mathematician J. Hadamard (1865–1963) and the Belgian mathematician C. J. de la Vallée Poussin (1866–1962) independently proved the PNT along the lines indicated by Riemann. (E. Landau later proved that the PNT was equivalent to $\zeta(s)$ having no zeros on the line Re $s = 1$.) The proof uses quite a bit of complex analysis, though some recent proofs have greatly simplified some of the more technical aspects. (There was a mathematical joke in the 1950s that proving the PNT was the surest way to immortality.)

Since the PNT is a statement about integers rather than complex numbers, number theorists sought an "elementary" proof of it, i.e., one that does not involve complex analysis. In 1949, Atle Selberg and Paul Erdös essentially working in tandem constructed such proofs. Their arguments are indeed intricate, however.

In this section we make a modest venture into this fascinating area. We begin with two important definitions:

Definition 9.7 (Chebyshev functions): Let $x > 0$ be a real number.
(a) $\vartheta(x) = \sum_{p \leq x} \log p$ where the sum is over all primes $p \leq x$.
(b) $\psi(x) = \sum_{p^m \leq x} \log p$ where the sum is over all prime powers $p^m \leq x$.

For example, $\vartheta(12) = \log 2 + \log 3 + \log 5 + \log 7 + \log 11 = \log 2310$ and $\psi(12) = 3 \log 2 + 2 \log 3 + \log 5 + \log 7 + \log 11 = \log 27720$. (In $\psi(12)$ there is a contribution of $\log 2$ from $n = 2, 4$, and 8 and a contribution of $\log 3$ from $n = 3$ and 9.) Recall the von Mangoldt lambda function, $\Lambda(n) = \log p$ if $n = p^m$ and $\Lambda(n) = 0$ otherwise (see Exercise 3.3.12). It readily follows that

$$\psi(x) = \sum_{n \leq x} \Lambda(n)$$

Next note that $p^m \leq x$ if and only if $p \leq x^{1/m}$. In addition, $x^{1/m} < 2$ whenever $m > \frac{\log x}{\log 2}$. Let $L_x = \log x / \log 2 = \log_2 x$. Hence, we have

$$\psi(x) = \sum_{1 \leq m \leq L_x} \vartheta\left(x^{1/m}\right) \tag{9.14}$$

Equivalently, there are $[\log x / \log p]$ powers of p less than x where $[\cdot]$ is the greatest integer function. Hence

$$\psi(x) = \sum_{p \leq x} [\log x / \log p] \log p \tag{9.14'}$$

Proposition 9.11: $\psi(x) = \vartheta(x) + O\left(\sqrt{x} \log^2 x\right)$.

Proof: By eq. (9.14), $\psi(x) = \vartheta(x) + \sum_{2 \leq m \leq L_x} \vartheta\left(x^{1/m}\right)$ where $L_x = \log x / \log 2$.

Now $\vartheta(t) < t \log t$ for all $t \geq 2$. Hence

$\vartheta\left(x^{1/m}\right) < x^{1/m} \log\left(x^{1/m}\right) < x^{1/m} \log x \leq \sqrt{x} \log x$ for $m \geq 2$. So
$\sum_{2 \leq m \leq L_x} \vartheta\left(x^{1/m}\right) < L_x \sqrt{x} \log x = O\left(\sqrt{x} \log^2 x\right)$. The result follows. ∎

Next, we show that $\psi(x)$ and $\vartheta(x)$ are of order exactly x. By Proposition 9.11, it suffices to obtain upper and lower bounds for $\psi(x)$ or $\vartheta(x)$ either of which is a constant times x. In fact, we will prove the following:

Theorem 9.12: (a) $\vartheta(n) < (2 \log 2)n$ for all $n \geq 1$.
(b) $\psi(n) > \left(\frac{\log 2}{3}\right)n$ for all $n \geq 2$.

Note that the constants in Theorem 9.12 are not at all sharp since by the PNT and some subsequent discussion there are better bounds for the Chebyshev functions. However, Theorem 9.12 is sufficient for our purposes.

Corollary 9.12.1: The Chebyshev functions $\psi(x)$ and $\vartheta(x)$ are of order x.

Proof of Theorem 9.12: (a) Let $k \geq 1$ and set $K = \binom{2k+1}{k}$. The quantity

$$2^{2k+1} = (1+1)^{2k+1} = \sum_{i=0}^{2k+1} \binom{2k+1}{i}$$

But $\binom{2k+1}{k} = \binom{2k+1}{k+1}$. Thus $K < 2^{2k}$.

If p is a prime with $k+1 < p \leq 2k+1$, then p divides the numerator of K but not the denominator. Thus p exactly divides K. Since K is an integer, the product of all such primes divides K. Let S be the set of integers in the interval $(k+1, 2k+1]$. It follows that

$$\vartheta(2k+1) - \vartheta(k+1) = \sum_{p \in S} \log p \leq \log K < \log 2^{2k} = (2\log 2)k \tag{9.15}$$

It is trivial to check that Theorem 9.12(a) holds for $n=1$ and $n=2$. Now assume that it holds for all $n \leq N$ for some $N > 2$. We proceed by induction to show that the result holds for $n = N+1$:

If $N+1$ is even, then $\vartheta(N+1) = \vartheta(N) < (2\log 2)N < (2\log 2)(N+1)$.
If $N+1$ is odd, then $N+1 = 2k+1$ for some integer $k \geq 1$. Then

$$\vartheta(N+1) = \vartheta(2k+1) = [\vartheta(2k+1) - \vartheta(k+1)] + \vartheta(k+1) < (2\log 2)k + (2\log 2)(k+1)$$

by inequality (9.15) and our inductive hypothesis. But

$$(2\log 2)k + (2\log 2)(k+1) = (2\log 2)(2k+1) = (2\log 2)(N+1)$$

Hence $\vartheta(N+1) < (2\log 2)(N+1)$ for $N+1$ odd too. This establishes (a).

(b) Let $L = \binom{2k}{k} = \frac{(2k)!}{k!k!}$. The exact power of a prime p dividing $n!$ is the sum $[n/p] + [n/p^2] + \cdots = \sum_{m \geq 1} [n/p^m]$. This is a finite sum since $[n/p^m] = 0$ for all $p^m > n$. It follows that

$$\text{if } L = \prod_{p \leq 2k} p^{r_p}, \text{ then } r_p = \sum_{m \geq 1} ([2k/p^m] - 2[k/p^m]) \tag{9.16}$$

Now $p^m > 2k$ if and only if $m > \frac{\log 2k}{\log p}$ and hence there are at most $[\log 2k / \log p]$ terms in the expression for r_p. Furthermore, if $[2k/p^m]$ is even, then the term $[2k/p^m] - 2[k/p^m]$

adds 0 to the sum. If $[2k/p^m]$ is odd, then the term $[2k/p^m] - 2[k/p^m])$ adds 1 to the sum (Exercise 9.4.7). Therefore,

$$r_p \leq \left\lceil \frac{\log 2k}{\log p} \right\rceil \tag{9.17}$$

So $\log L = \sum_{p \leq 2k} r_p \log p \leq \sum_{p \leq 2k} \left\lceil \frac{\log 2k}{\log p} \right\rceil \log p = \psi(2k)$ by (9.14').

But $L = \frac{k+1}{1} \cdot \frac{k+2}{2} \cdot \ldots \cdot \frac{2k}{k} \geq 2^k$. It follows that $\log L \geq (\log 2)k$ and so $\psi(2k) \geq (\log 2)k$. Set $k = [n/2]$ for $n \geq 2$. So $k \geq n/3$. Then

$$\psi(n) \geq \psi(2k) \geq (\log 2)k \geq \left(\frac{\log 2}{3} \right) n$$

as desired. ∎

Theorem 9.13 (Chebyshev's Theorem): The prime counting function $\pi(x)$ is of order $\frac{x}{\log x}$.

It follows that there are constants C_1 and C_2 such that

$C_1 \frac{x}{\log x} < \pi(x) < C_2 \frac{x}{\log x}$ for all $x \geq 2$.

Corollary 9.13.1: The nth prime p_n is of order $n \log n$.

Proof of Corollary: Since $\pi(p_n) = n$, p_n is the "inverse" of the prime counting function. ($\pi(x)$ is not a one-to-one function and so there is really no unique inverse function.) Let $y = \frac{x}{\log x}$. So $\pi(x)$ is of order y by Theorem 9.13. Then $\log y = \log x - \log \log x$ and so $\log x \sim \log y$ since $\log \log x = o(\log x)$. Hence $x = y \log x \sim y \log y$. So the inverse of y is asymptotic to $x \log x$. Similarly, the inverse of $\pi(x)$ is of order $x \log x$. Thus p_n is of order $n \log n$. ∎

Proof of Theorem 9.13: It is clear that

$$\vartheta(x) = \sum_{p \leq x} \log p \leq (\log x) \sum_{p \leq x} 1 = \pi(x) \log x$$

By Theorem 9.12, for some appropriate constant $k_1 > 0$,

$$\pi(x) \geq \frac{\vartheta(x)}{\log x} > \frac{k_1 x}{\log x} \tag{9.18}$$

In the other direction, if $0 < a < 1$, then let $S_a(x)$ be the set $(x^{1-a}, x]$. We have

$$\vartheta(x) \geq \sum_{p \in S_a(x)} \log p \geq \log x^{1-a} \sum_{p \in S_a(x)} 1$$

$$= (1 - a) \log x \left\{ \pi(x) - \pi(x^{1-a}) \right\}$$

$$\geq (1 - a) \log x \left\{ \pi(x) - x^{1-a} \right\}$$

So

$$\vartheta(x) \geq (1 - a) \log x \cdot \pi(x) - (1 - a) \log x \cdot x^{1-a}$$

Rearranging

$$\pi(x) \leq \frac{\vartheta(x) + (1 - a) \log x \cdot x^{1-a}}{(1 - a) \log x} = x^{1-a} + \frac{\vartheta(x)}{(1 - a) \log x} \qquad (9.19)$$

But for x sufficiently large, there is a constant $k_2 > 0$ such that

$$x^{1-a} < \frac{k_2 x}{2 \log x} \quad \text{and} \quad \frac{\vartheta(x)}{(1 - a) \log x} < \frac{k_2 x}{2 \log x}$$

by Theorem 9.12. Hence

$$k_1 \frac{x}{\log x} < \pi(x) < k_2 \frac{x}{\log x}$$

for x sufficiently large. ■

Of course, we can remove the condition "for x sufficiently large" by replacing k_1 and k_2 with new constants C_1 and C_2 having potentially a larger difference.

Chebyshev actually went beyond Theorem 9.13 by showing that

$$\liminf_{x \to \infty} \frac{\pi(x)}{x / \log x} \leq 1 \leq \limsup_{x \to \infty} \frac{\pi(x)}{x / \log x}$$

Hence if $\lim_{x \to \infty} \frac{\pi(x)}{x / \log x}$ exists, then it must be 1.

Next we relate the growth of $\pi(x)$ to that of the Chebyshev functions.

Theorem 9.14: $\pi(x) \sim \frac{\vartheta(x)}{\log x} \sim \frac{\psi(x)}{\log x}$ as $x \to \infty$.

Proof: By Theorem 9.12 and Proposition 9.11, it suffices to prove $\pi(x) \sim \frac{\vartheta(x)}{\log x}$. On the one hand, by (9.18)

$$\frac{\pi(x) \log x}{\vartheta(x)} \geq 1 \quad \text{for all } x \geq 2$$

On the other hand, by (9.19)

$$\frac{\pi(x) \log x}{\vartheta(x)} \leq \frac{x^{1-a} \log x}{\vartheta(x)} + \frac{1}{1 - a} \qquad (9.20)$$

For any $\beta > 0$ we can find an $\alpha = \alpha(\beta) > 0$ such that $\frac{1}{1-\alpha} < 1 + \frac{\beta}{2}$.

Choose $M = M(\beta)$ such that if $x > M$, then

$$\frac{x^{1-\alpha}\log x}{\vartheta(x)} < \frac{k\log x}{x^\alpha} < \frac{\beta}{2} \quad \text{for appropriate } k$$

This yields via eq. (9.20)

$$1 \le \frac{\pi(x)\log x}{\vartheta(x)} \le 1 + \beta \quad \text{for all } x > M$$

Since β is arbitrary, we obtain the desired result. ∎

By Theorem 9.14, it suffices to prove that $\psi(x) \sim x$ in order to prove the PNT. Unfortunately, it is a large leap from Corollary 9.12.1 to such a proof. What Hadamard and de la Vallée Poussin proved is that

$$\psi(x) = x + O\left(xe^{-c(\log x)^{1/14}}\right)$$

More recently, mathematicians have made gradual improvements to the error term. If the Riemann hypothesis is true, then

$$\psi(x) = x + O\left(x^{1/2+\varepsilon}\right) \text{ for arbitrary positive } \varepsilon$$

Quite remarkably, if the above expression can be proven, then the Riemann hypothesis would follow. Thus, the Riemann zeta function and the primes are inextricably linked.

We end this section with a short account on the gaps between successive primes. Number the primes in ascending order p_1, p_2, p_3, etc. and let $g_n = p_{n+1} - p_n$ be the nth *prime gap*. We have shown that g_n can be arbitrarily large but is also often equal to 2 (and may be so an infinite number of times if the Twin Prime Conjecture is true). The PNT implies that the average order of g_n approaches $\log n$ as $n \to \infty$. In 2005, D. Goldston, J. Pintz, and C. Yildirim used a weighted Selberg sieve to prove that

$$\liminf_{n\to\infty} \frac{g_n}{\log p_n} = 0$$

In 2007, they strengthened this to

$$\liminf_{n\to\infty} \frac{g_n}{\sqrt{\log p_n}(\log\log p_n)^2} = 0$$

A tremendous step forward was taken when Yitang Zhang (2014) established a specific finite bound that infinitely many prime gaps must satisfy. Specifically, he showed that

$$\liminf_{n\to\infty} g_n < 70{,}000{,}000$$

Subsequently, many mathematicians worked collaboratively via a Polymath Project with ideas due to T. Tao and J. Maynard. As of 2020, the value 70,000,000 has been reduced to 246.

Exercise 9.4

1. Show that a corollary to Theorem 8.13 is that the sum of the reciprocals of the primes diverges.
2. (a) Describe in words the expression $e^{\vartheta(x)}$.
   (b) Describe in words the expression $e^{\psi(x)}$.
3. Use the PNT to explain why it is said there are more primes than squares.
4. Use the PNT to show that the nth prime $p_n \sim n \log n$.
5. Use the PNT to give a heuristic argument why there ought to be infinitely many Mersenne primes.
6. Use the PNT to show that given $M > 0$, for n sufficiently large there are at least M primes between n and $2n$.
7. Show that

$$[2x] - 2[x] = \begin{cases} 1 \text{ if } [2x] \text{ is odd} \\ 0 \text{ if } [2x] \text{ is even} \end{cases}.$$

8. Let $\pi_2(x)$ represent the number of twin primes less than or equal to x.
   G.H. Hardy and J.E. Littlewood [54] conjectured that the following holds:

$$\pi_2(x) \sim 2C_2 \frac{x}{(\log x)^2} \text{ as } x \to \infty \text{ where } C_2 = \prod_{p>2}\left[1 - \frac{1}{(p-1)^2}\right]$$

Investigate the accuracy of this conjecture for $x = 100, x = 200, x = 1000, x = 2000$. (The twin prime constant C_2 is approximately 0.66016.)

9. Let $r_2(2n)$ represent the number of representations of $2n$ as a sum of two primes. Hardy and Littlewood conjectured that the following holds:

$$r_2(2n) \sim C_2 \prod_{p|n} \frac{p-1}{p-2} \cdot \frac{2n}{(\log 2n)^2} \text{ as } x \to \infty$$

where the product is over odd primes p dividing n. Investigate the accuracy of this conjecture for $50 \le n \le 500$.

9.5 Bertrand's postulate and applications

In 1845, the French mathematician J.L.F. Bertrand (1822–1900) conjectured that for every $n > 2$ there is a prime p with $n < p < 2n$. Bertrand was a child prodigy who began attending lectures at the École Polytechnique at age 11. By the age of 17 he had separate bachelor of arts and bachelor of science degrees as well as a doctor of science degree in thermodynamics. His mathematical work spanned such disciplines as differential geometry, mathematical analysis, the theory of symmetric groups, mathematical physics, differential equations, and probability. He also wrote some popular algebra textbooks for secondary schools.

Although Bertrand verified his conjecture up to $n = 3{,}000{,}000$, he was unable to exact a proof. In 1850, Chebyshev was able to produce a demonstration using analytical methods. In fact, Chebyshev actually proved that for $n \geq 25$, there is always a prime in the interval $[n, (1 + \varepsilon)n]$ for any $\varepsilon > 1/5$. Subsequently, other mathematicians were able to lower the value of ε somewhat. Of course, by the PNT, for any $\varepsilon > 0$ there will be a prime in $[n, (1 + \varepsilon)n]$ for n sufficiently large. Below we present a more transparent exposition partially based on a work of Paul Erdös (1932).

Theorem 9.15 (Bertrand's Postulate): If $n > 1$, then there is at least one prime p for which $n < p < 2n$.

Proof: Notice that the list of primes 2, 3, 5, 7, 13, 23, 43, 83, 163, 317, 631 establishes the result up to $n = 630$ since each is less than double its predecessor. Now assume for the sake of argument that there is some $n > 512 = 2^9$ for which Bertrand's postulate is false. We consider the binomial coefficient

$$L = \binom{2n}{n} = \frac{(2n)!}{n!n!}. \text{ Let } p \text{ be a prime with } p \mid L. \text{ By our hypothesis, } p \leq n.$$

We collect some previous results:

$$\text{if } L = \prod_{p \leq 2k} p^{r_p}, \text{then } r_p = \sum_{m \geq 1} ([2k/p^m] - 2[k/p^m]) \tag{9.16}$$

$$r_p \leq \left\lceil \frac{\log 2k}{\log p} \right\rceil \tag{9.17}$$

If $2n/3 < p \leq n$, then $2p \leq 2n < 3$. So $p^2 > 4n^2/9 > 2n$ for $n > 512$.

In this case (9.13) becomes $r_p = [2n/p] - 2[n/p] = 2 - 2(1) = 0$.

So $r_p = 0$ for $2n/3 < p \leq n$. (Alternatively, notice that if $2n/3 < p \leq n$ then $p^2 \| (2n)!$ and $p \| n!$. Thus $p \nmid L$.)

Hence $p \leq n/3$ for every $p \mid L$. By Theorem 9.12(a),

$$\sum_{p \mid L} \log p \leq \sum_{p \leq 2n/3} \log p = \vartheta(2n/3) \leq \frac{4 \log 2}{3} n$$

If $r_p \geq 2$, then $2 \leq \left\lceil \frac{\log 2n}{\log p} \right\rceil \leq \frac{\log 2n}{\log p}$ by eq. (9.14).

But this implies that $\log p^2 \leq \log 2n$ and so $p \leq \sqrt{2n}$.

Thus, there are at most $\sqrt{2n}$ such primes for which $r_p \geq 2$. In addition eq. (9.14) implies that $r_p \log p \leq \log 2n$. It follows that

$$\log L = \sum_{p \leq 2n} r_p \log p = \sum_{r_p = 1} \log p + \sum_{r_p \geq 2} r_p \log p$$

$$\leq \sum_{r_p = 1} \log p + \sqrt{2n} \log 2n$$

$$\leq \frac{4 \log 2}{3} n + \sqrt{2n} \log 2n$$

Now L is the largest term in the expansion of

$$2^{2n} = (1+1)^{2n} = 2 + \binom{2n}{1} + \cdots + \binom{2n}{2n-1}$$

Since there are $2n$ terms, $2^{2n} \leq 2\,nL$ and hence $(2\log 2)n \leq \log 2n + \log L$. By our previous discussion,

$$(2 \log 2)n \leq \log 2n + \frac{4 \log 2}{3} n + \sqrt{2n} \log 2n$$

Collecting terms,

$$\frac{2 \log 2}{3} n \leq \left(1 + \sqrt{2n}\right) \log 2n$$

Dividing both sides by $1 + \sqrt{2n}$,

$$\frac{\frac{2\log 2}{3} n \left(\sqrt{2n} - 1\right)}{2n - 1} \leq \log 2n \qquad (9.21)$$

We now show that eq. (9.21) is erroneous for n sufficiently large.

On the one hand,

$$\frac{\frac{2\log 2}{3} n \left(\sqrt{2n} - 1\right)}{2n - 1} > \frac{\log 2}{3} \left(\sqrt{2n} - 1\right) > \frac{6}{19} \sqrt{n} \quad \text{for } n \geq 512$$

Now if $n = 2^x$, then $\frac{6}{19} \sqrt{n} = \frac{6}{19} 2^{x/2}$.

On the other hand,

$$\text{if } n = 2^x, \text{ then } \log 2n = (x+1)\log 2 < \tfrac{7}{9} x \text{ for } x \geq 9.$$

But $\frac{6}{19} 2^{x/2} > \frac{7}{9} x$ if and only if $2^{x/2} > \frac{133}{54} x$.

However, if $x = 9$, then $2^{x/2} = 22.627 \cdots > 22.16\bar{6} \cdots = \frac{133 \cdot 9}{54}$. Furthermore,

if $f(x) = 2^{x/2}$ and $g(x) = \frac{133}{54} x$, then $f'(x) = \frac{\log 2}{2} 2^{x/2} > \frac{133}{54} = g'(x)$ for $x \geq 9$.

So formula (9.21) is invalid for all $n \geq 512$, thus contradicting our original hypothesis. Therefore, for all $n > 1$, there is at least one prime p for which $n < p < 2n$. ∎

We conclude this section with three interesting corollaries to Bertrand's postulate. The first corollary is due to Hans-Egon Richert (1949). It is noteworthy that the following corollary does not follow from the PNT. The advantage of Bertrand's postulate over the PNT in this context is that it guarantees the existence of primes between n and $2n$ for all $n > 1$, not just for n sufficiently large.

Corollary 9.15.1: Every positive integer beyond 6 is the sum of distinct primes.

Proof: Consider the 13 numbers 7, 8, and 19. Note that $7 = 7$, $8 = 5 + 3$, $9 = 7 + 2$, $10 = 7 + 3$, $11 = 11$, $12 = 7 + 5$, $13 = 11 + 2$, $14 = 11 + 3$, $15 = 7 + 5 + 3$, $16 = 11 + 5$, $17 = 7 + 5 + 3 + 2$, $18 = 11 + 7$, and $19 = 11 + 5 + 3$. So all are sums of distinct primes not exceeding 11. For our first step let $p_1 = 13$. By adding p_1 to each of our sums we can express the p_1 numbers $20 = 19 + 1$, $21 = 19 + 2$, …, $32 = 19 + p_1$ as the sum of distinct primes not exceeding 13. For example, $20 = 13 + 7$, $21 = 13 + 8 = 13 + 5 + 3$, …, $32 = 13 + 19 = 13 + 11 + 5 + 3$. By Bertrand's postulate, there is a prime p_2 with $p_1 < p_2 < 2p_1$. For the second step, we list the p_2 numbers $33 = 19 + p_1 + 1$, 34, …, $19 + p_1 + p_2$ as sums of distinct primes not exceeding p_2. In particular, $33 = (33 - p_2) + p_2$, $34 = (34 - p_2) + p_2$, …, $19 + p_1 + p_2 = (19 + p_1) + p_2$. Since $m - p_2 > m - 2p_1 = m - 26 \geq 7$ for $m \geq 33$, each expression in parentheses is the sum of distinct primes not exceeding p_1. In general, after the nth step, all integers $7, 8, \ldots, 19 + p_1 + p_2 + \cdots + p_n$ have been expressed as the sum of distinct primes not exceeding p_n. Here $p_r < p_{r+1} < 2p_r$ for $1 \leq r \leq n - 1$ as guaranteed by Bertrand's postulate. For $n \geq 1$, denote $19 + p_1 + p_2 + \cdots + p_n$ as P_n. Let p_{n+1} be a prime with $p_n < p_{n+1} < 2p_n$. The $(n+1)^{\text{st}}$ step involves listing the p_{n+1} numbers $P_n + 1, \ldots, P_n + p_{n+1}$ as the sum of distinct primes as follows: For $1 \leq m \leq p_{n+1}$, the number $P_n + m = p_{n+1} + (P_n + m - p_{n+1})$. But the number in parentheses,

$$P_n + m - p_{n+1} > P_{n-1} - p_n > P_{n-2} - p_{n-1} \cdots > 19 + p_1 - p_2 > 19 - p_1 = 6.$$

In addition, $P_n + m - p_{n+1} \leq P_n$. So $P_n + m - p_{n+1}$ has been expressed as the sum of distinct primes, none exceeding p_n, by the nth step. Adding p_{n+1} to each expression for $1 \leq m \leq p_{n+1}$ completes the $(n+1)$th step. In this way, all integers beyond 6 are expressed as the sum of distinct primes. ∎

Corollary 9.15.2: The decimal $\alpha = 0.2357111317\ldots$ consisting of juxtaposing all the primes in ascending order is irrational.

Proof: Suppose α is rational. Then its decimal expansion would either terminate or eventually repeat. Since there are infinitely many primes, it must be the case that α has an eventually periodic decimal expansion. Suppose that α has k initial digits followed by a repetitive pattern consist of r digits each. By Bertrand's postulate, for any number $n \geq 1$, there is at least one prime consisting of n digits (in fact, at least three such primes). For $n \geq 1$, let p_n denote some prime having n digits. Let $mr > k + r$. Then the prime p_{mr} appears in the decimal expansion of a and has a repetitive pattern of length r (since the prime preceding p_{mr} has at least $k + r - 1 > k$ digits). Let R be the number consisting of the r digits in p_{mr}. Then $R \mid p_{mr}$ and $R < p_{mr}$ since $m > 1$. This contradicts the primality of p_{mr}, thus establishing the result. ■

Note that the proof of Corollary 9.15.2 actually only utilized the fact that there is a prime between n and $10n$ for n sufficiently large. Hence, we did not need the full strength of Bertrand's postulate. Alternatively, we could have applied the PNT to prove Corollary 9.15.2. However, the PNT would be ineffective in the proofs of Corollary 9.15.1 and 9.15.3 since it only makes an assertion for primes sufficiently large.

Our final corollary contains an interesting application of Bertrand's postulate to partial sequences of consecutive squares. More recently, this line of inquiry has been greatly extended and still contains fertile areas of research. Our last result is due to Arnold, et al. (1985). We begin with a definition.

Definition 9.8: For $n \geq 1$, the *discriminator function* $D(n)$ is defined to be the minimum value of k such that the squares $1^2, 2^2, \ldots, n^2$ are incongruent modulo k.

So $D(n)$ is the smallest k for which the remainders upon division by k of the first n squares are all distinct. For example, $D(6) = 13$ since $1^2 \equiv 1, 2^2 \equiv 4, 3^2 \equiv 9, 4^2 \equiv 3, 5^2 \equiv 12$, and $6^2 \equiv 10 \pmod{13}$, while for all $k < 13$, there are $1 \leq \{a < b \leq 6$ for which $a^2 \equiv b^2 \pmod{k}$. Check it.

Here is a table of values of $D(n)$ for $1 \leq n \leq 18$:

Table 9.2: Discriminator function table.

n	1	2	3	4	5	6	7	8	9	10	11	12	13	14	15	16	17	18
$D(n)$	1	2	6	9	10	13	14	17	19	22	22	26	26	29	31	34	34	37

Corollary 9.15.3: For $n > 4$, the discriminator function, $D(n)$, is the smallest integer at least $2n$ that is equal to a prime or twice a prime.

We preface our proof with a helpful lemma.

Lemma 9.15.1: Let $f(x) = x + M/x$ where $M > 9$ is a constant. Let $S = [3, \sqrt{M})$. Then the maximum value of f on S is $f(3) = 3 + M/3$.

Proof: $f'(x) = 1 - M/x^2 < 0$ for all $x \in S$. Hence f is decreasing over S and the result follows. ■

Proof of Corollary 9.15.3: Let p be a prime with $p \geq 2n$. If $1 \leq a < b \leq n$ and $a^2 \equiv b^2$ $(\bmod\, p)$, then $p \mid (b+a)(b-a)$. By Euclid's lemma, either $p \mid (b+a)$ or $p \mid (b-a)$. But neither possibility is tenable since $p \geq 2n > b + a > b - a > 0$. So $D(n) \leq p$ for $p \geq 2n$.

Similarly, if $2p \geq 2n$ and there are a and b with $1 \leq a < b \leq n$ with $a^2 \equiv b^2$ $(\bmod\, 2p)$, then $2p \mid (b+a)(b-a)$. But $p \geq n$ implies that $p \mid (b+a)$.

Furthermore, $b + a < 2n$ and hence $p = b + a$. It follows that $2 \mid (b-a)$. But then $b + a$ and $b - a$ are of opposite parity, a contradiction. So $D(n) \leq 2p$ for $2p \geq 2n$.

Now let m be any integer less than the smallest prime or double of a prime exceeding $2n$. By Bertrand's postulate, $m < 4n$ since there must be a prime p between $2n$ and $4n$. If $m < 2n$, then there are a and b with $1 \leq a < b \leq n$ for which $m = a + b$. But then $a^2 \equiv b^2$ $(\bmod\, m)$. So $D(n) \geq 2n$.

We complete the proof by showing that if $2n \leq m < 4n$ with $m \neq p$ and $m \neq 2p$, then there are $1 \leq a < b \leq$ for which $a^2 \equiv b^2$ $(\bmod\, m)$. By Table 9.2, we may assume that $n > 18$. The number m may be written in at least one of the following forms:

(i) $m = s^2$ (m is a square);
(ii) $m = rs$ with $3 \leq r < s, r$ and s odd (m is odd but not a prime squared);
(iii) $m = 2s^2$ (m is twice a square);
(iv) $m = 2rs$ with $3 \leq r < s$, r and s odd (m is exactly divisible by 2, but not twice a prime squared);
(v) $m = 4rs$ with $2 \leq r < s$ (m is divisible by 4 and is not a square $4p^2$).

   (i) Let $a = s$ and $b = 2s$. Since $m = s^2 < 4n$, we have that $s < 2\sqrt{n}$. Hence $1 < a < b < 4\sqrt{n} \leq n$ for $n \geq 16$. In addition, $m \mid 3$ $s^2 = b^2 - a^2$. So $a^2 \equiv b^2 (\bmod\, m)$.

   (ii) Let $a = \frac{s-r}{2}$ and $b = \frac{s+r}{2}$. Since r and s are odd, a and b are integers. Further, $m = rs < 4n$ implies $s < 4n/r \leq 4n/3$. So $1 \leq a < s/2 \leq 2n/3 < n$. To bound b we use Lemma 9.15.1 with $M = m$. In this case $b = (s+r)/2 = \frac{r+m/r}{2} \leq \frac{3}{2} + \frac{m}{6} < 3/2 + 2n/3 < n$ for $n \geq 5$. In addition, $m \mid rs = b^2 - a^2$ and $a^2 \equiv b^2 (\bmod\, m)$.

   (iii) Let $a = s$ and $b = 3s$. Here $m = 2s^2 < 4n$ implies that $s < \sqrt{2n}$. Thus $1 < a < b = 3s = \sqrt{18n} \leq n$ for $n \geq 18$. Again $m \mid 8s^2 = b^2 - a^2$ and $a^2 \equiv b^2$ $(\bmod\, m)$.

   (iv) Let $a = s - r$ and $b = s + r$. Then $a < s = m/2r < 2n/r < 2n/3 < n$. To bound b apply Lemma 9.15.1 with $M = 2m$. Then $b = s + r = r + m/2r \leq 3 + m/6 < 3 + 2n/3 \leq n$ for $n \geq 9$. In this case, $m \mid 4rs = b^2 - a^2$ and hence $a^2 \equiv b^2 (\bmod\, m)$.

   (v) Let $a = s - r$ and $b = s + r$. Then $1 \leq a < b < 2s = m/2r < 2n/r \leq n$. As in case (iv), we have $a^2 \equiv b^2$ $(\bmod\, m)$.

This completes the proof. ■

A work by Bremser et al. (1990) extends Corollary 9.15.3 to a similar result for arbitrary exponents $j \geq 2$. In addition, the discriminator function has been analyzed for the so-called Dickson polynomials by Moree and Mullen. Furthermore, Moree has used finite field theory to obtain some general results for large classes of polynomials.

Despite the fact that there is no polynomial that generates only primes, the mathematician Mills [147] devised a more intricate function which does generate only primes. His construction is based on a much deeper result concerning the distribution of primes due to Ingham (1937). The demonstration of Ingham's result is well beyond our present concerns. However, as mentioned previously, the study of the gaps between successive primes continues to be an active research area. In fact, Ingham's result has been revised repeatedly. In 1986, Mozzochi proved that for n sufficiently large $p_{n+1} - p_n < p_n^{1051/1920}$.

Ingham's Theorem: Let p_n denote the nth prime. There is a constant k such that for all $n \geq 1, p_{n+1} - p_n < k p_n^{5/8}$.

Corollary: There is a constant M such that for all $m \geq M$ there is a prime between m^3 and $(m+1)^3 - 1$.

By Ingham's theorem, if $m > k^8$ and p_n is the largest prime less than m^3, then we have $p_n < m^3 < p_{n+1} < p_n + k p_n^{5/8} < m^3 + m^{1/8} \ m^{15/8} = m^3 + m^2 < (m+1)^3 - 1$. The result is immediate. ∎

It is interesting to note that the conjecture that there is always a prime between successive squares has not been proved. However, in a similar vein Friedlander and Ivaniec [38] proved the striking result that there are infinitely many primes of the form $x^2 + y^4$. Finally, we present a rather unusual prime producing function.

Theorem 9.16 (Mills' Theorem): There is a real number A such that $\left[A^{3^n}\right]$ is prime for all $n \geq 1$.

Proof: We construct a sequence of primes $q_1, q_2, \ldots$ as follows:
Let M be as in the corollary above and let $q_1 \geq M$. Let q_{n+1} be such that

$$q_n^3 < q_{n+1} < (q_n + 1)^3 - 1 \ for \ n \geq 1$$

The corollary guarantees the existence of q_n for all n. Now let

$$a_n = q_n^{3^{-n}} and \ b_n = (q_n + 1)^{3^{-n}} \ for \ n \geq 1 \qquad (9.22)$$

Since $a_{n+1} = (q_{n+1}^3)^{3^{-n-1}} > (q_n^3)^{3^{-n-1}} = q_n^{3^{-n}} = a_n$, the sequence $\{a_n\}$ is a monotone increasing sequence. Similarly,

$$b_{n+1} = (q_{n+1} + 1)^{3^{-n-1}} < \left((q_n + 1)^3\right)^{3^{-n-1}} = (q_n + 1)^{3^{-n}} = b_n$$

Hence the sequence $\{b_n\}$ is a monotone decreasing sequence.

By definition, $a_n < b_n$ for all n. So $a_n < b_1$ for all n. Analogously, $b_n > a_1$ for all n. Thus, the sequence $\{a_n\}$ is bounded above and the sequence $\{b_n\}$ is bounded below. By the monotone convergence theorem, the two sequences converge. Let $A = \lim_{n\to\infty} a_n$ and $B = \lim_{n\to\infty} b_n$. Then

$$a_n < A \le B < b_n \text{ for all } n$$

It follows that

$$a_n^{3^n} < A^{3^n} \le B^{3^n} < b_n^{3^n}$$

But by eq. (9.22), we get $q_n < A^{3^n} < q_n + 1$ for all n. Hence the prime $q_n = \left[A^{3^n}\right]$ for all n. ∎

Exercise 9.5

1. Show that Bertrand's postulate is equivalent to the following statement: If p_n is the nth prime, then $p_{n+1} < 2p_n$ for all n.

2. Use Bertrand's postulate to prove that $1 + 1/2 + 1/3 + \ldots + 1/n$ is never an integer for $n > 1$. (Can you prove it without assuming Bertrand's postulate?)

3. Express all integers $7 < n \le 100$ as the sum of distinct primes by utilizing the algorithm described in the proof of Corollary 8.15.1.

4. Show that $\binom{2n}{n} = \binom{2n-1}{n-1} + \binom{2n-1}{n}$. Conclude that $\binom{2n}{n}$ is even for all n.

5. (a) Show that $\binom{2n+2}{n+1} = \frac{2(2n+1)}{n+1}\binom{2n}{n}$ for all $n \ge 1$.

   (b) Show that if n is even and $2^m \mid\mid \binom{2n}{n}$, then $2^{m+1} \mid\mid \binom{2n+2}{n+1}$.

   (c) Use part (a) and Exercise 9.5.4 to show that if $n = 2^m - 1$, then 2^m divides $\binom{2n}{n}$. Hence $\binom{2n}{n}$ can contain arbitrarily high powers of 2.

6. Show that $2 \mid\mid \binom{2^{m+1}}{2^m}$ for all $m \ge 1$.

7. (a) Show that if $p > 2$, then $p^m \mid \binom{2p^m - 2}{p^m - 1}$. (Hint: Use the identity in Exercise 9.5.5(a).)

   (b) Show if $p > 2$, then $p \nmid \binom{2p^m}{p^m}$.

   (Hint: Let $N = \prod_{1 \le k \le p^m}(p^m + k)$, $D = \prod_{1 \le k \le p^m} k$, and $\Pi = \prod_{1 \le k \le p^{m-1}}(p^{m-1} + k)$.

   Then $\binom{2p^m}{p^m} = \frac{N/\Pi}{D/\Pi}$. Notice that the factors of N divisible by p are precisely the factors of Π. Now determine the exact powers of p dividing N/Π and D/Π.

   (c) Use parts (a) and (b) to conclude that $p^m \mid\mid \binom{2p^m - 2}{p^m - 1}$.

8. Let $e_p(n)$ denote the exponent e for which $p^e || \binom{2n}{n}$.
   Use – 7 to show the following:
   (a) $\lim \inf_{n\to\infty} e_2(n) = 1$.
   (b) $\lim \inf_{n\to\infty} e_p(n) = 0$ for all $p > 2$.
   (c) $\lim \sup_{n\to\infty} e_p(n) = \infty$ for all primes p.

   Conclude that $\lim_{n\to\infty} e_p(n)$ does not exist (nor equal ∞) for all primes p.

9. Show that $4 | \binom{2n}{n}$ for all n not a power of 2. Interestingly, it was not until 1985 that A. Sárközy proved that $\binom{2n}{n}$ is never square-free for $n > 4$ with at most finitely many exceptions. A. Granville and O. Ramaré completed the proof that there are no exceptions in 1993.

10. Show that for all $n > 1$, there is a prime p with $n < p < 2n$ for which $p | \binom{2n}{n}$. It follows that $\binom{2n}{n}$ is never square-full. (Compare with Exercise 1.1.25.)

11. Show that the decimal $\alpha = 0.12345678910111213\ldots$ is irrational. This number is known as the *Chapernowne constant* named after D. G. Chapernowne who, as a Cambridge University undergraduate in 1933, showed that it is a *normal* number (with all sequences of digits appearing the expected proportion of the time. (In fact, it is known to be transcendental – i.e., not the root of any polynomial with integral coefficients.)

12. Use Ingham's theorem to show that for any k there is an $N = N(k)$ such that if $n > N$, then there are at least k primes between n and $2n$.

13. What exponent $e < 5/8$ would be sufficient in Ingham's theorem to ensure the existence of a prime between n^2 and $(n + 1)^2$ for n sufficiently large?

14. Show that if $n > 1$, then $n!$ is never a perfect square, cube, etc.

15. Let the *discriminator* $D(3, n)$ be the minimum value of k for which the cubes 1^3, $2^3, \ldots, n^3$ are all incongruent (mod k). $D(3, 1) = 1$. For $n \geq 2$, $D(3, n)$ is the smallest $k \geq n$ for which k is square-free and has no prime divisors congruent to 1 (mod 3). Verify the result for $2 \leq n \leq 30$.

16. (a) For $n \geq 2$, let the *discriminator* $D_F(n)$ be the minimum value of k for which the Fibonacci numbers $F_2, F_3, \ldots, F_n$ are all incongruent (mod k). Determine $D_F(n)$.
    (b) For $n \geq 1$, let the *discriminator* $D_E(n)$ be the minimum value of k for which the even numbers $2, 4, \ldots, 2n$ are all incongruent (mod k). Find a formula for $D_E(n)$.

17. Find all integers n less than or equal to 30 with the property that if $1 < r < n$ and $\gcd(n, r) = 1$, then r is prime. It is known that 30 is the largest integer with this property. Can you prove it?

Chapter 10
Introduction to additive number theory

10.1 Waring's problem

In 1770, the British mathematician Edward Waring published his *Meditationes Arithmeticae*. In it he stated without proof that every positive integer is the sum of at most 4 squares, 9 cubes, 19 biquadrates, and so on. Posterity has interpreted Waring's statement to mean that for any integer $k > 1$, there is a constant $n(k)$ such that every positive integer can be expressed as the sum of at most $n(k)$ kth powers. Finding a complete proof to Waring's statement became known as Waring's problem. In the same year, Lagrange proved his celebrated result (Theorem 5.8) on the sum of 4 squares, lending some credence to Waring's conjecture. As remarked in Chapter 1, Hilbert (1909) settled Waring's problem in the affirmative by showing the existence of $n(k)$ for every $k \geq 2$.

Definition 10.1: Let $g(k)$ denote the minimal number for which all positive integers are expressible as the sum of $g(k)$ kth powers.

Hilbert's result is nonconstructive, but does guarantee that $g(k)$ is well-defined for all k. By Lagrange's result and the fact no integer congruent to 7 (mod 8) is the sum of 3 squares, $g(2) = 4$. There are still unanswered questions concerning the exact value of $g(k)$ for some k. In addition, there are other functions closely related to $g(k)$ which warrant greater scrutiny. We begin with a general result which gives a lower bound for $g(k)$ due to Johannes Albert Euler (1734–1800), a son of Leonhard Euler.

Theorem 10.1: For $k \geq 2$, $g(k) \geq [(3/2)^k] + 2^k - 2$.

Proof. For $k \geq 2$, let $3^k = q \cdot 2^k + r$ where $1 \leq r < 2^k$. Hence $q = [(3/2)^k]$. Let $m = q \cdot 2^k - 1 < 3^k$. Any expression for m as the sum of kth powers consists solely of terms of the form 1^k or 2^k. In fact, m is the sum of $q - 1$ terms 2^k and $2^k - 1$ terms 1^k. So m is the sum of $q + 2^k - 2 = [(3/2)^k] + 2^k - 2$ kth powers. All other representations of m would require at least $2^{k+1} - 1$ terms of 1^k which involves more summands.

By the proof of Theorem 10.1, the number 7 requires 4 squares, the number 23 requires 9 cubes, the number 79 requires 19 biquadrates, and the number 223 requires 37 fifth powers. Hence $g(2) \geq 4$, $g(3) \geq 9$, $g(4) \geq 19$, and $g(5) \geq 37$. Surprisingly, Theorem 10.1 seems to be fairly sharp. We have already commented that $g(2) = 4$. In 1909, A. Wieferich proved that $g(3) = 9$. In 1986, R. Balasubramanian, F. Dress, and J.M. Deshouillers proved that $g(4) = 19$. In addition, J.R. Chen proved that $g(5) = 37$ in 1964. In fact, all values of $g(k)$ are known for $k \leq 471600000$ by some deep results of the

https://doi.org/10.1515/9783111579283-010

Russian mathematician I.M. Vinogradov and extensive calculations due to R.M. Stemmler (1964) and M.C. Wunderlich and J.M. Kubina [143]. In all cases yet determined Theorem 9.1 has given the exact value of $g(k)$. Furthermore, Kurt Mahler (1957) has shown that the lower bound in Theorem 9.1 is the exact value of $g(k)$ with at most a finite number of exceptions.

The first demonstration of an upper bound on $g(k)$ for any k was due to Joseph Liouville. He proved that $g(4) \leq 53$. Below we modify his proof to get a mildly better result. Although Proposition 10.2 is far from optimal, it gives some of the flavor of techniques historically used in this area.

Proposition 10.2: $g(4) \leq 50$.

Proof. Let n be any positive integer. We may write $n = 6m + r$ where $0 \leq r \leq 5$. All positive integers are expressible as the sum of four squares (some possibly equal to zero). Let $m = A^2 + B^2 + C^2 + D^2$. Then

$$n = 6(A^2 + B^2 + C^2 + D^2) + r.$$

Now let $A = a_1{}^2 + a_2{}^2 + a_3{}^2 + a_4{}^2$. Then

$$6A^2 = \sum_{i<j} (a_i + a_j)^4 + \sum_{i<j} (a_i - a_j)^4. \tag{10.1}$$

Hence $6A^2$ is the sum of at most 12 biquadrates. Similarly, $6B^2, 6C^2$, and $6D^2$ are all the sum of at most 12 biquadrates. Hence $6m$ is the sum of at most 48 biquadrates and n is the sum of at most $48 + 5 = 53$ biquadrates.

To make a slight improvement, notice that if $n \geq 81$, then $n = 6m + r$ where m is a positive integer and $r = 0, 1, 2, 81, 16$, or 17 (check mod 6.) But $1 = 1^4, 2 = 1^4 + 1^4, 81 = 3^4, 16 = 2^4$, and $17 = 2^4 + 1^4$. Since $6m$ is the sum of at most 48 biquadrates, n is the sum of at most $48 + 2 = 50$ biquadrates. If $n \leq 50$, then n is the sum of at most 50 biquadrates (all terms 1^4). If $51 < n \leq 80$, then $n = 16 + 16 + 16 + (n - 48) \cdot 1^4$, a representation of n as the sum of at most $3 + (80 - 48) = 35 < 50$ biquadrates. ∎

Tables of sums of cubes supplied in 1851 by the lightning calculator Zachariah Dase suggested that, except for 23 and 239, all positive integers were the sum of at most 8 cubes. (Dase later tabulated all the primes between 6 million and 9 million.) Furthermore, all integers beyond 454 require but 7 cubes, and all numbers checked (up to 12,000) beyond 8,042 require just 6 cubes. Since small integers often possess unique peculiarities, it may be more natural to make the following definition:

Definition 10.2: Let $G(k)$ denote the minimal number for which all sufficiently large integers are expressible as the sum of $G(k)$ kth powers.

Although $g(3)$ was shown equal in 1909, that same year Edmund Landau proved $G(3) \le 8$. In 1943, Y.V. Linnik (1915–1972) demonstrated that $G(3) \le 7$. We will see by Theorem 10.3 that $G(3) \ge 4$. Presently, the possibilities of $4, 5, 6$, and 7 are still feasible for the value of $G(3)$.

By definition, $G(k) \le g(k)$ for all k. Clearly $G(2) = g(2) = 4$. The next result gives a lower bound for $G(k)$ due to E. Maillet (1895).

Theorem 10.3: $G(k) \ge k + 1$ for $k \ge 2$.

It is helpful to establish an auxiliary lemma before proceeding.

Lemma 10.3.1: Let R be a positive integer. Then $\sum_{r=0}^{R} \binom{m+r-1}{r} = \binom{m+R}{R}$.

Proof: (Induction on R) Lemma 10.3.1 checks for $R = 0$ and $R = 1$.

Assume that the lemma is true for all $R \le N - 1$. By hypothesis,

$$\sum_{r=0}^{N} \binom{m+r-1}{r} = \sum_{r=0}^{N-1} \binom{m+r-1}{r} + \binom{m+N-1}{N}$$

$$= \binom{m+N-1}{N-1} + \binom{m+N-1}{N}$$

$$= \frac{(m+N)(m+N-1)!}{N!m!} = \binom{m+N}{N}. \qquad \blacksquare$$

Proof of Theorem 10.3: Let $k \ge 2$ be given and let $W(N) =$ the number of positive integers not exceeding N that are expressible as the sum of at most k kth powers.

Let $A(N)$ equal the number of k-tuples $(a_1, \ldots, a_k)$ with

$$0 \le a_1 \le \cdots \le a_k \le \left[N^{1/k} \right] \tag{10.2}$$

such that

$$a_1^k + \cdots + a_k^k \le N$$

Then $W(N) \le A(N)$ for all N. We will complete the proof by showing that $A(N) < N$ for N sufficiently large. Thus for N sufficiently large, there are some integers $n < N$ which require at least $k + 1$ kth powers.

Since $0 \le a_1 \le a_2$, the number of possibilities for a_1 is $\sum_{a_1=0}^{a_2} 1 = a_2 + 1$. The number of possible 2-tuples (a_1, a_2) with $0 \le a_1 \le a_2 \le a_3$ is

$$\sum_{a_2=0}^{a_3} \sum_{a_1=0}^{a_2} 1 = \sum_{a_2=0}^{a_3} (a_2 + 1) = \frac{(a_3+1)(a_3+2)}{2!}$$

The number of possible 3-tuples (a_1, a_2, a_3) satisfying $0 \le a_1 \le a_2 \le a_3 \le a_4$ is

$$\sum_{a_3=0}^{a_4} \sum_{a_2=0}^{a_3} \sum_{a_1=0}^{a_2} 1 = \sum_{a_3=0}^{a_4} \frac{(a_3+1)(a_3+2)}{2!} = \frac{(a_4+1)(a_4+2)(a_4+3)}{3!}$$

Analogously, eq. (10.2) implies that we can express $A(N)$ as

$$A(N) = \sum_{a_k=0}^{[N^{1/k}]} \sum_{a_{k-1}=0}^{a_k} \cdots \sum_{a_1=0}^{a_2} 1.$$

Define $a_{k+1} = [N^{1/k}]$. In general, we wish to demonstrate that for $1 \le m \le k$,

$$\sum_{a_m=0}^{a_{m+1}} \frac{1}{(m-1)!} \prod_{r=1}^{m-1} (a_m + r) = \frac{1}{m!} \prod_{r=1}^{m} (a_{m+1} + r)$$

$$\sum_{a_m=0}^{a_{m+1}} \frac{1}{(m-1)!} \prod_{r=1}^{m-1} (a_m + r)$$

$$= \frac{(m-1)!}{(m-1)!} + \frac{2 \cdots m}{(m-1)!} + \frac{3 \cdots (m+1)}{(m-1)!} + \cdots + \frac{a_{m+1} \cdots (a_{m+1} + m - 1)}{(m-1)!}$$

$$= \sum_{r=0}^{a_m-1} \binom{m+r-1}{r}$$

Now apply Lemma 10.3.1 with $R = a_{m+1}$. So

$$\sum_{a_m=0}^{a_{m+1}} \frac{1}{(m-1)!} \prod_{r=1}^{m-1} (a_m + r) = \binom{m + a_{m+1}}{a_{m+1}} = \frac{1}{m!} \prod_{r=1}^{m} (a_{m+1} + r)$$

In particular, let $m = 1, 2, \ldots, k$ in turn. For $m = k$,

$$\sum_{a_k=0}^{\left[N^{\frac{1}{k}}\right]} \frac{1}{(k-1)!} \prod_{r=1}^{k-1} (a_k + r) = \frac{1}{k!} \prod_{r=1}^{k} \left(\left[N^{\frac{1}{k}}\right] + r \right)$$

Hence, $A(N) = \dfrac{1}{k!} \displaystyle\prod_{r=1}^{k} \left(\left[N^{1/k}\right] + r \right)$

As $N \to \infty$, $A(N) \sim \frac{N}{k!} < \left(\frac{1}{2} + \varepsilon\right)N$ for any $\varepsilon > 0$. It follows that for N sufficiently large, $A(N) < N$. ∎

Theorem 10.3 is about the best lower bound for $G(k)$ available for all k. However, for particular values of k, one can often make improvements. For example, $G(2)$ is actually 4 as we know from considering $n \equiv 7 \pmod 8$. The next proposition (due to A.J. Kempner – 1912) serves as another example.

Proposition 10.4: $G(4) \geq 16$.

Proof: Notice that $a^4 \equiv 0$ or $1 \pmod{16}$ for all a. For some n, if $16n$ is the sum of 15 or fewer biquadrates, then each of the biquadrates itself must be congruent to $0 \pmod{16}$. Hence if $16n = \sum_{k=1}^{15} a_k^4$, then $a_k^4 = 16 b_k^4$ for appropriate integers b_k. It follows that $n = \sum_{k=1}^{15} b_k^4$. But the number 31 is not the sum of 15 or fewer biquadrates. Therefore, $16^m \cdot 31$ is not the sum of 15 or fewer biquadrates for all $m \geq 1$. The result follows immediately. ∎

In fact, in 1939 the British mathematician Harold Davenport (1907–1969) proved that $G(4) = 16$. We complete our discussion of lower bounds for $G(k)$ by extending Proposition 10.4 to higher powers of 2.

Proposition 10.5: $G(2^m) \geq 2^{m+2}$ for all $m \geq 2$.

Proof: Proposition 10.4 establishes the result for $m = 2$. If $m > 2$, then let $k = 2^m$. If a is even, then $2^{m+2} \mid a^{2^m}$ since $2^m > m + 2$.

If a is odd, then $a = 2b + 1$ for some integer b. By the binomial theorem,

$$a^{2^m} = 1 + \binom{2^m}{1}(2b) + \binom{2^m}{2}(2b)^2 + \binom{2^m}{3}(2b)^3 + \cdots$$

$$\equiv 1 + 2^{m+1}b + 2^{m+1}(2^m - 1)b^2 \pmod{2^{m+2}}$$

because 2^{m+2} divides all other terms

$$\equiv 1 + 2^{m+1}(b - b^2) + 2^{2m+1}b^2 \pmod{2^{m+2}}$$

$$\equiv 1 - 2^{m+1}b(b-1) \pmod{2^{m+2}} \text{ since } 2m + 1 > m + 2$$

But $b(b-1)$ is even. Hence $a^{2^m} \equiv 1 \pmod{2^{m+2}}$ for an odd.

So $a^{2^m} \equiv 0$ or $1 \pmod{2^{m+2}}$ for all integers a.

Let n be any odd integer and suppose that $2^{m+2}n$ is the sum of at most $2^{m+2} - 1$ kth powers. By our observation above, each of the kth powers must itself be congruent to $0 \pmod{2^{m+2}}$. Hence if $2^{m+2}n = \sum_{i=1}^{2^{m+2}-1} a_i^k$, then $a_i^k = 2^{m+2}b_i$ for appropriate integers b_i. It follows that a_i is even for all i. But then 2^{k-m-2} divides n. Since $k = 2^m > m - 2$, n is even, a contradiction. Since $2^{m+2}n$ can be made arbitrarily large, the proof is complete. ∎

In the other direction, good upper bounds for $G(k)$ have been obtained by I.M Vinogradov (1947) and later by R.C. Vaughn (1989). The best result currently is due to T. Wooley (1997) who has shown that there is a constant C such that

$$G(k) \leq 2k \log k + 2k \log \log k + Ck \text{ for all } k \geq 2$$

We now turn our attention to the problem of representing natural numbers as the sum or difference of kth powers. Let us begin by introducing appropriate notation:

Definition 10.3: Let $v(k)$ denote the minimal number for which all positive integers are expressible as the sum or difference of at most $v(k)$ kth powers.

Clearly $v(k) \le g(k)$ and hence $v(k)$ exists for all k. Somewhat surprisingly, $v(k)$ is easier to analyze than $g(k)$ or $G(k)$. Theorem 10.6 will establish an upper bound for $v(k)$ for all $k \ge 2$, thus guaranteeing the existence of $v(k)$ independent of Hilbert's solution of the Waring problem for $g(k)$.

Let us evaluate $v(2)$. If n is even, then $n = 2a$ for some integer a and $n = (a+1)^2 - a^2 - 1^2$. If n is odd, then $n = 2a + 1$ for some integer a and $n = (a+1)^2 - a^2$. So $v(2) \le 3$. However, the number 6 cannot be expressed as the sum of two squares. Furthermore, suppose $6 = x^2 - y^2 = (x+y)(x-y)$ for some integers x and y. Since the parity of $x+y$ and $x-y$ are the same, either 6 is odd or 6 is divisible by 4. In either case we have a contradiction.

Therefore, $v(2) = 3$.

Theorem 10.6: For $k \ge 2$, $v(k) \le 2^{k-1} + \frac{k!}{2}$.

The proof relies on a lemma concerning successive differences of kth powers, a separate topic fun to investigate on your own.

Lemma 10.6.1: For fixed $k \ge 2$, let $P(x) = x^k$. Define $P_1(x) = (x+1)^k - x^k$ and $P_{m+1}(x) = P_m(x+1) - P_m(x)$ for $m \ge 1$. Then $P_{k-1}(x) = k! \, x + d$ where d is an integer (independent of x.)

Proof of Lemma: By the binomial theorem, $P_1(x)$ is a polynomial of degree $k - 1$ with leading coefficient k. $P_2(x) = P_1(x+1) - P_1(x)$ is a polynomial of degree $k - 2$ with leading coefficient $k(k-1)$. In general, if $P_m(x)$ is a polynomial of degree $k - m$ with leading coefficient a_m, then $P_{m+1}(x)$ is a polynomial of degree $k - (m+1)$ with leading coefficient $(k - m)a_m$. It follows that $P_{k-1}(x)$ is a polynomial of degree 1 with leading coefficient $k!$. Furthermore, $P_k(x)$ is a constant and so the constant term d in $P_{k-1}(x)$ must be independent of x. ∎

Proof of Theorem: Let n be a positive integer. Choose the integer x such that $k! x + d$ is as close to n as possible. (Note that n may be larger or smaller than $k! x + d$.) So $n = k! x + d + m$ where $|m| \le \frac{k!}{2}$. By definition, $P_{k-1}(x)$ is the sum and difference of at most 2^{k-1} kth powers. But by the lemma, $P_{k-1}(x) = k! x + d$. Since m can be expressed as the sum of at most $\frac{k!}{2}$ 1's or −1's, the result follows. ∎

Theorem 10.6 gives the best upper bound for $k = 2$, but is quite a bit too large in general.

For $k = 3$, Theorem 10.6 implies that $v(3) \leq 7$. Elementary considerations allow us to improve that bound. Notice that for all $n > 1$,

$$n^3 - n = (n-1)n(n+1) \equiv 0 \pmod{6}$$

So there is a positive integer a for which $n = n^3 - 6a$. But

$$6a = (a+1)^3 + (a-1)^3 - 2a^3$$

Hence $n = n^3 - (a+1)^3 - (a-1)^3 + a^3 + a^3$, a sum of five cubes. Therefore, $v(3) \leq 5$.

In the other direction, $a^3 \equiv -1, 0,$ or $1 \pmod 9$ for all integers a. It follows that if $n \equiv \pm 4 \pmod 9$, then n requires the sum or difference of at least four cubes. Therefore, $v(3) \geq 4$. No one has been able to prove whether $v(3)$ is 4 or 5 (though 4 is the more popular conjecture.)

A related result due to L.J. Mordell (1936) makes clever utilization of Pell's equation to establish the following:

Let the integer n be representable as $a^3 + b^3 + c^3 + d^3$ for integers $a, b, c,$ and d where for some ordering $-(a+b)(c+d) > 0$ is not a perfect square and either $a \neq b$ or $c \neq d$. Then n has infinitely many representations as a sum of 4 integral cubes. (Note that some of a, b, c, d can be negative. So there is no need to say "sum or difference".)

Our final result relates $v(k)$ to $G(k)$ and gives an upper bound for $v(k)$ which is usually superior to that given in Theorem 10.6.

Proposition 10.7: For all $k \geq 2, v(k) \leq G(k) + 1$.

Proof: By the definition of $G(k)$, there is an integer $M = M(k)$ such that if m is any integer exceeding M, then m is the sum of $G(k)$ kth powers. Now let n be any natural number. Let x be such that $n + x^k > M$. Then n is the sum or difference of at most $G(k) + 1$ kth powers. ∎

We finish the section with some comments of historical interest. Gauss used the theory of ternary quadratic forms to prove that every number is the sum of at most three triangular numbers. This exciting discovery is cited in his mathematical diary dated July 10, 1796. From this he was able to specify the set of integers expressible as the sum of at most three squares (and later to give the number of such representations). In 1815, Cauchy proved the more general result that all natural numbers are the sum of at most k k-gonal numbers, originally conjectured by Fermat. In fact, Cauchy showed that all but four of the k-gonal numbers could be taken to be 0 or 1. Cauchy's result was subsequently strengthened by other mathematicians. For example, Legendre proved that if $n > 28(k-2)^2$ and k is odd, then n is the sum of at most four k-gonal numbers. Similarly, if $n > 7(k-2)^3$ and k is even, then n is the sum of at most five k-gonal numbers one of which is 0 or 1.

Exercise 10.1

1. Use the proof of Theorem 10.1 to determine the minimal number of sixth powers necessary to represent the number 703.
2. Verify identity eq. (10.1).
3. Show that the method of proof in Proposition 10.2 cannot be strengthened to prove that $g(4) \leq 49$ by considering $a^4 \pmod 6$ for $0 \leq a \leq 5$.
4. Show directly that $v(4) \geq 8$ by considering 4th powers modulo 16. (In fact, $v(4)$ is known to be either 9 or 10.)
5. Characterize all natural numbers that require precisely the sum or difference of two squares.
6. (a) Show that every integer $n \equiv 3 \pmod 6$ is the sum of four integral cubes by verifying the identity (H.W. Richmond – 1922)

    $$k^3 + (-k+4)^3 + (-2k+4)^3 + (2k-5)^3 = 6k+3$$

    (b) Express 3, 9, 15, and 21 as the sum of four integral cubes.
    (c) Show that every integer $n \equiv 7 \pmod{18}$ is the sum of four integral cubes by verifying the identity

    $$(k+2)^3 + (6k-1)^3 + (8k-2)^3 + (-9k+2)^3 = 18k+7$$

    (d) Express −11, 7, 25, and 331 as the sum of four integral cubes.
7. With the notation of Lemma 10.6.1, show that $P_k(x) = k!$ for all x.
8. Let $P(x) = x^k$, $R_1(x) = (x+r)^k - x^k$, and $R_{m+1}(x) = R_m(x+r) - R_m(x)$. Find a formula for $R_k(x)$.
9. (a) Verify that $d = 1$ if $k = 2$, $d = 6$ if $k = 3$, and $d = 36$ if $k = 4$ in Lemma 9.6.1.
    (b) Show that $d = (k-1)\frac{k!}{2}$ for all k.
10. Compare Theorem 10.6 and Proposition 10.7 for $2 \leq k \leq 4$ utilizing all known results for $G(k)$ in this range.
11. (a) Show that there are infinitely many ways to express the number 1 as a sum of three integral cubes by verifying the following identity:

    $$1 = (9n^4)^3 + (1 - 9n^3)^3 + (3n)^3(1 - 3n^3)^3$$

(b) It has been shown that 2 can be expressed as the sum of three cubes in infinitely many ways. However, $3 = 1^3 + 1^3 + 1^3 = 4^3 + 4^3 + (-5)^3$ are the only known representations for the number 3. Verify that there are no more such representations of $3 = a^3 + b^3 + c^3$ for $|a|, |b|, |c| < 20$.

12. (a) Verify that all natural numbers less than 100 are the sum of at most three triangular numbers.

(b) Verify that all natural numbers less than 100 are the sum of at most five pentagonal numbers.

13. Verify that 8042 requires the sum of seven cubes. It is the largest known number requiring seven cubes.

10.2 Schnirelmann density and the $\alpha + \beta$ theorem

In this section we view additive number theory from the viewpoint of set density. The density of a set of natural numbers is a measure of its prevalence versus the entire set **N**. What properties can we infer about a set by simply knowing its density? Does the density of a set tell us anything about the representation of integers as sums of terms from the set? These are the sort of questions typically addressed in this area.

Let A denote a set of nonnegative integers. We will assume throughout this section that $0 \in A$ for all sets considered. We think of A as a sequence of integers written in ascending order. Let $A(n) = A \cap \{0, 1, \ldots, n\}$ be the set of all elements of A not exceeding n and let $|A(n)|$ be the number of positive integers in $A(n)$. For example, if $A = \{0, 3, 6, 9, 12, \ldots\}$ and $n = 10$, then $A(10) = \{0, 3, 6, 9\}$ and $|A(10)| = 3$. Below we define one form of density due to the Russian mathematician L.G. Schnirelmann (1905–1938).

Definition 10.4: The *(Schnirelmann) density* of the set A is defined to be

$$d(A) = \inf_{n \geq 1} \frac{|A(n)|}{n}$$

Notice that the Schnirelmann density is very sensitive to the beginning of the sequence A. In particular, if $1 \notin A$, then $d(A) = 0$. If $r \notin A$, then $d(A) \leq \frac{r-1}{r}$. For example, if $O = \{0, 1, 3, 5, 7, \ldots\}$ is the set of odd numbers, then $d(O) = 1/2$. If $S = \{0, 1, 4, 9, 16, \ldots\}$ is the set of squares, then $d(S) = 0$. In addition, $d(A) = 1$ if and only if A contains all natural numbers.

If A and B are two sets of nonnegative integers, then the set $A + B$ is defined as $A + B = \{a + b : a \in A \text{ and } b \in B\}$. Note that since $0 \in A \cap B$, $A \cup B \subset A + B$. For example, if $E = \{0, 2, 4, 6, \ldots\}$ is the set of all nonnegative even integers and $F = \{0, 5, 10, 15, \ldots\}$ is the set of nonnegative integers divisible by 5, then $E + F$ consists of all nonnegative integers except for 1 and 3. Furthermore, we define $2A$ as $A + A$, and in general, $kA = (k-1)A + A$ for $k > 2$.

Much of Section 10.1 could be recast in this new light. Let $N_0 = N \cup \{0\}$ denote the set of nonnegative integers. If S is the set of squares defined above, then Lagrange's theorem (Theorem 5.8) establishes that $4S = N_0$. Of course, $kS = N_0$ for any $k \geq 4$. Similarly, since $g(3) = 9$, $9C = N_0$ where $C = \{0, 1, 8, 27, 64, \ldots\}$ is the set of cubes

Definition 10.5: The set A is a *basis of order* k for N_0 if $kA = N_0$.

It follows that the squares form a basis of order 4 for the nonnegative integers. Similarly, the cubes form a basis of order 9. Hilbert's theorem implies that for any $k \geq 2$, the set of kth powers forms a basis of N_0. Note that the order of a basis A is not defined uniquely since we do not require k to be the minimal value for which $kA = N_0$.

A natural consequence of the definition of density is the following elegant formula due to Schnirelmann.

Proposition 10.8: If A and B are sets of nonnegative integers with $0 \in A \cap B$, then

$$d(A + B) \geq d(A) + d(B) - d(A)d(B) \tag{10.3}$$

The form of formula (10.3) should be reminiscent of other formulas you know. For example, if A and B are finite sets with $|S|$ denoting the number of elements in the set S, then $|A \cup B| = |A| + |B| - |A \cap B|$. Analogously, if A and B are independent events and $p(E)$ represents the probability that E occurs, then $p(A \text{ or } B) = p(A) + p(B) - p(A)p(B)$. Here is the proof of Proposition 10.8.

Proof: Let $d(A) = \alpha$, $d(B) = \beta$, $A + B = C$, and $d(C) = \gamma$. Let n be a positive integer and let $A(n) = A \cap \{0, 1, \ldots, n\}$ as defined previously. All $|A(n)|$ members of A also belong to C. Let a_k and a_{k+1} be successive numbers in $A(n)$. Here $a_0 = 0$ and we append $a_K = n + 1$ where $K = |A(n)| + 1$. There are $a_{k+1} - a_k - 1 = r_k$ numbers between a_k and a_{k+1}, none of which belongs to A. If $b \in B$ with $1 \leq b \leq r_k$, then $a_k + b \in C$. So the segment between a_k and a_{k+1} contains at least $|B(r_k)|$ numbers that belong to C. It follows that

$$|C(n)| \geq |A(n)| + \sum_{k=0}^{K-1} |B(r_k)|$$

as k runs through all segments between numbers of A. (If $a_{k+1} = a_k + 1$ for some k, then the segment is the empty set and $r_k = 0 = |B(0)|$.) By the definition of density, $|B(r_k)| \geq \beta r_k$ for all r_k. Hence

$$|C(n)| \geq |A(n)| + \beta(n - |A(n)|)$$

since

$$\sum_{k=0}^{K-1} r_k = n - |A(n)|$$

But $|A(n)| \geq an$. Thus

$$|C(n)| \geq |A(n)|(1-\beta) + \beta n \geq an(1-\beta) + \beta n$$

Thus,

$$\frac{C(n)}{n} \geq \alpha + \beta - \alpha\beta$$

Since n is arbitrary and $\gamma \geq \frac{|C(n)|}{n}$ for all n,

$$d(A+B) \geq d(A) + d(B) - d(A)d(B) \qquad \blacksquare$$

By induction, Proposition 10.8 can readily be extended to a sum of any finite number of sets.

Corollary 10.8.1: Let $A_1, A_2, \ldots, A_n$ be a collection of sets of nonnegative integers with $0 \in A_k$ for all $k = 1, \ldots, n$. Then

$$d\left(\sum_{k=1}^{n} A_k\right) \geq 1 - \prod_{k=1}^{n}(1 - d(A_k))$$

An important application of Proposition 10.8 is the following result, due to Schnirelmann.

Theorem 10.9: If A is a set of nonnegative integers with positive density, then A is a basis of N_0.

Proof: Let A be a set with $d(A) = \alpha > 0$. By Corollary 10.8.1,

$$d(kA) \geq 1 - (1 - \alpha)^k$$

But $1 - \alpha < 1$ implies that there exists a K sufficiently large such that $d(KA) > 1/2$. Hence for any positive integer m,

$$|(KA)(m)| > m/2 \qquad (10.4)$$

We will show that A forms a basis of order $2K$.

Let n be a positive integer. We wish to show that $n \in KA + KA = 2\,KA$. If $n \in KA$, then $n \in 2\,KA$. Suppose that $n \notin KA$. Let $r = |KA(n)|$. By inequality (10.4), $|(KA(n)| > n/2$.

Let $0 < a_1 < \cdots < a_r \leq n - 1$ be all the elements of KA between 1 and n, inclusive. Consider the set $S = \{a_1, \ldots, a_r, n - a_1, \ldots, n - a_r\}$. Since there are more than n elements of S, by the Pigeonhole Principle there must be an i and j such that $a_i = n - a_j$. But then $n = a_i + a_j$. Hence, $n \in KA + KA = 2\,KA$. Since n is arbitrary, A forms a basis of N_0. ∎

Schnirelmann (1930) showed that the set of primes $P = \{0, 1, 2, 3, 5, 7, 11, \ldots\}$ is such that $2P$ has positive density. It follows from Theorem 10.9 that P forms a basis for N_0.

What is the order of the basis P? If Goldbach's Conjecture is true (all even integers greater than 2 are the sum of two primes), then P forms a basis of order 3. In fact, using different analytical methods, the Russian mathematician I.M. Vinogradov (1937) showed that every sufficiently large integer is the sum of at most three primes. We say that the primes form an *asymptotic basis* of order 3. The theorem of Vinogradov is an astounding result even though "sufficiently large" is not effectively computable. In 1966, the Chinese mathematician J.R. Chen proved that every sufficiently large integer is the sum of a prime and an integer having at most two prime factors. Chen's proof involves the large sieve inequality and deep analysis. Unfortunately, it appears that Chen's methods cannot be directly extended to prove Goldbach's Conjecture. More recently, A. Perelli and J. Pintz (1993) proved that if N is sufficiently large, almost all even integers in the interval $(N - K, N]$ are expressible as the sum of two primes as long as $K > N^{\frac{7}{36} + \varepsilon}$.

In 1931, Schnirelmann and Landau noticed that for all known examples, that in fact, $d(A + B) \geq d(A) + d(B)$ as long as the right side is at most 1. After many partial proofs of this conjecture, the American H. B Mann established the full result (1942).

Theorem 10.10 (Mann's $a + \beta$ Theorem): If A and B are sets of nonnegative integers with $0 \in A \cap B$ and $C = A + B$, then

$$d(C) \geq \min\{1, \, d(A) + d(B)\} \tag{10.5}$$

Notice that Theorem 10.10 completely supplants Proposition 10.8.

Currently there are several different proofs of Theorem 10.10. We follow one presented by I. Niven, H. S. Zuckerman, and H.L. Montgomery [1991] with some modifications. Let us begin with two useful lemmas.

Lemma 10.10.1: Let n be a positive integer and A and B be sets of nonnegative integers with $B(n) \subset A(n), 0 \in A$, and $|B(n)| \geq 1$. Let there be a constant θ with $0 < \theta \leq 1$ for which

$$|A(k)| + |B(k)| \geq \theta k \text{ for } k = 1, \ldots, n \tag{10.6}$$

Define a_0 to be the smallest integer in $A(n)$ such that there exists $ab \in B(n)$ with $a_0 + b \notin A(n)$.
(i) Suppose there are integers b and d for which $b \in B(n)$ and $d - a_0 < b \leq d \leq n$. Then for all $a \in A(n)$ satisfying $1 \leq a \leq d - b$, $a + b \in A(n)$ and $|A(d)| \geq |A(b)| + |A(d - b)|$.
(ii) If $1 \leq m \leq a_0$, then $|A(m)| \geq \theta m$.

Proof: The existence of a_0 is guaranteed by the fact $|B(n)| \geq 1$; for the sum of any positive integer in $B(n)$ with the largest element of $A(n)$ is not in $A(n)$.

(i) Notice that $d - b < a_0$ and hence if $1 \leq a \leq d - b$, then $a < a_0$. Hence $a + b \in A(n)$ for all such a by the minimality of a_0. There are $|A(d-b)|$ positive integers $a \in A$ with $a \leq d - b$. For each such $a, a + b \in A(n)$ by the above and $b < a + b \leq d$. Note that the number of elements of $a \in A$ with $b < a + b \leq d$ is $|A(d)| - |A(b)|$. Therefore,

$$|A(d)| - |A(b)| \geq |A(d-b)|$$

(ii) Let r be the smallest integer not exceeding n such that $|A(r)| < \theta r$ (if there is such an r). Condition (10.6) implies that $|B(r)| > 0$. Hence there exists a $b \in B$ with $0 < b \leq r \leq n$.

If $r \leq a_0$, then we may let $d = r$ in part (i) obtaining

$$|A(r)| \geq |A(b)| + |A(r-b)| \tag{10.7}$$

By hypothesis, $B(n) \subset A(n)$. Hence $|A(b)| = |A(b-1)| + 1$. But $|A(b-1)| \geq \theta(b-1)$ since $b - 1 < r$. Also $|A(r-b)| \geq \theta(r-b)$ since $b > 0$. By eq. (10.7), we get $|A(r)| \geq \theta(b-1) + 1 + \theta(r-b) = \theta r + 1 - \theta \geq \theta r$, a contradiction. Thus $r > a_0$ and the result follows. ∎

Lemma 10.10.2: Let n be a positive integer and A and B be sets of nonnegative integers with $0 \in A \cap B$ and $|B(n)| \geq 1$. Set $C = A + B$. If there exists a constant θ with $0 < \theta \leq 1$ satisfying inequalities (10.6), then there are sets A' and B' with (i) $A' + B' \subset C$, (iii) $|B'(n)| < |B(n)|$, and (iii) $|A'(k)| + |B'(k)| \geq \theta k$ for $k = 1, \ldots, n$.

Proof: Assume $B(n) \not\subset A(n)$. In this case let $A' = A \cup B$ and $B' = A \cap B$. (i) Let $a' \in A'$ and $b' \in B'$. So $b' \in A$ and $b' \in B$. If $a' \in A$, then $a' + b' \in A + B$. If $a' \in B$, then $a' + b' \in B + A = A + B$. So $A' + B' \subset C$.

Furthermore, $|B'(n)| < |B(n)|$ since $B(n) \not\subset A(n)$. Hence (ii) follows.
Finally, for $1 \leq k \leq n$, $|A'(k)| = |A(k)| + |B(k)| - |A(k) \cap B(k)|$ and $|B'(k)| = |A(k) \cap B(k)|$. Hence $|A'(k)| + |B'(k)| = |A(k)| + |B(k)| \geq \theta k$, which establishes (iii).

Now assume that $B(n) \subset A(n)$. Let a_0 be as defined in Lemma 10.10.1.
Let $B' = \{b \in B(n): a_0 + b \in A(n)\}$ and let $A^* = \{a_0 + b: b \in B(n), b \notin B', \text{ and } a_0 + b \leq n\}$.
Certainly $0 \in B'$. It is possible that $A^*(n)$ is empty, but in any event, $A(n)$ and $A^*(n)$ are disjoint. Now let $A' = A \cup A^*$.

(i) If $c \in A' + B'$, then $c = a + b$ where $a \in A'$ and $b \in B'$. If $a \in A$ then $c \in A + B$ since $B' \subset B$. If $a \in A^*(n)$, then $a = a_0 + b_0$ with $b_0 \in B(n)$ but $b_0 \notin B'$. Hence $c = (a_0 + b) + b_0$. But $a_0 + b \in A(n)$ since $b \in B'$. So $A' + B' \subset C$.

Since there is a value $b > 0$ with $b \in B(n)$ for which $a_0 + b \notin A(n)$, it is the case that $B(n)$ contains at least one positive integer that is not in B'.
Therefore, $|B'(n)| < |B(n)|$, which establishes (ii).

We now establish part (iii). Let $B^* = \{b \in B(n):a_0 + b \notin A(n)\}$ be the set difference $B(n) - B'$. Notice that for $k = 1, \ldots, n$, $|A'(k)| = |A(k)| + |A^*(k)|$ and $|B'(k)| = |B(k)| - |B^*(k)|$. Further, $|A^*(k)| = |B^*(k - a_0)|$ since there is a one-to-one correspondence between elements of $A^*(k)$ and $B^*(k - a_0)$. Hence,

$$|A'(k)| + |B'(k)| = |A(k)| + |A^*(k)| + |B(k)| - |B^*(k)|$$
$$= |A(k)| + |B(k)| - (|B^*(k)| - |B^*(k - a_0)|)$$

It follows that (iii) holds for all k for which $|B^*(k)| = |B^*(k - a_0)|$. Now let $k \leq n$ be such that $|B^*(k)| > |B^*(k - a_0)|$. Then $B(k) - B(k - a_0) \geq 1$. Let b_s be the smallest element of B such that $k - a_0 < b_s \leq k$. Since

$$|B(k)| - |B(k - a_0)| \geq |B^*(k)| - |B^*(k - a_0)|$$

it follows that

$$|A'(k)| + |B'(k)| \geq |A(k)| + |B(k)| - (|B(k)| - |B(k - a_0)|)$$
$$= |A(k)| + |B(k - a_0)|$$

But $B(k - a_0) = B(b_s - 1)$ by the definition of b_s. Hence

$$|A'(k)| + |B'(k)| \geq |A(k)| + |B(b_s - 1)| \tag{10.8}$$

But $k - a_0 < b_s \leq k \leq n$. Apply Lemma 10.10.1(i) with $d = k$ and $b = b_s$. We get $|A(k)| \geq |A(b_s)| + |A(k - b_s)|$. In addition, $k - b_s < a_0$. By Lemma 10.10.1(ii), $|A(k - b_s)| \geq \theta(k - b_s)$. So eq. (10.8) becomes

$$|A'(k)| + |B'(k)| \geq |A(b_s)| + \theta(k - b_s) + |B(b_s - 1)|$$

But $b_s \in B(n) \subset A(n)$. Hence $|A(b_s)| = |A(b_s - 1)| + 1$. So

$$|A'(k)| + |B'(k)| \geq \theta(k - b_s) + (|A(b_s - 1)| + |B(b_s - 1)|) + 1$$
$$\geq \theta(k - b_s) + \theta(b_s - 1) + 1 = \theta k + 1 - \theta \geq \theta k$$

This establishes the lemma. ∎

Proof of Theorem 10.10: Let n be a nonnegative integer. Let $d(A) = \alpha$ and $d(B) = \beta$. By definition of density, $|A(k)| \geq ak$ and $|B(k)| \geq \beta k$ for all $k \geq 0$. Let $\theta = min\{1, \alpha + \beta\}$. We will show that

$$|C(n)| \geq \theta n \tag{10.9}$$

Since n is arbitrary, it will follow that $\frac{|C(n)|}{n} \geq \theta$ for all n. Hence eq. (10.5) holds and the theorem is proved.

Notice that inequality (10.6) holds for A and B with $k = 1, \ldots, n$ by the definition of θ. Let $|B(n)| = m \leq n$. We will prove that inequality (10.9) holds by induction on m. If $m = 0$, then $B(n) = \{0\}$ and $C(n) = A(n)$. Since $|A(n)| = |A(n)| + |B(n)| \geq \theta n$, (10.9) holds as well.

Now suppose that $M \geq 1$ and that $|(Q + B)(n)| \geq \theta n$ holds for any Q and B satisfying (10.6) for which $|B(n)| = m < M$. Let $|B(n)| = M$. By Lemma 10.10.2, there exist sets A' and B' with $A' + B' \subset C, |B'(n)| < |B(n)|$, and $|A'(k)| + |B'(k)| \geq \theta k$ for all $k \leq n$. Since $|B'(n)| < M$, by the inductive hypothesis $|(A' + B')(n)| \geq \theta n$. But $A' + B' \subset C$ implies that $|C(n)| \geq \theta n$. ∎

The $\alpha + \beta$ theorem implies that if $0 < d(A) < 1$ and $d(B) > 0$, then $d(A + B) > d(A)$. There are also some sets B with $d(B) = 0$ such that $d(A + B) > d(A)$ for any set A with $0 < d(A) < 1$.

Definition 10.6: If B is a set of nonnegative integers for which $d(A + B) > d(A)$ for any set A with $0 < d(A) < 1$, then B is called an *essential component*.

An interesting result of P. Erdös (1936) states that every basis is an essential component. In particular, the set of squares form an essential component as do the set of primes. In addition, sets that are essential components but not bases have also been constructed.

There are many interesting extensions to Mann's $\alpha + \beta$ theorem. For example, Luther Cheo (1952) proved that if $\alpha, \beta > 0$, and $\alpha + \beta \leq \gamma \leq 1$, then there are sets A and B of density α and β respectively such that $d(A + B) = \gamma$. There is still much uncharted territory here.

It should be mentioned that there are other definitions of density that have proven useful. In Definition 8.6, we defined the natural density of the set A to be $\lim_{n \to \infty} \frac{|A(n)|}{n}$. Of course, not every set has a natural density since the limit may not exist. However, when the natural density exists, it provides a reasonable interpretation for the probability of a random natural number being in the set. One other form of density is asymptotic density.

Definition 10.7: Let A be a set of nonnegative integers. The *asymptotic density* of A is defined as

$$\delta(A) = \lim_{n \to \infty} \inf \frac{|A(n)|}{n}$$

Asymptotic density is not as sensitive as Schnirelmann density to the initial terms of a sequence but still applies in many situations where there is no natural density. If A has a natural density, $\delta_n(A)$, then $\delta(A) = \delta_n(A)$. Furthermore, it is always the case that $0 \leq d(A) \leq \delta(A) \leq 1$. Here are two examples:

If $E = \{0, 2, 4, 6, \ldots\}$ is the set of even integers in N_0, then $d(E) = 0$ while $\delta(E) = \delta_n(E) = 1/2$. If A is the set with $0 \in A$ and $a > 0$ in A if and only if there is a $k \geq 0$ such that $3 \cdot 2^{k-1} < a \leq 2^{k+1}$, then $d(A) = 0$, $\delta(A) = 1/3$, and $\delta_n(A)$ does not exist (Exercise 10.2.2).

Some final comments are in order. In 1974, in a splendid tour de force E. Szemerédi proved the long-standing conjecture that if $\lim_{n \to \infty} \sup \frac{|A(n)|}{n} > 0$, then the set A contains arbitrarily long arithmetic progressions. Previously, K. Roth (1953) had shown that A must contain a 3-term arithmetic progression, and E. Szemerédi (1967) had shown that A must contain a 4-term arithmetic progression. By a spectacular extension of Szemerédi's theorem, Ben Green and Terrence Tao (2008) proved the long-held belief that the set of primes contains arbitrarily long arithmetic progressions! In particular, they proved that if A is a subset of the primes for which

$$\underset{N \to \infty}{\text{limsup}} \frac{|A(n)|}{\pi(n)} > 0$$

where $\pi(n)$ is the prime counting function, then for all positive integers k, the set A contains infinitely many arithmetic progressions of length k. Since then, similar theorems have been proven dealing with Gaussian primes and with various polynomial progressions.

There is a perpetual search for long arithmetic strings of primes. The 21-term prime arithmetic progression $142072321123 + 1419763024680k$ for $0 \leq k \leq 20$ was found by Andrew Moran and Paul Pritchard (1995). This was extended by P. Pritchard et al who discovered a 22-term prime arithmetic progression $11410337850553 + 4609098694200k$ for $0 \leq k \leq 21$.) The current record for constructing a long sequence of consecutive primes is due to Rob Gahn and a distributed search group called Prime-Grid. In 2019, they announced that the 27 numbers

$$224, 584, 605, 939, 537, 911 + 81, 292, 139 \cdot 23\# \cdot n$$

for $n = 0, 1, \ldots, 26$ are all prime where p# denotes the product of all primes up to p (called p *primorial.*)

Exercise 10.2

1. Determine the Schnirelmann densities of the following sets:
   (a) $H = \{0, 1, 4, 7, 10, 13, 16, 19 \ldots\}$, numbers congruent to 1 (mod 3).
   (b) $S = \{0, 1, 3, 10, 17, 24, 31, 38 \ldots\}$, numbers congruent to 3 (mod 7).
   (c) $F = \{0, 1, 2, 3, 5, 8, 13, 21, \ldots\}$, the set of Fibonacci numbers.
   (d) $T = \{0, 1, 3, 6, 10, 15, 21, 28 \ldots\}$, the set of triangular numbers.

2. Show that if $= \{a \geq 1:$ there is a $k \geq 0$ with $3 \cdot 2^{k-1} < a \leq 2^{k+1}\}$, then $d(A) = 0$, $\delta(A) = 1/3$, and $\delta_n(A)$ does not exist.

3. Prove Corollary 10.8.1. (Hint: Rewrite Proposition 10.8 as
    $1 - d(A + B) \leq (1 - d(A))(1 - d(B))$ and then use induction.)

4. Use Corollary 9.13.1 to show that if $P = \{0, 1, 2, 3, 5, 7, 11, \ldots\}$ is the set of primes, then $d(P) = 0$.

5. Use Corollary 9.10.5 to show that every positive integer is either square-free or the sum of two square-free integers.

6. Let a_n be the nth positive integer in A. Show that is A has a natural density $\delta_n(A)$, then $\delta_n(A) = \liminf_{n \to \infty} \frac{n}{a_n}$.

7. A set S is said to be *complete* if every positive integer can be expressed as a sum of distinct elements of S.

    (a) Show that $S = \{0, 1, 2, 4, 8, 16, \ldots\}$ is complete. Is S a basis for N_0?

    (b) Let $F = \{0, 1, 1, 2, 3, 5, 8, 13, \ldots\}$ be the sequence of Fibonacci numbers. Let $F - \{1\} = \{0, 1, 2, 3, 5, 8, 13, \ldots\}$. Show that $F - \{1\}$ is complete. Is $F - \{1\}$ a basis for N_0?

    (c) Show that $F - \{F_k\}$ where F_k is the kth Fibonacci number is complete.

    (d) Show that $F - \{F_j, F_k\}$ with $j < k$ is not complete.

    (e) Prove Zeckendorf's Theorem (1951): If n is a positive integer, then n can be written as sum of distinct elements of $F - \{1\}$ no two being consecutive.

8. A set S is *asymptotically complete* if every sufficiently large integer is the sum of distinct elements of S. Explain why the following sets are asymptotically complete:

    (a) The set of triangular numbers (see Exercise 5.4.11).

    (b) The set of squares (see Exercise 5.4.7).

    (c) The set of primes (see Corollary 9.15.1).

9. Determine all integers n less than or equal to 210 for which if $n/2 < p < n$ is prime, then $n - p$ is prime. (J.-M. Deshouillers, A. Granville, W. Narkiewicz, and C. Pomerance have proven that 210 is the largest integer with this property – 1993.)

10.3 Van der Waerden's theorem

The Dutch mathematician P.J.H. Baudet conjectured that if $\mathbf{N}$ is partitioned in any way into two classes, then at least one of the classes must contain an arithmetic progression of ℓ terms for arbitrarily large ℓ. Many mathematicians tried unsuccessfully to prove the conjecture. In 1926, another Dutch mathematician B.L. van der Waerden proved the result (with some assistance from Emil Artin and Otto Schreier.) In fact, he proved somewhat more by showing that (i) instead of $\mathbf{N}$ an appropriate finite subset of consecutive integers would do, and (ii) the subset could be partitioned into an arbitrary number of classes without disturbing the conclusion. As is often the case, the generalization of a mathematical statement can be more readily proved than can the original statement.

Theorem 10.11 (van der Waerden's Theorem): Let $k, l \epsilon N$. Then there is an $n = n(k, l)$ such that if a sequence of n consecutive integers is partitioned into at most k classes, then at least one of the classes contains an arithmetic progression of length l.

Let $\mathbf{N}(n) = \{1, 2, 3, \ldots, n\}$. Clearly it is sufficient to prove the theorem for $N(n)$ since all other sequences of n consecutive integers are simply a linear translation of $\mathbf{N}(n)$. If a and b are in the same class we will write $a \sim b$. If $S = \{a+1, \ldots, a+t\}$ and $S' = \{b+1, \ldots, b+t\}$ are sequences of equal length with $a+i \sim b+i$ for $1 \le i \le t$, then we write $S \sim S'$ and say S and S' are of the same type. In addition, we say that sequences $S_1, \ldots, S_m$ of equal length form an arithmetic progression if their initial elements (or any corresponding element for that matter) form an arithmetic progression. Below we present a proof of van der Waerden's theorem due to M.A. Lukomskaya as beautifully described by Khinchin [68].

Proof: (Induction on) If $l = 2$, then for any k let $n = n(k, l) = k + 1$. By the pigeonhole principle, there is a class with at least two elements. Those two elements are necessarily in arithmetic progression.

Now assume the theorem is true for some $l \ge 2$ and all k. We will show that it must be true for $l' = l + 1$ and all k by actually constructing an $n = n(k, l+1)$ for any given k. To this end define $r_0, r_1, r_2, \ldots$ and $n_0, n_1, n_2, \ldots$ recursively

$$r_0 = 1, n_0 = n(k, l), r_m = 2n_{m-1}r_{m-1}, n_m = n(k^{r_m}, l) \text{ for } m = 1, \ldots, k \qquad (10.10)$$

Notice that n_m is well defined for all m by our inductive hypothesis. It will be shown that r_k suffices for $n(k, +l)$.

Let $n = r_k$ and partition $\mathbf{N}(n)$ into at most k classes (equivalently, into k classes where we allow empty classes.) View $\mathbf{N}(n)$ as being made up of $2n_{k-1}$ segments of length r_{k-1}. The left half of $\mathbf{N}(n)$, namely $\mathbf{N}(n/2)$, is made up of n_{k-1} segments of length r_{k-1}. There are $k^{r_{k-1}}$ possible types of segments of length r_{k-1} since each number must be from one of k possible classes. By eq. (10.10), $n_{k-1} = n(k^{r_{k-1}}, l)$. Hence the left half of $\mathbf{N}(n)$ contains an arithmetic progression of l segments $S_1, \ldots, S_l$ of the same type. Let d_1 be the difference between the first term of S_1 and S_2 (or equivalently between corresponding terms of any consecutive segments.)

If $S_l = \{a_l + 1, \ldots, a_l + r_{k-1}\}$, then define $S_{l'}$ as $\{a_l + 1 + d_1, \ldots, a_l + r_{k-1} + d_1\}$. The segments $S_1, \ldots, S_{l'}$ form an arithmetic progression of $l+1$ segments all lying within $\mathbf{N}(n)$, although $S_{l'}$ may well be of another type.

Consider S_1. S_1 is of length $r_{k-1} = 2n_{k-2}r_{k-2}$ by eq. (10.10). The left half of S_1 consists of n_{k-2} segments of length r_{k-2}. There are $k^{r_{k-2}}$ possible types of segments of length r_{k-2}. But $n_{k-2} = n(k^{r_{k-2}}, l)$. Hence the left half of S_1 contains an arithmetic progression of l segments $S_{11}, \ldots, S_{1l}$ of the same type. Let $d_2 = a_{12} - a_{11}$ be the difference between the first term of S_{11} and S_{12}. If $S_{1l} = \{a_{1l} + 1, \ldots, a_{1l} + r_{k-2}\}$, then define $S_{1l'} = \{a_{1l} + 1 + d_2, \ldots, a_{1l} + r_{k-2} + d_2\}$. The segments $S_{11}, \ldots, S_{1l'}$ form an arithmetic progression of $l+1$ segments all lying within S_1. Again we make no assertions about the type of the segment $S_{1l'}$.

Next carry over the construction congruently to each S_m for $m = 2, \ldots, l'$. For example, $S_{21} = \{a_{11} + 1 + d_1, \ldots, a_{11} + r_{k-2} + d_1\}$. The segments $S_{m1}, \ldots, S_{ml'}$ form an arithmetic progression of $l+1$ segments all lying within S_m for $1 \le m \le l'$. At this point we have constructed $l+1$ sequences of $l+1$ segments of length r_{k-2}.

Now consider S_{11}. S_{11} is of length r_{k-2}. The left half of S_{11} consists of n_{k-3} segments of length r_{k-3}. Similar to what we observed before, the left half of S_{11} contains an arithmetic progression of l segments $S_{111}, \ldots, S_{11\ell}$ of the same type. Let d_3 be the difference between the first term of S_{111} and S_{112}. If $S_{11\ell} = \{a_{11\ell} + 1 + d_3, \ldots, a_{11\ell} + r_{k-3} + d_3\}$, then the segment $S_{11\ell'} = \{a_{11\ell} + 1 + d_3, \ldots, a_{11\ell} + r_{k-3} + d_3\}$ forms an arithmetic progression of length $l+1$ lying entirely within S_{11}. Again, carry over this construction congruently to each $S_{m_1 m_2}$ with $1 \le m_1, m_2 \le l+1$. At this point we have $(l+1)^2$ sequences of $l+1$ segments of length r_{k-3}.

Repeat this procedure k times. In the last step we obtain an arithmetic progression of $l+1$ segments $S_{11\ldots 11} \ldots, S_{11\ldots 1l}$ of length $r_0 = 1$, the first l terms being of the same type (i.e., in the same class.) Finally carry out this procedure congruently to all previous segments obtaining $(l+1)^k$ sequences of $l+1$ segments of length 1. Furthermore, for $1 \le j \le k$ and $1 \le m_1, \ldots, m_j, m'_1, \ldots, m'_j \le l$,

$$S_{m_1 \ldots m_j} \sim S_{m'_1 \ldots m'_j} \tag{10.11}$$

In eq. (10.11), if $j < k$ and $m_{j+1}, \ldots, m_{j+k}$ are any indices from among $1, \ldots, l'$, then the number $S_{m_1 \ldots m_j m_{j+1} \ldots m_{j+k}}$ appears in the same position in the segment $S_{m_1 \ldots m_j}$ as does the number $S_{m'_1 \ldots m'_j m_{j+1} \ldots m_k}$ in the segment $S_{m'_1 \ldots m'_j}$. By eq. (10.11), these segments are of the same type. Hence for $1 \le m_1, \ldots, m_j, m'_1, \ldots, m'_j \le l$ and $1 \le m_{j+1}, \ldots, m_k \le l+1$,

$$S_{m_1 \ldots m_j m_{j+1} \ldots m_k} \sim S_{m'_1 \ldots m'_j m_{j+1} \ldots m_k} \tag{10.12}$$

If $j \le k$ and $m'_j = : m_j + 1$, then $S_{m_1 \ldots m_{j-1} m'_j}$ and $S_{m_1 \ldots m_{j-1} m_j}$ appear in the same position in successive segments. It follows that

$$S_{m'_1 \ldots m'_j m_{j+1} \ldots m_k} - S_{m_1 \ldots m_j m_{j+1} \ldots m_k} = d_j \tag{10.13}$$

Now consider the $k+1$ numbers

$$a_0 = S_{11 \ldots 111}, a_1 = S_{11 \ldots 11l'}, a_2 = S_{11 \ldots 1l'l'}, \ldots, a_k = S_{l'l' \ldots l'l'}$$

where the subscript of S in a_m contains $k - m$ 1's followed by m l's.

By the pigeonhole principle, there are two numbers, say a_r and a_s where $a_r \sim a_s$ and $0 \le r < s \le k$ (hence $a_r < a_s$).

Next we consider the $l+1$ numbers

$$n_1 = \underbrace{S_{1\ldots1}}_{k-s} \underbrace{1\ldots1}_{s-r} \underbrace{\ell\ldots\ell}_{r}, n_2 = \underbrace{S_{1\ldots1}}_{k-s} \underbrace{2\ldots2}_{s-r} \underbrace{\ell\ldots\ell}_{r} \cdots, n_{\ell'} = \underbrace{S_{1\ldots1}}_{k-s} \underbrace{\ell\ldots\ell}_{s-r} \underbrace{\ell\ldots\ell}_{r}.$$

Notice that $n_1 = a_r$ and $n_{\ell'} = a_s$. Furthermore, by eq. (10.12), $n_1 \sim n_2 \sim \ldots \sim n_l$.

(We cannot guarantee that n_l is of the same type from eq. (10.12) since it may be that $s = k$.) But $a_r \sim a_s$. Hence all $l+1$ numbers n_i ($1 \le i \le l+1$) are from the same class. It remains to show that $n_1, \ldots, n_l$ are in arithmetic progression.

Let $i' = i+1$ and let $a_{i,m} = \underbrace{S_{1\ldots1}}_{k-s} \underbrace{i\ldots i}_{s-r-m} \underbrace{i\ldots i}_{m} \underbrace{\ell\ldots\ell}_{r}$ for $0 \le m \le s-r$.

Then $a_{i,0} = n_i$ and $a_{i,s-r} = n_{i'}$ for any $i = 1, \ldots, l$. We have that

$$n_{i'} - n_i = \sum_{m=1}^{s-r} (a_{i,m} - a_{i,m-1})$$

By eq. (10.13), $a_{i,m} - a_{i,m-1} = d_{k-r-m+1}$ since all indices are identical in $a_{i,m}$ and $a_{i,m-1}$ save for a difference of 1 in the $k-r-m+1$ position. It follows that

$$n_{i'} - n_i = d_{k-s+1} + \cdots + d_{k-r-1} + d_{k-r}$$

Since the sum on the right side is independent of i, $n_{i'} - n_i$ is constant for all $i = 1, \ldots, \ell$ and so $n_1, \ldots, n_{\ell'}$ are in arithmetic progression. ∎

Van der Waerden's theorem admits of great generalization. In fact, an entire area of study called Ramsey theory has been one outgrowth of a group of related theorems. An outstanding conjecture of Erdös states that if A is any sequence of positive integers for which $\sum_{a \in A} \frac{1}{a}$ diverges, then A must contain arbitrarily long arithmetic progressions. In fact, Erdös had offered \$3000 to anyone who can settle his conjecture! Note that if the conjecture is true, then the sequence of primes would contain arbitrarily long arithmetic progressions.

One area of interest is to find the smallest possible values of $n(k, \ell)$. Although our proof of Theorem 9.11 was constructive, the value we obtain is far from optimal. Let $W = W(k, \ell)$ denote the smallest number such that if $\mathbf{N}(W)$ is partitioned into at most k classes, then at least one class contains an arithmetic progression of length ℓ. $W(k, \ell)$ is called the van der Waerden function. We noted that $W(k, 2) = k+1$. In addition, it is fairly easy to check that $W(2, 3) = 9$ (exercise). The numbers $W(2, 4) = 18$, $W(2, 5) = 22$, $W(2, 6) = 32$, and $W(2, 7) = 46$ were computed by V. Chvátal (1969). Many other values of $W(k, \ell)$ are now known.

Exercise 10.3

1. Let $W(k, \ell)$ denote the van der Waerden function.
   (a) Verify that $W(2, 3) = 9$.
   (b) Show that $W(2, 4) \geq 18$ by constructing a partition of $N(17)$ into two classes with neither class containing an arithmetic progression of length four.
2. It is known that $W(3, 3) = 27$. Construct a partition of $N(n)$ into three classes with no class containing an arithmetic progression of length three for n as large as you can.
3. (a) Let M be a positive integer and let $S = \{a_n\}_{n=1}^{\infty}$ where $a_1 < a_2 < a_3 < \dots$ be any infinite sequence of integers for which $a_i - a_{i-1} \leq M$ for all i. Show that S contains arithmetic progressions of arbitrary length (J.R. Rabung – 1975). (Hint: Let $S_0 = S$ and $S_i = \{a_n + i\}_{n=1}^{\infty} \cap (S_0 \cup \dots \cup S_{i-1})^c$ for $1 \leq i \leq M - 1$ where A^c is the complement of the set A. Apply van der Waerden's theorem to the sets $S_0, \dots, S_{M-1}$.)
   (b) Show that if the natural numbers are partitioned into two classes, then either (i) one class contains arbitrarily long strings of consecutive integers or (ii) both classes contain arithmetic progressions of arbitrary length.
4. Show that there is a constant M and set S as in Exercise 10.3.3 for which S has no arithmetic progression of infinite length.
5. Partition the set of positive integers into two sets so that each set contains arithmetic progressions of arbitrary finite length but neither contains an infinite arithmetic progression.
6. Verify that $199 + 210k$ for $0 \leq k \leq 9$ is an arithmetic progression containing 10 primes.

10.4 Introduction to the theory of partitions

Suppose you wished to mail an n-cent letter and had an unlimited supply of stamps of all denominations. How many choices would you have in selecting the stamps? The arrangement on the letter itself is irrelevant. For example, if the letter were a 4-cent letter you could choose a 4-cent stamp, or a 3-cent and a 1-cent stamp, or two 2-cent stamps, or a 2-cent and two 1-cent stamps, or four 1-cent steps. (We ignore the fact the letter may be returned to you for insufficient postage!) There are five choices in all. We can list the possibilities as $4, 3 + 1, 2 + 2, 2 + 1 + 1, 1 + 1 + 1 + 1$. Each of these expressions is called a partition of 4 and the terms are called the parts of the partition. Hence there are 5 (unrestricted) partitions of the number 4. Here is a more formal definition:

Definition 10.8: Let n be an integer. A *partition* of n is a representation of n as a sum of positive integers. The total number of essentially distinct partitions is denoted by $p(n)$.

Here are all the partitions of $n = 7$: 7, $6 + 1$, $5 + 2$, $5 + 1 + 1$, $4 + 3$, $4 + 2 + 1$, $4 + 1 + 1 + 1$, $3 + 3 + 1$, $3 + 2 + 2$, $3 + 2 + 1 + 1$, $3 + 1 + 1 + 1 + 1$, $2 + 2 + 2 + 1$, $2 + 2 + 1 + 1 + 1$, $2 + 1 + 1 + 1 + 1 + 1$, $1 + 1 + 1 + 1 + 1 + 1 + 1$.

Hence $p(7) = 15$.

Notice that repetitions are allowed and that the order of terms does not matter. Hence it is often convenient to list the terms of a partition in descending (actually nonascending) order as we have done above. It is convenient to define $p(0) = 1$ and $p(n) = 0$ for negative integers n.

It is instructive to work up some examples for yourself and determine p(n) for a few small values of n. Check that $p(3) = 3, p(5) = 7$, and $p(6) = 11$. The values of $p(n)$ grow very rapidly with n. For example, $p(10) = 42, p(22) = 1002, p(50) = 204226, p(100) = 190,569,292$, and $p(500) = 2,300,165,032,574,323,995,027$.

The partition function was first studied by G.W. Leibniz (1646–1716) who discussed its difficulty and importance in a letter to Johann Bernoulli. The partition function and its many variants have since been studied by many prominent mathematicians, the most notable being Euler. In fact, all results in this section are due to him unless specified otherwise. There have also been significant contributions from Gauss, Cauchy, Jacobi, Sylvester, Hardy, Ramanujan, Rademacher, F. Dyson, A.O.L. Atkin, G. Andrews, K. Ono, and many others. Interestingly the theory of partitions has found applications in group theory, combinatorics, nonparametric statistics, and even particle physics.

Euler derived several recursive formulas for calculating $p(n)$, one of which we study in this section. Unfortunately, there is no simple closed formula for $p(n)$. However, a very complicated convergent infinite series for $p(n)$ has been derived in the twentieth century. We will say a few words about it at the end of this section.

There is another useful way to view partitions, namely as strips of dots. These are sometimes called Ferrers' graphs, named after N.M. Ferrers (1829–1908), a Senior Wrangler at Cambridge (the top honor in the Tripos exam), who used them in a communication with J.J. Sylvester (1853). For example, we can view the partition $5 + 3 + 3 + 2$ of 13 as a series of dots as in Figure 10.1:

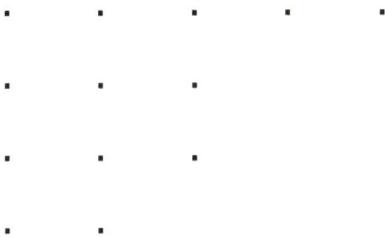

Figure 10.1: Ferrers' graph for one partition of 13.

Each part is listed horizontally. Alternately, we can view the dots vertically to obtain the conjugate partition $4 + 4 + 3 + 1 + 1$ of 13. The following is an obvious consequence of Ferrers' graphs.

Proposition 10.12: The number of partitions of n into k parts is equal to the number of partitions of n with largest part k.

Proof: The proof is immediate. ∎

Corollary 10.12.1: The number of partitions of n into at most k parts is equal to the number of partitions of n with largest part less than or equal to k.

Proof: For a given n, apply Proposition 10.12 with largest part k, $k - 1$, . . ., 1 in succession. The numbers may now be summed since the sets of partitions considered are disjoint. ∎

Corollary 10.12.2: The number of partitions of n into an even (alternately, odd) number of parts is equal to the number of partitions of n with largest part an even (alternately, odd) number.

Proof: The proof is left as Exercise 10.4.2(b). ∎

Before proceeding, it is helpful to introduce notation for the number of partitions of n with various restrictions. Here are some useful ones:

$p_e(n)$ = the number of partitions of n into an even number of parts.
$p_0(n)$ = the number of partitions of n into an odd number or parts.
$p_u(n)$ = the number of partitions of n into unequal parts.
$p_k(n)$ = the number of partitions of n having exactly k parts.
$p(n, o)$ = the number of partitions of n having odd parts only.
$p(n, e)$ = the number of partitions of n having even parts only.
$p(n, k)$ = the number of partitions of n with largest part being at most k.

We will call partitions of n with an even number of parts *even partitions*. Analogously, *odd partitions* are those with an odd number of parts. For example, if we look at the partitions of $n = 7$, we find that $p_e(7) = 7$, $p_o(7) = 8$, $p_u(7) = 5$, $p_2(7) = 3$, $p_3(7) = 4$, $p(7, o) = 5$, $p(7, e) = 0$, $p(7, 2) = 4$, and $p(7, 3) = 8$. Check your understanding by verifying that (i) $p(n) = p_e(n) + p_o(n)$, (ii) $p(n, e) = 0$ if n is odd, (iii) $\sum_{k=1}^{n} p_k(n) = p(n)$, and (iv) $p(n, m) = p(n)$ for $m \geq n$. Corollary 9.12.1 establishes that $p(n, k)$ equals the number of partitions of n into at most k parts.

Proposition 10.13: The number of partitions of n containing part k at least m times is

$$p(n - mk)$$

Proof: If p is a partition of n with part k appearing at least m times, then the partition with m k's deleted is a partition of $n - mk$. Conversely, if P is a partition of $n - mk$, then the partition with m k's appended is a partition of n containing part k at least m times. ∎

For example, the number of partitions of 7 containing part 1 at least three times is $p(7 - 3 \cdot 1) = 5$.

Corollary 10.13.1:
(a)   The number of partitions of n containing part k exactly m times is $p(n - mk) - p(n - (m + 1)k$.
(b)   The number of partitions of n with no part k is $p(n) - p(n - k)$.

Proof: Exercise 10.4.3(a). ∎

Next we introduce the concept of *generating functions,* an important topic throughout much of mathematics. Recall that the geometric series $1 + x^k + x^{2k} + \ldots$ converges to $\frac{1}{1-x^k}$ for $|x| < 1$. If we multiply the series for $k = 1$ together with that for $k = 2$, we get

$$\frac{1}{(1-x)(1-x^2)} = \left(1 + x + x^2 + x^3 + \cdots\right)\left(1 + x^2 + x^4 + x^6 + \cdots\right)$$

valid for $|x| < 1$. Collecting like terms on the right-hand side, we obtain

$$\frac{1}{(1-x)(1-x^2)} = 1 + x + 2x^2 + 2x^3 + 2x^4 + 3x^5 + 4x^6 + \cdots$$

It is valid to rearrange terms since each series converges absolutely. For $n \geq 1$, the coefficient of x^n on the right-hand side is equal to $p(n, 2)$. This is a natural consequence of the law of exponents: $x^a x^b = x^{a+b}$. For example, the partition $1 + 2 + 2$ of 5 corresponds to the product $x \cdot x^4 = x^1 \cdot x^{2+2}$. Similarly, $1 + 1 + 1 + 2$ corresponds to $x^3 \cdot x^2 = x^{1+1+1} \cdot x^2$ and $1 + 1 + 1 + 1 + 1$ corresponds to $x^5 \cdot 1 = x^{1+1+1+1} \cdot 1$. Thus, $p(5, 2) = 3$. Similarly,

$$\prod_{k=1}^{m} \frac{1}{(1-x^k)} = \sum_{n=0}^{\infty} p(n, m)x^n \tag{10.14}$$

In eq. (10.14) we define $p(0, m) = 1$. The next theorem allows us to extend eq. (10.14) as $m \to \infty$.

Theorem 10.14: Let $P(x) = \prod_{k=1}^{\infty} \frac{1}{(1-x^k)}$. Then $P(x) = \sum_{n=0}^{\infty} p(n)x^n$ for $0 < x < 1$.

Hence, $P(x)$ is the generating function for the partition function $p(n)$.

Proof: Let $P_m(x) = \prod_{k=1}^{m} \frac{1}{(1-x^k)}$. By (10.14), $P_m(x) = \sum_{n=0}^{\infty} p(n, m)x^n$. Rewrite this as

$$P_m(x) = \sum_{n=0}^{m} p(n, m)x^n + \sum_{n=m+1}^{\infty} p(n, m)x^n$$

Notice that $p(n, m) = p(n)$ if $m \geq n$. It follows that

$$P_m(x) = \sum_{n=0}^{m} p(n)x^n + \sum_{n=m+1}^{\infty} p(n,m)x^n$$

But $p(n,m) \le p(n)$ for all m, n and $\lim_{m \to \infty} p(n,m) = p(n)$ for all n. Hence, $\lim_{m \to \infty} P_m(x) = P(x)$. It follows that

$$\sum_{n=0}^{m} p(n)x^n < P_m(x) < P(x) \text{ for all } m$$

So $\sum_{n=0}^{\infty} p(n)x^n$ is convergent for $0 < x < 1$. Since $\lim_{m \to \infty} p(n,m) = p(n)$ for all n, we have

$$P(x) = \lim_{m \to \infty} P_m(x) = \lim_{m \to \infty} \sum_{n=0}^{\infty} p(n,m)x^n = \sum_{n=0}^{\infty} p(n)x^n \qquad \blacksquare$$

Although our proof holds for $0 < x < 1$, it can be extended so that Theorem 10.14 is valid for $|x| < 1$. In any event, we will usually treat generating functions as formal identities for which questions of convergence are moot. The following examples of generating functions for various restricted partition functions should then be considered as formal identities, although rigorous proofs of their convergence for appropriate x can be given along the lines of the previous proof.

We define $p_u(n,o)$ as the number of partitions of n into odd and unequal parts and $p_u(n,e)$ as the number of partitions of n into even and unequal parts. For example, $p_u(7,o) = 1$ since 7 is the only such partition of 7. Similarly, $p_u(8,o) = 2$ since $7+1$ and $5+3$ are the only allowable partitions of 8 with this restriction. Additionally, $p_e(7,o) = 0$ and $p_e(8,o) = 2$.

Furthermore, let $p(0,o) = p(0,e) = p_u(0) = p_u(0) = p_u(0,o) = p_u(0,e) = 1$.

Proposition 10.15:

(a) $\prod_{k=1}^{\infty} \frac{1}{1-x^{2k-1}} = \sum_{n=0}^{\infty} p(n,o)x^n$.

(b) $\prod_{k=1}^{\infty} \frac{1}{1-x^{2k}} = \sum_{n=0}^{\infty} p(n,e)x^n$.

(c) $\prod_{k=1}^{\infty} \left(1+x^k\right) = \sum_{n=0}^{\infty} p_u(n)x^n$.

(d) $\prod_{k=1}^{\infty} \left(1+x^{2k-1}\right) = \sum_{n=0}^{\infty} p_u(n,o)x^n$.

(e) $\prod_{k=1}^{\infty} \left(1+x^{2k}\right) = \sum_{n=0}^{\infty} p_u(n,e)x^n$.

Proof: The proof is left as Exercise 10.4.8.

We are now in a position to prove a rather striking result.

Theorem 10.16: For all $n \ge 0$, $p_u(n) = p(n,o)$.

Proof: By Proposition 10.15, it suffices to show that

$$\prod_{k=1}^{\infty} \left(1+x^k\right) = \prod_{k=1}^{\infty} \frac{1}{1-x^{2k-1}}$$

But $1 + x^k = \frac{1-x^{2k}}{1-x^k}$. So

$$\prod_{k=1}^{\infty}\left(1+x^k\right) = \prod_{k=1}^{\infty}\frac{1-x^{2k}}{1-x^k} = \prod_{k=1}^{\infty}\frac{1}{1-x^{2k-1}}$$

since all factors with even exponents divide out. ∎

The following result is representative of some of the beautiful identities discovered by Euler:

Theorem 10.17:

(a) $\sum_{n=0}^{\infty} p_{uu}(n,0)x^n = 1 + \frac{x}{1-x^2} + \frac{x^4}{\left(1-x^2\right)\left(1-x^4\right)} + \frac{x^9}{\left(1-x^2\right)\left(1-x^4\right)\left(1-x^6\right)} + \cdots$

(b) $\sum_{n=0}^{\infty} p_u(n,e)x^n = 1 + \frac{x^2}{1-x^2} + \frac{x^6}{\left(1-x^2\right)\left(1-x^4\right)} + \frac{x^{12}}{\left(1-x^2\right)\left(1-x^4\right)\left(1-x^6\right)} + \cdots$

In part (a), the exponents in the numerators are successive squares. In part (b), the exponents in the numerators are of the form $m(m+1)$. The method of proof involves introducing a parameter r and then proving a more general identity from which both (a) and (b) follow as special cases.

Proof: Let $F(r,x) = \prod_{k=1}^{\infty}\left(1+rx^{2k-1}\right) =: \sum_{n=0}^{\infty} a_n r^n$ where $a_n = a_n(x)$ and $a_0 = 1$.

$$F(rx^2, x) = \left(1+rx^3\right)\left(1+rx^5\right)\left(1+rx^7\right)\cdots \qquad (10.15)$$

Expanding (10.15),

$$1 + a_1 r + a_2 r^2 + a_3 r^3 + \cdots$$

$$= 1 + \left(a_1 x^2 + x\right)r + \left(a_2 x^4 + a_1 x^3\right)r^2 + \left(a_3 x^6 + a_2 x^5\right)r^3 + \cdots$$

Hence $a_n = \frac{x^{2n-1}}{1-x^{2n}}a_{n-1}$ for $n \geq 1$. In particular, $a_1 = \frac{x}{1-x^2}$, $a_2 = \frac{x^4}{\left(1-x^2\right)\left(1-x^4\right)}$, and

$a_n = \frac{x^{n^2}}{\left(1-x^2\right)\cdots\left(1-x^{2n}\right)}$ for $n \geq 1$ by induction.

It follows that

$$\prod_{k=1}^{\infty}\left(1+rx^{2k-1}\right) = 1 + \sum_{n=1}^{\infty}\frac{x^{n^2}}{\left(1-x^2\right)\cdots\left(1-x^{2n}\right)}r^n$$

Part (a) follows by letting $r=1$ and applying Proposition 10.15(d). Part (b) follows by letting $r = x$ and applying Proposition 10.15(e). ∎

We now turn our attention to a rather remarkable result of Euler's. Recall that the pentagonal numbers $f_5(n) = 1 + 4 + 7 + \cdots + (3n-2)$ were defined as the number of dots in a pentagonal array. In particular, $f_5(n) = \frac{n(3n-1)}{2}$ (see Exercise 1.2.4(a).) If we extend our definition of pentagonal numbers to include negative arguments, then $f_5(-n) = \frac{n(3n+1)}{2}$ are also pentagonal numbers. Thus, the sequence of pentagonal numbers begins $\{1, 2, 5, 7, 12, 15, 22, 26, 35, 40, \ldots\}$. Notice that the pentagonal numbers generated from positive arguments are disjoint from those with negative arguments (Exercise 10.4.20).

Let us define two more functions:

$E(n)$ = the number of partitions of n into an even number of unequal parts,

$O(n)$ = the number of partitions of n into an odd number of unequal parts.

For example, the partitions of 6 into an even number of unequal parts are $5 + 1$ and $4 + 2$, while the partitions into an odd number of unequal parts are 6 and $3 + 2 + 1$. Hence $E(6) = 2 = O(6)$. For $n = 7$ we obtain $E(7) = 3$ and $O(7) = 2$. Of course, $E(n) + O(n) = p_u(n)$ for all n. What is initially far from obvious is that the difference between $E(n)$ and $O(n)$ is always $-1, 0$, or 1. Furthermore, $E(n) = O(n)$ if and only if n is not a pentagonal number. The following theorem was announced by Euler in a letter to Nikolaus Bernoulli (Johann's nephew – not his father, brother, or son) dated November 10, 1742. Nikolaus Bernoulli (1687–1759) was the creator of the Saint Petersburg paradox in probability.

Theorem 10.18 (Pentagonal Number Theorem): $E(n) = O(n)$ unless $n = \frac{k(3k \pm 1)}{2}$ is a pentagonal number, in which case $E(n) - O(n) = (-1)^k$.

Proof (F. Franklin – 1881): Let $n \geq 1$ and let p be a partition of n into unequal parts. Let s be the smallest part of p and let $\ell_1 > \ell_2 > \ldots$ be the largest parts in descending order. Let k be the maximum integer for which $\ell_1 = \ell_2 + 1 = \ldots = \ell_k + (k-1)$. So k is the number of consecutive numbers in p beginning with the largest part. There are two possibilities:

(i) $s \leq k$ or (ii) $s > k$.

(i) If $s \leq k$, then transform p to another partition p' of n by removing part s and adding one to each of $\ell_1, \ell_2, \ldots, \ell_s$. Define s' and k' analogously for p'. The partition p' has only unequal parts with $s' > s = k'$. The number of parts of p' is one less than the number of parts of p. Hence the contribution from p and p' to $E(n)$ and $O(n)$ balances.

There is one exception, however, where the transformation from p to p' described cannot be carried out. This is the case where $s = k$ and $s = \ell_k$. Here the part ℓ_k is missing once we remove part s. In this situation, $n = k + (k+1) + \cdots + (2k-1) = \frac{k(3k-1)}{2}$.

(ii) If $s > k$, then form the partition p' of n by subtracting 1 from each of $\ell_1, \ldots, \ell_k$ and adding an additional part k (which is the smallest part of p'). Notice that p' has only unequal parts and $s' = k \leq k'$. Furthermore, the number of parts of p' is one more than the number of parts of p; hence one of p and p' is an even partition, the other odd.

There is one exceptional situation, namely that when $s = k+1$ and $s = \ell_k$. Here the part s would be altered by subtracting 1 from ℓ_k and p' would have two parts $s - 1$. In this situation, $n = (k+1) + (k+2) + \cdots + 2k = \frac{k(3k+1)}{2}$.

Since $(p')' = p$ in all but the exceptional cases, barring them there is a one-to-one correspondence between unequal partitions of n into an even and odd number of parts respectively. In the two exceptional cases k is the number of parts of p. Hence there is an excess of one even partition when k is even. Analogously, there is an excess of one odd partition when k is odd. Note that there is no value of n having both exceptions as noted previously.

Therefore, $E(n) = O(n)$ unless $n = \frac{k(3k \pm 1)}{2}$, in which case $E(n) - O(n) = (-1)^k$. ∎

We can use the pentagonal number theorem to characterize pentagonal numbers in a variety of ways. Here is one such example.

Corollary 10.18.1: The number n is pentagonal if and only if $p(n, \mathrm{o})$ is odd.

Proof: For all n, $p_u(n) = E(n) + O(n)$. By Theorem 10.16, $p_u(n) = p(n, \mathrm{o})$. By the Pentagonal Number Theorem, $p_u(n)$ is odd or even, respectively, depending on whether or not n is a pentagonal number. ∎

Let $d(n) = E(n) - O(n)$ for $n \geq 1$ and let $d(0) = 1$. The function $d(n)$ has a particularly simple generating function:

$$\prod_{k=1}^{\infty} (1 - x^k) = \sum_{n=0}^{\infty} d(n)x^n \tag{10.16}$$

Notice that it is the reciprocal of the generating function for $p(n)$ given by Theorem 10.14. It follows that

$$\left(\sum_{n=0}^{\infty} d(n)x^n \right) \left(\sum_{n=0}^{\infty} p(n)x^n \right) = 1$$

By equating coefficients of x^n, we get $d(0)p(0) = 1$ and

$$\sum_{k=0}^{n} d(k)p(n-k) = 0 \text{ for } n \geq 1$$

By the Pentagonal Number Theorem, we get *Euler's identity*

$$p(n) = p(n-1) + p(n-2) - p(n-5) - p(n-7) + \cdots$$
$$+ (-1)^{k+1} \left\{ p\left(n - \frac{k(3k-1)}{2} \right) + p\left(n - \frac{k(3k+1)}{2} \right) \right\} + \cdots \tag{10.17}$$

where the sum is extended over all positive values of k.

Example 10.1: Calculate $p(10)$.

Solution: The only pentagonal numbers less than or equal to 10 are $1, 2, 5$, and 7. By Euler's identity, $p(10) = p(9) + p(8) - p(5) - p(3)$. Assuming we know $p(0) = 1, p(1) = 1$, $p(2) = 2, p(3) = 3, p(4) = 5$, and $p(5) = 7$; we still need to compute $p(8)$ and $p(9)$. We apply Euler's identity as often as needed:

$$p(8) = p(7) + p(6) - p(3) - p(1) = p(7) + p(6) - 4.$$
$$p(9) = p(8) + p(7) - p(4) - p(2) = p(8) + p(7) - 7. \text{ So}$$
$$p(10) = p(9) + p(8) - 10 = \{p(8) + p(7) - 7\} + p(8) - 10$$
$$= 2\{p(7) + p(6) - 4\} + p(7) - 17 = 3p(7) + 2p(6) - 25.$$
$$\text{But } p(7) = p(6) + p(5) - p(2) - p(0) = p(6) + 4.$$
$$\text{Hence } p(10) = 3\{p(6) + 4\} + 2p(6) - 25 = 5p(6) - 13.$$
$$\text{Finally, } p(6) = p(5) + p(4) - p(1) = 11.$$
$$\text{Hence } p(10) = 5 \cdot 11 - 13 = 42$$

There is no simple formula for $p(n)$. However, in 1918 G.H. Hardy and S. Ramanujan (1887–1920) proved that $p(n) \sim \frac{\exp\left(\pi\sqrt{2n/3}\right)}{4n\sqrt{3}}$ as $n -- > \infty$. In fact, using sophisticated analytical methods including the celebrated circle method, they derived an asymptotic formula for $p(n)$. In theory, the formula can be used to derive the exact value of $p(n)$ when the neglected terms can be shown to be bounded by 1/2. In 1937, Hans Rademacher (1892–1969) extended the analysis to establish a rapidly convergent series representation for $p(n)$. The formula involves the evaluation of elliptic modular functions, the theory of which lies outside our study here. By the way, Ramanujan was a self-taught genius from India whose results and insights have continued to mystify and delight mathematicians right up to the present. Hardy called him the "most romantic figure in the recent history of mathematics."

An area where there is still much to be learned is that of the arithmetic properties of $p(n)$. For example, Ramanujan proved that $p(5k + 4) \equiv 0 \pmod{5}$, $p(7k + 5) \equiv 0 \pmod{7}$, and $p(11k + 6) \equiv 0 \pmod{11}$, interesting discoveries that are difficult to establish. Although Ramanujan and others established similar results for powers of 5, 7, and 11, no analogous results have been discovered for moduli 2 or 3. In 1957, O. Kolberg proved in an essentially nonconstructive manner that $p(n)$ is odd infinitely often and even infinitely often. Best of luck in your investigations!

Exercise 10.4

1. Calculate $p(8)$ and $p(9)$ directly by listing all appropriate partitions.
2. (a) Show that the number of partitions of n into at least k parts is equal to the number of partitions of n with largest part greater than or equal to k.
   (b) Prove Corollary 10.12.2.
3. (a) Prove Corollary 10.13.1.
   (b) Show that the number of partitions of n containing part k is the same as the number of partitions of n containing part one at least k times.
   (c) Show that the number of partitions of n with smallest part at least 2 is $p(n) - p(n-1)$.
4. (a) Let $c(n)$ denote the number of partitions of n where order counts. For example, $c(3) = 4$ since $3, 2+1, 1+2, 1+1+1$ are the four ordered partitions of the number 3. Compute $c(4)$ and $c(5)$.
   (b) Derive a formula for $c(n)$, the number of *compositions* of n, valid for all n. (Hint: Think of ordered partitions with $k+1$ parts in terms of placing k slashes between n dots.)
   (c) Find the number of compositions of n with exactly m parts, $0 \le m \le n$.
   (d) Show that the number of compositions of n with no part 1 is F_{n-1}, the $(n-1)^{st}$ Fibonacci number.
5. Use Ferrers' graph to show that the number of partitions of n with smallest part k equals the number of partitions of n with largest part repeated exactly k times.
6. (a) Investigate under what conditions a partition and its conjugate are identical (so-called *self-conjugate* partitions.)
   (b) How many self-conjugate partitions of 7 are there? Note that it is the same as $pu_u(7,0)$. Investigate this phenomenon for other small values. In fact, the number of self-conjugate partitions of order n equals $p_u(n,o)$. Can you set up a one-to-one correspondence between them?
7. (W.P. Durfee – 1882)
   (a) Show that if p is a self-conjugate partition of n, then the Ferrer's graph of p contains a $k \times k$ square in its northwest corner with n and k the same parity.
   (b) Show that the number of self-conjugate partitions of n is given by $\sum_{0<k\le\sqrt{n}} p\big((n-k^2)/2, k\big)$ where the summation is over k of the same parity as n.
8. Establish Proposition 10.15.
9. Show that $d(n) = p(n) - p(n-1)$ is a monotone increasing function for $n \ge 7$.
10. (a) Use generating functions to show that $p(n) = \sum_{k=0}^{n} p(k,o) \cdot p(n-k,e)$.
    (b) Show that $p_u(n) = \sum_{k=0}^{n} p_u(k,o) \cdot p_u(n-k,e)$.
11. Verify the generating function (10.16).
12. Prove that $p_k(n) = p(n-k, k)$.
13. Show that the number of partitions of n containing part k at least m times is the same as the number of partitions of n containing part m at least k times.

14. Show that $p(n, e) = p(n/2)$ for n even.

15. Calculate $p(11)$, $p(12)$, and $p(13)$ using Euler's identity, eq. (10.17).

16. (a) Make a chart of $p_e(n)$ and $p_o(n)$ for $1 \le n \le 12$.

    (b) Can you make any conjectures based on (a)? Not much is known here save for an interesting result of Euler's that $|p_e(n) - p_o(n)| = p_u(n, o)$. Check it for $1 \le n \le 12$.

    (c) Use the result in Exercise 10.4.6 to show that there is always an even number of nonself-conjugate partitions of n.

17. (a) Show that the number of partitions of n with an even part repeated equals the number of partitions of n with some part repeated at least four times.

    (b) Show that the number of partitions of n with an even part repeated equals the number of partitions of n with some part divisible by 4.

18. Let $\Pr(p)$ denote the product of the parts of a partition p. Investigate the function $M(n)$, the maximum of $\Pr(p)$ as p ranges over all partitions of n.

19. Prove Theorem 10.16 directly by setting up a one-to-one correspondence between partitions with unequal parts and those having only odd parts (J. J. Sylvester). (Hint: Let p be a partition of n with k odd parts m. Write k as a sum of powers of 2 and then multiply m by each power of two, obtaining k distinct numbers, etc.)

20. Show that the pentagonal numbers generated from positive arguments are disjoint from those with negative arguments.

21. Compare the value $p(100) = 190, 569, 292$ with the Hardy-Ramanujan asymptotic formula.

22. Show that there are approximately $2\sqrt{2n/3}$ terms in Euler's identity for $p(n)$.

23. How many ways can change be made for a quarter using pennies, nickels, and/or dimes?

24. (a) Establish that $p(n, k) = p(n, k - 1) + p(n - k, k)$.

    (b) Use the recurrence relation in (a) to find $p(n, k)$ for $5 \le n \le 7, 3 \le k \le 5$. (Euler used (a) to find $p(n, k)$ for $1 \le n \le 70, 1 \le k \le 20$ and $k = n$.)

25. (a) Show directly that $p(n) > 2^m$ where $m = \left[\sqrt{n}\right]$ for $n > 1$. (Hint: consider the number of ways to write $n = a_1 + \cdots + a_k + (n - a_1 - \cdots - a_k)$ where $1 \le a_1 < \cdots < a_k \le m$.)

    (b) Show that $p_u(n) > 2^r$ where $r = \left[n^{1/3}\right]$ for $n > 4$.

26. (a) The number of partitions of n in which the difference between any 2 parts is at least two equals the number of partitions of n into parts $\equiv 1$ or $4 \pmod 5$ (**L.J. Rogers – S. Ramanujan Identities**). Verify this for $n = 6$ and $n = 7$.

    (b) The number of partitions of n in which the difference between any two parts is at least 2 and with no part one equals the number of partitions of n into parts $\equiv 2$ or $3 \pmod 5$. Verify this for $n = 6$ and $n = 7$.

27. Use generating functions to show that the number of partitions of n into parts congruent to either 1 or 5 (mod 6) equals the number of partitions of n into unequal parts congruent to either one or 2 (mod 3).

Hints and answers to selected exercises

Exercise 1.1

1. 84.
2. 60, 80.
3. (a) $3/7 = 1/3 + 1/11 + 1/231$; (b) $11/13 = 1/2 + 1/3 + 1/78$.
5. 18.
7. $(n-1)n/2 + n(n+1)/2 = n^2$.
13. (a) Consider $n = 8k + 7$; (b) 33; (c) 128 is not the sum of distinct squares.
17. $6{,}561 = 3^8$.
19. Note that $6 = 3!$.
23. You may want to take a quick look at Section 7.4.

Exercise 1.2

2. (b) $(n^2 - n + 1) + (n^2 - n + 3) + \cdots + (n^2 + n - 1) = n^3$.
3. (b) $g_n = a(ar^{n+1} - 1)/(r - 1)$.
7. How many lines does the kth line cross for $1 \le k \le n$?
11. Φ.
13. Show that $\Phi^{n+2} = \Phi^{n+1} + \Phi^n$, $\Phi'^{n+2} = \Phi'^{n+1} + \Phi'^n$, $\Phi + \Phi' = 1$, and $\Phi^2 + \Phi'^2 = 3$.
17. Note that $F_n < F_{n+2}/2$.
19. (a) $21 = 6 \cdot 7/2$, $2{,}211 = 66 \cdot 67/2$, $222111 = 666 \cdot 667/2$, and so on.
    (b) $55 = 10 \cdot 11/2$, $5{,}050 = 100 \cdot 101/2$, $500500 = 1{,}000 \cdot 1001/2$, and so on.
23. The set S does not consist entirely of positive integers.

Exercise 1.3

3. (a) For all k: $10^k \equiv 1 \pmod 9$.
    (b) $10^{2k} \equiv 1 \pmod{11}$, $10^{2k-1} \equiv -1 \pmod{11}$.
5. S_1 and S_3.
7. (a) Use induction on $t_{3n}, t_{3n+1}, t_{3n+2}$. (b) Reason modulo 3.
11. 200 elements are divisible by 3, 60 are divisible by 5.
13. The left side is divisible by 3.
17. Factor $n^5 - n$.
20. Reason modulo 3.
23. There is an m such that $F_1 \equiv F_1 + m \pmod n$ and $F_2 \equiv F_2 + m \pmod n$. $F_0 = F_2 - F_1$.

https://doi.org/10.1515/9783111579283-011

Exercise 1.4

3. $13!/2!^3 3!$, $8 \cdot 12!/2!^3 3!$

5. (a) For each subset, a given element is either in or out.
   (b) Let $a = 1$, $b = -1$ in the binomial theorem.

7. For $1 \le n \le 1001$, let a_n be the chosen integers and consider $a_n - 9$.

11. Consider possible parities for the three coordinates of a lattice point.

13. (a) $2^n - 1$. (b) Use induction.

17. Consider $\binom{p}{q}$ where q is the smallest prime factor of p.

19. Let $I(n) = \int_0^1 (1 - x^2)^n \, dx$. On the left side, use the binomial theorem to evaluate $I(n)$. On the right side, use integration by parts to establish $I(n + 1) = \frac{2n+2}{2n+3} I(n)$.

23. Let $P_1, \ldots, P_{n+1}$ be the players with corresponding number of wins and losses P_{wi} and P_{li} player i's number of wins and losses, respectively. Find $\sum_{i=1}^{n+1} P_{wi}$ and note $L_{iw} = n - P_{iw}$.

Exercise 2.1

2. (a) $(462, 2002) = 154$.   (b) $-4(462) + 1(2002) = 154$.

3. (a) $(1234, 5{,}678) = 2$.   (b) $704(1234) + -153(5{,}678) = 2$.

5. (a) $(2002, 2{,}600) = 26$.   (b) $(13 + 100n)(2002) + (-10 - 77n)(2{,}600) = 26$.

7. Show that F_n and F_{n+1} are relatively prime.

11. (a) $(21, 81, 120) = 3$.   (b) $[21, 81, 120] = 2^3 \cdot 3^4 \cdot 5 \cdot 7 = 22{,}680$.

13. Consider $\gcd(6, 10, 15)$.

17. S must contain two consecutive integers.

19. (a) $(x, y) = (9 + 2n, -1 - 3n)$.   (b) $(x, y) = (1 + 9n, -22 - 2n)$.
   (c) No solution.   (d) $(x, y) = (-237 + 286n, 87 - 105n)$.

23. Note that $\gcd(a, b, c) = \gcd(\gcd(a, b), c)$. Apply Corollary 2.1.2 twice.

Exercise 2.2

2. $1{,}001^* \equiv 1113 \pmod{2048}$.

3. $4{,}821^* \equiv -10{,}419 \pmod{10{,}000}$.

5. (a) No solution.   (b) $x \equiv 6 \pmod{11}$.

7. (a) $3 \nmid 3{,}100$.   (b) $11 \nmid -3{,}000$.

11. 119.

17. $46 + 420n$.

19. $(x, y) = (11, 25), (16, 11), (31, 32)$.

Exercise 2.3

2. (a) 3, 7, 31, 211, 2,311 are all prime.
3. (a) Consider $N = 3p_1 \cdots p_s - 1$.   (b) Consider $N = N = 6p_1 \cdots p_s - 1$.
   (c) All sufficiently large odd primes are of the form $2k + 101$.
7. 7, 11, 13, 17, 19, 23, 29 are all prime.
11. Consider numbers of the form $N = p_1 \cdots p_k$ where $p_1, \ldots, p_{k-1} \equiv 1 \pmod{b}$ and $p_k \equiv a \pmod{b}$.
13. (a) For all n, consider $2d, \ldots, (n + 1)d$.
   (b) Let p be the least element of the arithmetic progression. Reason modulo p.
17. (a) Assume $\sqrt{2} = a/b$ with $(a, b) = 1$, and so on.
   (b) If $\log_3 10 = a/b$, then $3^a = 10^b$, and so on.
19. (a) The identity $x^{km} - y^{km} = (x^k - y^k)(x^{k(m-1)} + x^{k(m-2)}y + \cdots + y^{k(m-1)})$ is helpful with $x = \Phi$ and $y = \Phi'$. Next pair up appropriate terms and use the fact that $\Phi \Phi' = -1$.
23. 17 and 73 are multiplicative primes (as is 2475989, the 181440th prime).

Exercise 2.4

2. (a) $8{,}137 = 79 \cdot 103$,   (b) $9{,}919 = 7 \cdot 13 \cdot 109$,   (c) $20{,}711 = 139 \cdot 149$.
3. Consider $n \equiv 7 \pmod{10}$.
5. Mimic the proof of Proposition 2.16.
7. 2, 5, 13, 89, 233, and 1597 are all prime. $F_{19} = 4{,}181 = 37 \cdot 113$.
11. (b) All odd primes except 23, 47, and 83.

Exercise 2.5

2. (a) S_1,   (b) S_2, S_3.
3. (a) $3^{96} \equiv 1 \pmod{97}$, $8^{102} \equiv 9 \pmod{11}$, $(-5)^{12002} \equiv 12 \pmod{13}$.
   (b) $7^{1234} \equiv 9 \pmod{10}$, $5^{1111} \equiv 5 \pmod{12}$, $3^{4000} \equiv 1 \pmod{20}$.
5. (a) $2^{3011} \equiv 2 \pmod{31}$,   (b) $3^{52009} \equiv 20 \pmod{53}$.
7. Note that $3! \equiv 2 \pmod{4}$ and $(n - 1)! \equiv 0 \pmod{n}$ for all composite $n > 4$.
11. Apply Fermat's little theorem and Wilson's theorem.
13. $D \equiv C \equiv B \equiv A = 97^{9797} \equiv 79^{9797} = 7^{1632 \cdot 6 + 5} \equiv 7^5 \equiv (-2)^5 \equiv 4 \pmod{9}$.
   $B < 9(10{,}000) = 90{,}000$, $C < 9(5) = 45$. Hence $C = 4, 13, 22, 31$, or 40 and $D = 4$.
19. (a) $x^5 - 5x^3 + 4x = (x - 2)(x - 1)x(x + 1)(x + 2)$.
   (b) $2x^5 - 20x^3 + 18x = 2(x - 3)(x - 1)x(x + 1)(x + 3)$.
23. Use Wilson's theorem.

Exercise 2.6

2. $x^6 + 98x^5 + 35x^4 + 84x^3 + 21x^2 + 133x + 1 \equiv x^6 + 1 \pmod 7$. But $x \equiv 0 \pmod 7$ implies. $x^6 \equiv 0 \pmod 7$, and $x \not\equiv 0 \pmod{+7}$ implies $x^6 \equiv 1 \pmod 7$.

3. Reason modulo 3.

5. By Lagrange's Theorem, $f(x)$ has at most m roots (mod p), say $a_1, \ldots, a_r$ where $r \le m$. Let $g(x) = (x - a_1) \ldots (x - a_r)$.

7. $x = 3$.

Exercise 3.1

2. (a) $\tau(4) = 3$, $\tau(27) = 4$, $\tau(6) = 4$, $\tau(18) = 6$, $\tau(108) = 12$.
   (b) $\sigma(4) = 7$, $\sigma(27) = 40$, $\sigma(6) = 12$, $\sigma(18) = 39$, $\sigma(108) = 280$.

3. (a) $\sigma_1(10) = 18$, $\sigma_2(10) = 130$, $\sigma_3(10) = 1334$, $\sigma_4(10) = 10{,}642$.
   (b) $\mu(17) = -1$, $\mu(105) = -1$, $\mu(277) = -1$, $\mu(1234567890) = 0$.

5. (a) $\omega(n) = \Omega(n)$ if and only if n is square-free.
   (b) Let $n = \prod_{i=1}^{t} p_i^{a_i}$. $2\omega(n) = \Omega(n)$ if and only if $\omega(n) = \Omega(n/p_1 \ldots p_t)$.

7. Consider $n = p^{m-1}$ as p runs through the set of primes.

11. (a) $\tau(n)\sigma(n) = 12$ implies $n = 5$.
    (d) $\tau(n) + \sigma(n) = 22$ implies $n = 10$ or $n = 19$.

13. (a) If n is prime, then $\tau(n) \phi(n) + 2 = 2n$.
    (b) If n is prime, then $\tau(n) \sigma(n) - 2 = 2n$. Note that $10 \mid \tau(10) \tau(10) - 2$.

Exercise 3.2

2. $\sigma_r(n) = \prod_{i=1}^{t} \dfrac{p_i^{r(a_i+1)} - 1}{p_i^r - 1}$.

3. $f(n) = f(1)f(n)$.

5. Yes.

7. If $(m, n) = 1$, $\dfrac{f}{g}(mn) = \dfrac{f(mn)}{g(mn)} = \dfrac{f(m)f(n)}{g(m)g(n)} = \dfrac{f}{g}(m)\dfrac{f}{g}(n)$.

11. Note that $\sigma(n) > n$.

13. $10000 - 100 = 9900$.

17. $F(p_1 \ldots p_t) = 3^t$.

19. First show that $\sigma(n)$ prime implies $n = p^a$. Next show that if $m \mid (a + 1)$, then $(p^m - 1) \mid \sigma(n)$.

Exercise 3.3

2. Apply Möbius inversion and Theorem 3.3(b).
3. $p|a \Rightarrow p|b$ means if $a = \prod_{i=1}^{t} p_i^{a_i}$, then $b = \prod_{i=1}^{t} p_i^{b_i}$ where $a_i, b_i \geq 1$ for $i = 1, \ldots t$. Now apply formula (3.9) to $n = ab$.
5. (a) $\phi(n) = 6$ implies $n = 7, 9, 14,$ or 18.
   (b) $\phi(n) = 12$ implies $n = 13, 21, 26, 28, 36,$ or 42.
7. $g\left(\prod_{i=1}^{t} p_i^{a_i}\right) = \prod_{i=1}^{t} p_i^{2a_i-2}(p_i^2 - 1)$.
11. $|\mu(d)| = 1$ if and only if d is square-free. By Exercise 1.4.5(a), there are $2^{\omega(d)}$ square-free divisors of n.
13. (a) 343, (b) 0063.
17. (a) For $1 \leq r \leq n$, count how many integers $d \leq n$ satisfy $r|d$.
    (b) Apply Theorem 3.4 to (a) with $f(r) = \mu(r)$.
    (c) Apply Proposition 3.6 to (a) with $f(r) = \phi(r)$.
19. (b) Assume otherwise. Then $\phi(n/2) = \phi(n)/2 = \phi(m)$ for some $m \neq n/2$. $\phi(2m) = \phi(n)$. If m is odd, then $\phi(4m) = \phi(n)$.

Exercise 3.4

2. (a) Apply Theorem 3.8. (b) Let $n = 10$.
3. Note that $d|n$ if and only if $\frac{n}{d}|n$.
5. Proposition checks for $n = 6$. If $n > 6$, then by Theorem 3.8, $n = 4^r (2 \cdot 4^r - 1)$. But $4^r \equiv 4 \pmod{10}$ implies $2 \cdot 4^r - 1 \equiv 7 \pmod{10}$, whereas $4^r \equiv 6 \pmod{10}$ implies $2 \cdot 4^r - 1 \equiv 1 \pmod{10}$.
11. Show that $\sigma(kn) > k\,\sigma(n)$ for $k > 1$.
13. 12, 18, 20, 24.
17. (b) $34 \rightarrow 20 \rightarrow 22 \rightarrow 14 \rightarrow 10 \rightarrow 8 \rightarrow 7 \rightarrow 1$.
19. (d) $10744 = 2^3 \cdot 17 \cdot 79$, $10856 = 2^3 \cdot 23 \cdot 59$.
23. Use the fact that $n = \sigma(m) - m$ to obtain $p = 1199936447$.

Exercise 4.1

2. (a) 2, 6, 7, 8; (b) 3, 5, 6, 7, 10, 11, 12, 14.
3. (a) 6, (b) 5.
11. (a) $x = 3$ or 4, (b) $x = 6, 7, 10,$ or 11.
13. $x = 7$.
17. 8, 8.
19. Consider $g^1, g^2, \ldots, g^{p-1}$.
23. Note that $1 + 2 + \cdots + \mathrm{ord}_p a = \mathrm{ord}_p a(1 + \mathrm{ord}_p a)/2$.
29. (a) $\mathrm{ord}_{15} 2 = 4$, (b) $\mathrm{ord}_{53} 2 = 52$.

Exercise 4.2

2. Use Proposition 4.10(a).
3. Let $r_1 = 2^2 + 1$ and $r_{n+1} = (r_1 \ldots r_n)^2 + 1$ for $n \geq 1$.
7. (a) $x^6 \equiv 5 \pmod 7$ not solvable implies no solutions to $x^3 \equiv 5 \pmod{77}$.
    (b) $x^3 \equiv 6 \pmod 7$ has three solutions by Proposition 4.8 and $x^3 \equiv 6 \pmod{11}$ has one solution. By the CRT there are three solutions to $x^3 \equiv 6 \pmod{77}$.
11. (b), (c), and (d) are solvable.
13. (c) is solvable.

Exercise 4.3

2. (b) and (c) are solvable.
3. (a), (b), and (d) are solvable.
5. $\left(\frac{-a^2}{p}\right) = 1 \Leftrightarrow \left(\frac{-1}{p}\right) = 1$.
7. even + even = even, even + odd = odd, odd + odd = even.
11. (a) $\left(\frac{3}{23}\right) = 1$ (b) $\left(\frac{-2}{23}\right) = -1$ (c) $\left(\frac{5}{23}\right) = -1$.
13. $\left(\frac{5}{73}\right) = -1$.
17. No, there is a maximum of four consecutive quadratic residues (mod 19) corresponding to 4, 5, 6, 7 (mod 19).
19. (a) Use Corollary 4.13.1.
    (b) Pair up distinct quadratic residues as in (a).
    (c) Apply Wilson's theorem.

Exercise 4.4

2. (a) $\left(\frac{1776}{1511}\right) = -1$ (b) $\left(\frac{-65}{1949}\right) = 1$ (c) $\left(\frac{103}{1999}\right) = -1$ (d) $\left(\frac{15}{10007}\right) = -1$.
3. (a) $p \equiv 1, 7 \pmod{12}$ (b) $p \equiv 1, 3, 7, 9 \pmod{20}$.
    (c) $p \equiv 1, 3, 9, 19, 25, 27 \pmod{28}$,   (d) $p \equiv 1, 5, 7, 9, 25, 35, 37, 39, 43 \pmod{44}$.
11. (a) 2, 3, 5, and 7 are quadratic nonresidues (mod 43).
    (b) No, since $\left(\frac{6}{p}\right) = \left(\frac{2}{p}\right)\left(\frac{3}{p}\right)$.
13. (a) $1^2 + 2^2 + \cdots + \left(\frac{p-1}{2}\right)^2 = p(p^2 - 1)/24$ and $24 \mid (p^2 - 1)$.
    (b) Consider $1 + 2 + \cdots + (p - 1)$ and use part (a).

Exercise 5.1

2. Reason modulo 4.
3. (a) $x = 2, 3$.

5. 2, 5, 13, 17, 29, 37, 41, 53, 61, 73, 89, 97.
7. 2, 3, 11, 17, 19, 41, 43, 59, 67, 73, 83, 89, 97.
11. (a) 7, (b) 23.
13. Let $s^2 \leq c^3 \leq (s + 1)^2$. Then $(s + 1)^2 = s^2 + 2s + 1 \leq c^3 + 2c^{3/2} + 1 < (c + 1)^3$.

Exercise 5.2

3. $65^2 = 25^2 + 60^2 = 39^2 + 52^2 = 16^2 + 63^2 = 33^2 + 56^2$.
5. (a) $5^2 + 14^2 = 221 = 10^2 + 11^2$. (b) $13^2 + 18^2 = 493 = 3^2 + 22^2$.
    (c) $10^2 + 23^2 = 629 = 2^2 + 25^2$. (d) $7^2 + 32^2 = 1073 = 17^2 + 28^2$.
    (e) $10^2 + 49^2 = 2501 = 1^2 + 50^2$.
7. $14^2 + 7^2$ is the only such representation of 245. It is not the case that composites $n \equiv$ 1 (mod 4) are necessarily expressible as the sum of two squares in more than one way.
11. (a) $41 = 4^2 + 5^2$, (b) $101 = 1^2 + 10^2$,
    (c) $181 = 9^2 + 10^2$, (d) $1693 = 18^2 + 37^2$.
13. (a) If $p^2 = a^2 + b^2$, then $(2ab)^2 + (a^2 - b^2)^2 = p^4 = (ap)^2 + (bp)^2$.
    (b) Use induction.
17. There exist a and b for which $2ab(a^2 - b^2)(a^2 + b^2)\,|\,xyz$. Now consider the factors of the left side modulo 2, 3, and 5.
19. There are 17 Pythagorean triangles with perimeter less than or equal to 100, namely the 16 listed in Example 5.4 plus the (9, 40, 41) triangle.
23. (a) Show that if r is the inradius of triangle (x, y, z), then $r = xy/(x + y + z)$. Now use Theorem 5.3.
    (b) By the solution to (a), $r = kb(a - b)$. Letting $k = 1$, $b = r$, and $a = r + 1$ shows that there is at least one *primitive* Pythagorean triangle with inradius r.
    (c) (30, 40, 50), (28, 45, 53), (25, 60, 65), (24, 70, 74), (22, 120, 122), (21, 220, 221).
29. (b) $12^4 + 15^4 + 20^4 = 231631 = 481^2$.
37. The case $p = 2$ is trivial. Let p be an odd prime and suppose there is a k with all $n \equiv$ k (mod p) not expressible as the sum of two squares. (We may assume that $0 \leq k \leq p - 1$ without loss of generality.) Then k is neither a quadratic residue nor congruent to the sum of two quadratic residues (mod p). Let $R = \{0 = r_0, r_1, \ldots, r_{(p-1)/2}\}$ be all the quadratic residues (plus zero) (mod p). Then $k - r_i \notin R$ for $i = 0, \ldots, (p - 1)/2$ and $k - r_i \not\equiv k - r_j$ (mod p) for $i \neq j$. But there are only $(p - 1)/2$ quadratic nonresidues (mod p) as opposed to $(p + 1)/2$ values of $k - r_i$, a contradiction.
41. Primitive Pythagorean triangles with square perimeter are characterized by Theorem 5.3 with $a = 2r^2$, $b = s^2 - 2r^2$, s odd, $\sqrt{2}r < s < 2r$ and $(a, b) = 1$. The two with smallest perimeter are (16, 63, 65) and (252, 275, 373) corresponding to $(r, s) = (2, 3)$ and (3, 5), respectively.

Exercise 5.3

2. Use Theorem 5.5 to obtain $(x, y, z, w) = (35, 10, 14, 39)$.
3. (a) Face diagonals are 125, 244, 267. Body diagonal is $\sqrt{73225}$.
    (b) Integral face diagonals of 533, 765. Body diagonal of 925.
5. If $k = \sum_{i=1}^{3} n_i(n_i+1)/2$, then $8k + 3 = \sum_{i=1}^{3}(2n_i+1)^2$.
7. (b) $(x_1^2 + ny_1^2)(x_2^2 + ny_2^2) = (x_1x_2 + ny_1y_2)^2 + n(x_1y_2 - x_2y_1)^2$.
11. $27 = 5^2 + 2\cdot 1^2$, $211 = 7^2 + 2\cdot 9^2$, $507 = 13^2 + 2\cdot 13^2$, $1353 = 31^2 + 2\cdot 14^2$, $10450 = 100^2 + 2\cdot 15^2$.
13. (b) Compare (a) with Theorem 5.3.
17. Consider $X = 3x$, $Y = 3y$, $Z = 3z$.

Exercise 5.4

2. Being a competent math detective involves some leg work.
3. (a) $420 = 19^2 + 7^2 + 3^2 + 1^2$.   (b) $1457 = 38^2 + 3^2 + 2^2 + 0^2$.
7. Use Theorem 5.2.
11. Note that x_1^2 is a square and that x_{k+1}^2 equals a square plus x_k^2 for $k \geq 1$.
13. Verify that if $34 \leq n \leq 78$, then n is the sum of distinct triangular numbers less than or equal to 36. Now add 36 to 43, 44, . . . , 78 to get expressions for n with $79 \leq n \leq$ as the sum of distinct triangular numbers. Add 45 to 70, 71, . . . , 114 to get expressions for n with $115 \leq n \leq 159$ as the sum of distinct triangular numbers, and so on. This process extends indefinitely since each triangular number ≥ 6 is less than twice the one before (compare with Corollary 9.15.1).

Exercise 5.5

2. If, say $p\,|\,x$ and $p\,|\,z$, then $p^2\,|\,by^2$ and either $p\,|\,y$ or $p^2\,|\,b$, and so on.
3. (a) $x = 2rsk$, $y = (2r^2 - s^2)k$, $z = (2r^2 + s^2)k$.
    (b) (2, 1, 3), (4, 2, 6), (4, 7, 9), (6, 3, 9), (6, 7, 11), (8, 4, 12), (8, 14, 18), (8, 31, 33).
5. (a) $\left(\frac{2}{7}\right) = \left(\frac{7}{2}\right) = 1$,   (b) (1, 1, 3),   (c) (−2, 5, 3).
7. (a) $\left(\frac{7}{37}\right) = \left(\frac{37}{7}\right) = 1$.
11. (a) $\left(\frac{473}{7}\right) = \left(\frac{301}{11}\right) = \left(\frac{-77}{43}\right) = 1$,   (b) (7, 2, 3).
13. (1, 3, 2), (4, 1, 5).

Exercise 6.1

2. (a) 1, 1/2, 2/3, 3/5, 5/8, 8/13, 13/21, 21/34, 34/55, 55/89.
    (d) −4, −3, −121/4, −245/81.
    (e) 2, 3, 8/3, 11/4, 19/7, 87/32, 106/39, 1359/500.

3. r must be positive, that is, a_0 must be nonnegative.
5. See comments preceding Proposition 6.2.
7. Apply Corollary 6.3.1, Corollary 6.4.2(a), and Theorem 6.5.
11. (a) $46/133 = [0; 2, 1, 8, 5] = [0; 3, -9, -5]$
      $-56/27 = [-3; 1, 12, 2] = [-2; -13, -2] = [-2; -14, 2]$.
17. (a) $5/7 = [1; 4, 2]_n$, $17/11 = [2; 3, 4]_n$, $55/89 = [1; 3, 3, 3, 3, 2]_n$.
    (b) $(n + 1)/n = [2; 2, 2, \ldots, 2]_n$ with n 2's.
    (c) If $p_{-2} = q_{-1} = 0$ and $p_{-1} = q_{-2} = 1$, then $p_i = a_i p_{i-1} - p_{i-2}$ and $q_i = a_i q_{i-1} - q_{i-2}$, $p_i q_{i-1} - p_{i-1} q_i = -1$,
      $r = r_n < r_{n-1} < \ldots < r_2 < r_1$.

Exercise 6.2

3. (a) $p/q = 5/7$,   (c) $p/q = 7/5$,   (e) $p/q = 37/5$.
5. (a) $ad - bc = -1$ implies $\gcd(b, d) = 1$.
    (b) Consider the three terms with $1/2$ in the middle.
7. Apply Theorem 6.9. Note that $q^2 < q(n + 1)$.
11. $2/3$.

Exercise 6.3

2. (a) $\sqrt{2} = [1; \overline{2}]$,   (b) $\sqrt{3} = [1; \overline{1, 2}]$,   (c) $\sqrt{11} = [3; \overline{3, 6}$,
    (d) $2 + 3\sqrt{7} = [9; \overline{1, 14}]$,   (e) $1 + 4\sqrt{3} = [7; \overline{1, 12}]$.
3. (a) Compare with Exercise 6.2.7.
    (b) Note that $\left|\frac{a}{b} - \frac{p}{q}\right| \geq 1/q^2$.
    (c) $3/2$, $7/5$, $17/12$, $41/29$, $99/70$.
5. (a) $\sqrt{8} = [2; \overline{1, 4}]$,   (b) $\sqrt{15} = [3; \overline{1, 6}]$,   (c) $24 = [4; \overline{1, 8}]$.
    $\sqrt{k^2 - 1} = [k - 1; \overline{2k - 2}]$.
7. (a) $1 + \sqrt{2}$,   (b) $k + \sqrt{k^2 + 1}$.
11. (a) $\log_{10} 2 = [3, 3, 9, 2, 2, \ldots]$,   (b) $\log_2 3 = [1; 1, 1, 2, 2, 3, \ldots]$.

Exercise 6.4

2. (a) $2, 3, 8/3, 11/4, 19/7, 87/32, 106/39, 193/71, 1264/465.\ 1457/536, 2721/1001, 23225/8544.$
    (b) $1457/536$.
7. $1146408/364913$.

Exercise 6.5

3. (a) $x/y = \sqrt{2+1/y^2}$, (b) $(x, y) = (3, 2), (17, 12), (99, 70)$.
5. (a) $(x, y) = (9, 4) = (161, 72)$, (b) $(x, y) = (161, 24) = (51841, 7728)$,
    (c) Use the result of (a) to obtain $(x, y) = (170, 76)$.
7. If $x_0^2 - dy_0^2 = 1$, then $x = n\,x_0$ and $y = n\,y_0$.
11. (a) Mimic the proof of Theorem 6.18.
    (b) 3/1 is not a convergent for $\sqrt{6}$.
13. (a) $(x, y) = (2, 1)$, (b) $(x, y) = (18, 8)$, (c) $(x, y) = (27, 12)$.
17. (b) Use (a) with $x_1 = y_1 = 1$ and $x_2^2 - dy_2^2 = 1$.

Exercise 6.6

2. (a) $e \approx 2721/1001$ implies $\sqrt{e} \approx 1.64872124$.
    (b) $\sqrt{e} \approx 1908/1157 \approx 1.649092481$ (actual value of $\sqrt{e}$ is $1.64872127\ldots$).

Exercise 6.7

3. $r = \frac{\log p}{\log q}$ implies $p = q^r$, and so on.
5. Consider prime powers.

Exercise 7.1

2. (a) $2^{76} \equiv 9 \pmod{77}$, (d) $2^{2478} \equiv 1935 \pmod{2479}$.
3. (b) $3^{252} \equiv 31 \pmod{253}$, (c) $3^{342} \equiv 337 \pmod{343}$.
5. $L = 133$, $A = 25$, $y \equiv 5, 33, 100, 128 \pmod{133}$.
7. (a) $50629 = 197 \cdot 257$.
    (b) $n^4 + 4m^4 = (n^2 + 2mn + 2m^2)(n^2 - 2mn + 2m^2)$. Note that $949 = 5^4 + 4 \cdot 3^4$.
    (c) $4194305 = 5 \cdot 397 \cdot 2113$.

Exercise 7.2

2. (a) $2^{1104} \equiv 1 \pmod{1105}$. Note that $3 \mid 561$.
    (c) 91 is the least psp(3).
5. Check that $18k \mid (n - 1)$. Then apply Carmichael's theorem. Let $k = 6$.
7. If a is not a primitive root $\pmod{n}$, then $a^m \equiv 1 \pmod{n}$ for some $m \mid (n - 1)$ with $m < n - 1$. Thus there is a prime $p \mid (n - 1)$ with $m \mid (n - 1)/p$. Hence $a^{(n-1)/p} \equiv 1 \pmod{n}$.

11. (a) $41041 = 7 \cdot 11 \cdot 13 \cdot$ Check that $2^3 \cdot 3 \cdot 5 | (n-1)$.
    (b) Check that $2^4 \cdot 3^2 | (n-1)$.
13. (a) If $p^2 | n$, then $p | \phi(n)$, but $p \nmid (n-1)$.

Exercise 7.3

2. (a) $3^{45} \equiv 27 \pmod{91}$, (b) $3^{840} \equiv 436 \pmod{841}$, (c) $3^{1026} \equiv 482 \pmod{1027}$,
    (d) $3^{552} \equiv 781 \pmod{1105}$, (e) $3^{5698} \equiv 3002 \pmod{5699}$.
3. $2^{1373652/2} \equiv -1 \pmod{1373653}$ and $3^{1373652/4} \equiv 1 \pmod{1373653}$.
5. $n = 15841$ is a strong pseudoprime since $2^{(n-1)/32} \equiv 1 \pmod{n}$. To see that n is a Carmichael number, note that $2^3 \cdot 3^2 \cdot 5 | (n-1)$.
7. (b) $\left(\frac{2}{561}\right) = 1 = 2^{280} \pmod{561}$.
    (c) $\left(\frac{2}{1105}\right) = 1 = 2^{552} \pmod{1105}$.
    (e) $\left(\frac{2}{341}\right) = -1 \neq 1 = 2^{170} \pmod{341}$.

Exercise 7.4

2. Note that Proposition 7.8 gives a sufficient condition for Germain primes.
3. (a) $M_{37} = 223 \cdot 616318177$, (b) $M_{43} = 431 \cdot 9719 \cdot 2099863$.
5. (b) Use the CRT. In fact, $M_{31} \equiv 63 \pmod{248}$ is prime.
    (c) $M_{47} = 2351 \cdot 4513 \cdot 13264529$.
11. $(15, 125) = 5$. So $(M_{15}, M_{125}) = M_5 = 31$.

Exercise 7.5

3. (a) Check $5^{48} \equiv -1 \pmod{97}$, (b) $5^{96} \equiv -1 \pmod{193}$, (f) $3^{608} \equiv -1 \pmod{257}$.
5. Check $3^{128} \equiv -1 \pmod{257}$.
7. Use the division algorithm.
11. Let $r = (f_n - 1)/2$. Then $(f_n - g)^r \equiv g^r \neq 1 \pmod{f_n}$.

Exercise 7.6

2. $3799 = 29 \cdot 131$.
3. $9943 = 61 \cdot 163$.
5. $7811 = 73 \cdot 107$.
7. $74104 = 2^3 \cdot 59 \cdot 157$.
11. $n! + 1$ is prime for $n = 1, 2, 3$, and 11. (The next such value is $n = 27$.).
13. (a) Apply formula (6.4) repeatedly.

Exercise 7.7

2. $694449 = 3^2 \cdot 73 \cdot 151$.
3. $17201 = 103 \cdot 167$.

Exercise 8.1

2. (a) You are very good at deciphering.
   (b) It always seems impossible until it is done.
3. Glad you could come.
5. (a) Is now a good time?   (b) The end of time is near.

Exercise 8.2

2. P = 09270715202701142701, C = 14481320254801494801, $\phi(m)$ = 40, d = 27 (P is broken up into blocks of length 2).
3. P = 09271215220527220051813151420, C = 81273871293127293144137114476, $\phi(m)$ = 72, d = 29 (P is broken up into blocks of length 2).
5. (a) $p = 571, q = 311$,   (b) $p = 709, q = 179$.
11. (a) Study number theory and you will always be in your prime.
    (b) The proof is in the pudding.
13. Chance favors the prepared mind.

Exercise 8.3

2. (a) $X_i = 1, 7, 19, 11, 27, 27, 27, \ldots$.
   (b) $X_i = 2, 17, 62, 5, 26, 25, 22, 11, 44, 15, 56, 51, 36, 55, 48, 27, 28, 41, 6, 29, 34, 49, 30, 37, 58, 57, 54, 45, 18, 1, 14, 53, 42, 9, 38, 61, 2, 17, \ldots$
3. (a) $\pi_2 = 3, \pi_3 = 8, \pi_4 = 6, \pi_5 = 20, \pi_6 = 24, \pi_7 = 16, \pi_8 = 12, \pi_9 = 24, \pi_{10} = 60$, and so on.
   (b) $\pi_{625} = 2500, \pi_{1000} = 6000, \pi_{1050} = 76800, \pi_{2700} = 43200$.

Exercise 9.1

2. (a) This is just a geometric series.   (b) Integrate (a) term by term.
3. These are geometric series.
5. (a) If $f_i(x) = O(g(x))$ for $i = 1, 2$, then there exists K_i such that $\frac{|f_i(x)|}{g(x)} < K_i$ for $x > a$. But $|f_1(x) + f_2(x)| < |f_1(x)| + |f_2(x)| \Rightarrow \frac{|f_1(x) + f_2(x)|}{g(x)} < K_1 + K_2$.

7. See Figure 9.1.

11. From the Euler product, $\log \zeta(s) = -\sum_p \log(1 - p^{-s})$. Apply Exercise 9.1.2(b) to obtain.

   $\log \zeta(s) = \sum_p \sum_{k=1}^{\infty} \frac{1}{k} p^{-ks} = \sum_{n=1}^{\infty} \frac{\Lambda(n)}{\log n} n^{-s}$ after rearrangement. Now differentiate term by term with respect to s.

13. Integrate by parts.

Exercise 9.2

5. For $n = 5$, $S = 6$. Note that $5 = 2^2 + 1^2 + 0^2 + 0^2$. There are four choices of position for the number 2, three choices of position for the number 1 once 2 is fixed, and four choices of sign for the numbers 1 and 2. 8 $S = (4)(3)(4)$.

Exercise 9.3

2. $B_2 = -1/2$, $B_3 = 0$, $B_4 = -1/30$, $B_5 = 0$, $B_6 = 1/42$, $B_7 = 0$, $B_8 = -1/30$.

3. $\zeta(2) = \pi^2/6$, $\zeta(4) = \pi^4/90$, $\pi(6) = \pi^6/945$, $\zeta(8) = \pi^8/9450$.

5. $\sum_{n=1}^{\infty} \frac{1}{(2n)^2} = \pi^2/24$.

7. $p^n \nmid m \Leftrightarrow m \not\equiv 0 \pmod{p^n}$ is $1 - 1/p^n$ and $\prod_p \left(1 - \frac{1}{p^n}\right) = 1/\zeta(n)$.

11. $1/(2n + 1)^2$.

Exercise 9.4

2. (a) The product of all primes less than or equal to x.

   (b) the product of all largest prime powers less than or equal to x.

3. $\pi(x) \sim x/\log x$, but the number of squares less than or equal to x is asymptotic to $\sqrt{x}$.

5. The probability that a random n is prime is approximately $1/\log n$. So the probability that M_p is prime should be about $1/p \log 2$. The number of Mersenne primes is then $\sum_p \frac{1}{p \log 2}$ which diverges. Hence there should be infinitely many Mersenne primes.

8. $\pi_2(100) = 15$, $\pi_2(200) = 29$, $\pi_2 1,000) = 35$, $\pi_2(2000) = 61$. The conjecture is not very accurate for such small values of x. The "predicted" values are $\pi_2(100) \approx 6.2257$, $\pi_2(200) \approx 9.4066$, $\pi_2(1,000) \approx 27.6698$, $\pi_2(2000) \approx 45.7066$.

Exercise 9.5

2. Get a common denominator and note that there is a prime p with $n/2 < p < n$. Note that all but one term in the numerator is divisible by p. (Without Bertrand's Postulate, check powers of 2.).

5. (a) Use Definition 1.8 directly.

   (c) Rewrite (a) as $\binom{2n}{n} = \frac{n+1}{2(2n+1)} \binom{2n+2}{n+1}$.

7. (a) Let $n = p^m - 1$.

   (b) $p^r \mid\mid (N/\Pi)$ where $r = p^{m-1}$. But $r = (p-1)\frac{p^{n-1}-1}{p-1} + 1$, which is the exact power of p dividing D/Π.

11. Note that a has arbitrarily many consecutive zeros.

13. Any exponent $e < 1/2$ would suffice. (Note that any exponent $e < 2/3$ is sufficient to prove the corollary to Ingham's theorem).

17. 3, 4, 6, 8, 12, 18, 24, 30.

Exercise 10.1

3. If $n = 6m + r$ with $r = 2$ or 5, then n is not a perfect fourth power.

5. The integers not expressible as the sum or difference of two squares are precisely the negative integers $\equiv 2 \pmod 4$ and the positive integers that are $\equiv 2 \pmod 4$ and are divisible by an odd power of a prime $\equiv 3 \pmod 4$.

7. By the proof of Lemma 10.6.1, $P_{k-1}(x) = k!x + d$ for all x. Hence $P_k(x) = k!(x + 1) + d - (k!x + d) = k!$

17. $8{,}042 = 17^3 + 10^3 + 10^3 + 10^3 + 4^3 + 4^3 + 1^3$. It is of interest that the greedy algorithm fails.

Exercise 10.2

2. $A = \{2, 4, 7, 8, 13, \ldots, 16, 25, \ldots, 32, \ldots\}$. $d(A) = 0$ since $1 \notin A$. Since $|A(3 \cdot 2^{k-1})|/3 \cdot 2^{k-1} = 1/3$. The natural density $\delta_n(A)$ does not exist since $|A(n)/n|$ equals both 1/2 and 1/3 infinitely often.

5. Let S be the set of square-free integers. By Corollary 9.10.5, $\delta_n(S) = 6/\pi^2 = d(S)$. By Mann's $\alpha + \beta$ theorem, $d(2S) \geq \min\{1, 2d(S)\} = 1$. Hence $2S$ contains all positive integers. (Note that Proposition 10.8 is not strong enough here).

7. (a) S is complete since every positive integer has a binary representation given by successively subtracting the largest power of 2. The greedy algorithm works since each term is at most twice the preceding term. S is not basic since we need n terms for $2^n - 1$ where n is arbitrary.

   (b) Same argument as in (a).

(c) The only issue is the ability to write $F_k, \ldots, F_{k+1} - 1$ as a sum of distinct Fibonacci numbers not including F_k. The fact that $F_1 = F_2 = 1$ is helpful.

(d) Consider the representation of $F_{k+1} - 1$.

(e) The greedy algorithm works once again.

Exercise 10.3

2. For example, $A = \{1, 5, 7, 11, 14, 16, 19\}$, $B = \{2, 3, 6, 8, 12, 17, 20\}$, and $C = \{4, 9, 10, 13, 18, 21\}$ works for $n = 21$.

3. (a) Let ℓ be given. The sets $S_0, \ldots, S_{M-1}$ partition **N**. By van der Waerden's Theorem, some class (say S_d) contains an arithmetic progression of length ℓ. But then so does S.

   (b) If (i) does not hold in either class, then (a) applies to both classes.

5. Let $S = \{1, 4, 5, 6, 11, 12, 13, 14, 15, \ldots\}$ and $T = \{2, 3, 7, 8, 9, 10, 16, 17, 18, 19, 20, 21, \ldots\}$.

Exercise 10.4

2. (a), (b) Consider the conjugates of the partitions.

3. (b) Let $m = 1$ in Proposition 10.13.

7. W.E. Durfee taught at Hobart College for 45 years and was acting president four times!

   (a) All partitions have a square in the northwest corner with $k \geq 1$. If it's a self-conjugate partition, then the number of dots to the right of the square must equal the number of dots below it.

   (b) If p is a self-conjugate partition of n with a $k \times k$ Durfee square, note the amount of freedom in placing half the remaining dots to the right of the square.

13. Both are equal to $p(n - mk)$.

17. (a) If p is a partition of n with parts $2k + 2k$, then so is the partition p' formed by replacing $2k + 2k$ with $k + k + k + k$ and vice versa.

    (b) Replace $2k + 2k$ by $4k$ and vice versa.

19. If p is a partition of n with odd parts only, use the hint for each odd part m. By the Fundamental Theorem of Arithmetic, $2^a m_1 \neq 2^b m_2$ for odd $m_1 \neq m_2$. If p' is a partition of n into unequal parts, write each part r as $r = 2^k m$ where m is odd. Now form the partition p containing 2^k copies of m, and so on.

23. 12.

Bibliography

[1]   Adams, W.A. and Goldstein, L.J. *Introduction to Number Theory* (1976). Englewood Cliffs, NJ: Prentice – Hall.

[2]   Agraval, M., Kayal, N. and Saxena, N. "Primes in P," *Annals of Mathematics* 160(2), (2004), 781–793.

[3]   Alaoglu, L. and Erdos, P. "On Highly Composite and Similar Numbers," *Transactions AMS* 56(3), (1944), 448–469.

[4]   Alford, W.R., Granville, A. and Pomerance, C. "There are Infinitely Many Carmichael Numbers," *Annals of Mathematics* 140 (1994), 703–722.

[5]   Alford, W.R., Granville, A. and Pomerance, C. ""On the Difficulty of Finding Reliable Witnesses.", *Algorithmic Number Theory Proceedings (ANTS-1)*." In: Adelaman, L.M. and Huang, M.D. (Eds.). Lecture Notes in Comp. Sci Vol. 877 (1994). Boston: Springer Verlag, 1–16.

[6]   Al-khazin, A.J. "Muhammad ibn Al-Hasan Al-Khurasani." In: Dold-Samponius, Y. Dictionary of Scientific Biography (1981). New York: Charles Scribner Sons.

[7]   Andrews, G.E. *The Theory of Partitions* (1984). Cambridge: Cambridge U. Press.

[8]   Andrew, G.E. and Eriksson, K. *Integer Partitions* (2004). Cambridge: Cambridge U. Press.

[9]   Anglin, W.S. *Mathematics: A Concise History and Philosophy* (1994). New York: Springer-Verlag.

[10]   Ansorge, R. ""Lothar Collatz (6 July 1910–26 September 1990)," Mitteilungen der Gesellschaft für Angewandte Mathematik und Mechanik (1991), 4–9.

[11]   Apostol, T.M. *Introduction to Analytic Number Theory* (1976). New York: Springer-Verlag.

[12]   Archibald, R.C. "Unpublished Letters of James Joseph Sylvester and Other New Information Concerning his Life and Work" (Jan. 1936). Vol. 1, 85–154, Osiris, U. of Chicago Press.

[13]   Arnold, L.K., Benkoski, S.J. and McCabe, B.J. "The Discriminator (a simple application of Bertrand's Postulate)," *The American Mathematical Monthly* 92 (1985), 275–277.

[14]   Ayoub, R. "Euler and the Zeta Function," *American Mathematical Monthly* (December 1974), 1067–1086.

[15]   Balasubramanian, R. "On Waring's Problem: g(4) ≤ 21," *Hardy-Ramanujan Journal* 2 (1979), 1–32.

[16]   Banks, W., Pollack, P. and Pomerance, C. "Symmetric Pairs Revisited," *Integers* 19(#A54), (2019), 7.

[17]   Barina, D. "Improved verification limit for the convergence of the Collatz conjecture," *Journal of Supercomputing* 81(810), (2025), 1–14.

[18]   Bedocci, E. "Nota ad una congettura sui numeri prima", Rivista Mathematica Universita," Parma 11 (1985), 229–236.

[19]   Beeger, N.G.W.H. "Ona New Case of the Congruence 2p-1 ≡ 1 (mod p2)," *Messenger of Math* 51 (1922), 149–150.

[20]   Bell, E.T. "An Arithmetical Theory of Certain Numerical Functions," Department o [Physics, University of Washington, Seattle, Washington 98195, USA 1(1), (1915).

[21]   Bell, E.T. "Exponential Numbers," *American Math. Monthly* 41 (1934), 411–419.

[22]   Beutelspacher, A. *Cryptology* (1994). Washington, D.C: Mathematical Association of America.

[23]   Binet, J.P. "Memoire sur l'integration des equations lineaires aux differences finies, d'un ordre quelconques a coefficients variables, Comptes Rendus des Seances de l'Academie des Sciences," 17 (1843), 559–567.

[24]   Blum, L., Blum, M. and Shub, M. "Comparison of Two Pseudo-Random Number Generators." In: *Advances in Cryptology* (1983). Boston: Springer-Verlag.

[25]   Boyer, C. and Merzbach, U.C. *A History of Mathematics* 2d ed. (1989). New York: Wiley.

[26]   Brauer, A. "On a Property of k Consecutive Intgers," Bulletin AMS 47(4), (1941), 328–331.

[27]   Bremser, P.S., Schumer, P.D. and Washington, L.C. "A Note on the Incongruence of Consecutive Integers to a Fixed Power," *Journal of Number Theory* 35(1) (May 1990), 105–108.

[28]   Bressoud, D.M. *Factorization and Primality Testing* (1990). New York: Springer-Verlag.

https://doi.org/10.1515/9783111579283-012

[29] Brezinski, C. *History of Continued Fractions and Pade Approximants* (1991). Berlin and Heidelberg: Springer-Verlag.

[30] Brillhart, J. "Note on Representing a Prime as the Sum of Two Squares," *Mathematics of Computation* (120), (October 1972), 1011–1013.

[31] Brillhart, J., Lehmer, D.H., Selfridge, J.L., Tuckerman, B. and Wagstaff, S.S. *Factorizations of* $b^n \pm 1$ (1983). American Mathematical Society.

[32] Bruce, J.W. "A Really Trivial Proof of the Lucas-Lehmer Primality Test," *American Mathematical Monthly* 100(4) (April 1993), 370–371.

[33] Burton, D.M. *The History of Mathematics: An Introduction* 2d ed. (1991). Dubuque, IA: William C. Brown.

[34] Chandrasekharan, K. *An Introduction to Analytic NumberTheory* (1968). Berlin and Heidelberg: Springer-Verlag.

[35] Chein, E.Z. An Odd Perfect Number has at least Eight Prime Factors (1979). Penn State University thesis.

[36] Chernick, J. "On Fermat's Simple Theorem," *Bulletin AMS* 45 (1939), 269–274.

[37] Chowla, S. "There Exists an Infinity of 3-Combinations of Primes in Arithmetic Progression," Lahore of Pre Historic Era 6 (1944), 15–16.

[38] Chvatal, V. "Some Unknown van der Waerden Numbers." In: Graham, R., et al. (Ed.) *Combinatorial Structures and their Applications*, Proceedings of Calgary International Conference, Gordon and Prasch (1969). 31–33.

[39] Cohen, H. *A Course in Computational Analytic Number Theory* (1993). Berling and Heidelberg: Spring-Verlag.

[40] Conrey, J.B. "More than Two-fifths of the Zeros of the Riemann Zeta Function Are on the Critical Line," *Journal für die reine und angewandte Mathematik* 399 (1989), 1–26.

[41] Cornacchia, G. "Su di un metodo per la rizoluzione in numeri interie dell' equazione h=0nChxn-hyh = Pn," *Giornale di Matematiche de Buttaglini* 46 (1908), 33–90.

[42] Cox, D.A. *Primes of the Form* $x^2 + ny^2$ (1989). New York: Wiley.

[43] Crandall, R., Doenias, J., Norrie, C. and Young, J. "The Twenty-Second Fermat Number is Composite," *Mathematics of Computation* 64(210) (April 1995), 863–868.

[44] Damianou, P. and Schumer, P.D. "A Theorem Involving the Denominators of Bernoulli Numbers," *Mathematics Magazine* 76(3) (June 2003), 219–224.

[45] Da Silva, D.A. "Propriedades geraes e resolucao ao estudo da Theoria dos numeros." In: *Memorias da Academia das Ciencias (New Series)* I (1) (Impresa National) (1854). 1–16. Lisbon.

[46] Davenport, H. and Heilbronn, H. "On Waring's Problem for Fourth Powers," *Proc. of London Math. Society* 1 (1936), 143–150.

[47] De Koninck, J.-M. and Doyon, N. *The Life of Primes in 37 Episodes* (2020). Providence, RI: AMS.

[48] DeMoivre, A. "De Fractionibus Algebraicus Radicalitate Immunibus ad Fractiones Simpliciores Reducendis, Deque Summandis Terminus Quarumdam Aequeli Intervalla a se Distantibus," *Phil. Trans.of the Royal Society* 32 (1722), 162–178.

[49] Deshouillers, J.-M., Granville, A., Narkiewicz, W. and Pomerance, C. "An Upper Bound in Goldbach's Problem," *Mathematics of Computation* 61(203) (July 1993), 209–213.

[50] Dickson, L.E. *Introduction to the Theory of Numbers* (1929). Chicago: Univ. of Chicago press.

[51] Dickson, L.E. *History of the Theory of Numbers* (1966). Vol. 3, New York: Chelsea.

[52] Diffie, W. and Hellman, M. "New Directions in Cryptography," *IEEE Transactions in Information Theory* 22 (1976), 644–655.

[53] Dubouis, E. "L'intermediare des math," 18 (1911), 55–56,224–225.

[54] Dudley, U. "History of a Formula for Primes," *American Mathematical Monthly* 76 (January 1969), 23–28.

[55] Edwards, H.M. *Riemann's Zeta Function* (1974). New York and London: Academic Press.

[56] Elkies, N.D. "On $A^4 + B^4 + C^4 = D^4$," *Mathematics of Computation* 51(184) (October 1988), 825–835.

[57] ErdoEs, P. "Beweis eines Satzes von Tschebyshef," *Acta Scientiarum Mathematicarum (Szeged)* 5 (1930–1932), 194–198.

[58] ErdoEs, P. and Selfridge, J.L. "The Product of Consecutive Integers is Never a Power," Illinois Journal of Mathematics 19(2), (1975), 292–301.

[59] Euclid *The Thirteen Books of Euclid's Elements* Translated by Heiberg, J.L. with commentary by Heath (1956). T.L. New York: Dover.

[60] Euler, L. *"Opera Omnia,"* Societatis Scientiarum Naturalium Helveticae (1911).

[61] Euler, L. "Theorematum quorandum ad numeros primos spectantium demonstration." In: (E54) Opera Omnia (1736).

[62] Euler, L. "Variae observations circa series infinitas," Commentarii Scientiarum Petropolitanae 9 (1737), 160–188.

[63] Euler, L. "Theoremata circa divisores numerorum in hac forma paa ± qbb contentorum" (E164," Opera Omnia (for the years 1744–1746) I.2 194–222.

[64] Euler, L. "Theoremata circa divisors numerorum," Novi Commentarii Academiae Scientiarum Petropolitanae (for 1747–1748, pub. 1750), 20–48.

[65] Euler, L. "Analytical Observations." (E326) (1763) Transl. Huffman, C. *Euleriana* (2023)3 (1), 3–22

[66] Euler, L. "Nouveaux Memoires de l'Academie royale des Sciences" (1772). Berlin.

[67] Faltings, G. "Endlichkeitssatze fur abelsche Varietaten uber Zahlkorpern," *Inventiones mathematicae* 73 (1983), 349–366.

[68] Fermat, P. in Oeuvres de Fermat Tannery, P. and Henry, C. ed. (1894). 2, 212–217, Paris.

[69] Fibonacci, L.P. *The Book of Squares* English ed. (1987).translated by Sigler, L.E. Boston: Academic Press.

[70] Friedlander, J. and Ivaniec, H. "The Polynomial $X^2 + Y^4$ Captures its Primes," *Annals of Mathematics* 2 (148), (1998), 945–1030.

[71] Gahan, R. in Online Encyclopedia of Integer Sequences (A327760).

[72] Gauss, C.F. *Disquisitiones Arithmeticae* English ed. (1986). Translated by Clark, A.A. and Waterhouse, W.C. New York: Springer-Verlag.

[73] Gelfond, A.O. and Linnik, Y.V. *Elementary Methods in Analytic Number Theory* (1965). Chicago: Rand McNally.

[74] Gentle, J.E. *Random Number Generation and Monte Carlo Methods* 2d ed. (2003). New York: Springer.

[75] Giblin, P. *Primes and Programming* (1993). Cambridge: Cambridge University Press.

[76] Gillispie, C.C. ed. *Dictionary of Scientific Biography*. 16 vols (1970). New York: Scribners.

[77] Girard, A. "in L'Arith. de Simon Stevin ... annotations by Girard, A., Leide, 1625, 622, Oeuvres Math," de Simon Stevin par Albert Girard (1634), 156.

[78] Goldberg, K. "A Table of Wilson Quotients and the Third Wilson Prime," *Journal of the London Mathematical Society* 28 (1953), 252–256.

[79] Goldstein, C. "Frenicle." In: Foisneau, L. (Ed.). Dictionary of 17th (2001). London: Century French Philosophers

[80] Goldstein, L.J. "A History of the Prime Number Theorem," *American Mathematical Monthly* 80(6) (June-July 1973), 599–615.

[81] Goldston, D.A., Pintz, J. and Yidirim, C.Y. "Primes in Tuples 1," *Annals of Mathematics* 170(2), (2009), 819–862.

[82] Goto, T. and Ohno, Y. "Odd Perfect Numbers have a Prime Factor Exceeding 10^8," *Mathematics of Computation* 77(263), (2008), 1859–1868.

[83] Graham, R.L., Knuth, D.E. and Patashnik, O. *Concrete Mathematics* (1989). Reading, MA: Addison-Wesley.

[84] Granville, A. *Number Theory Revisited: A Masterclass* (2019). Providence, RI: AMS.

[85] Green, B. and Tao, T. "The Primes Contain Arbitrarily Long Arithmetic Progressions," *Annals of Mathematics* 2(167), (2008), 481–547.

[86] Grosswald, E. *Representations of Integers as Sums of Squares* (1985). New York: Springer-Verlag.

[87] Guth, L. and Maynard, J. "New Large Value Estimates for Dirichlet Polynomials," In arXiv:2405.20552 [math.NT ((May 31, 2024)), 48.

[88] Guy, R.K., (Ed.). *Reviews in Number Theory 1973–1983* (1984). 6 vols, American Mathematical Society.

[89] Guy, R.K. "Every Number is Expressible as the Sum of How Many Polygonal Numbers?," *American Mathematical Monthly* 101(2) (February 1994), 161–172.

[90] Guy, R.K. *Unsolved Problems in Number Theory* 2d ed. (1994). New York: Springer-Verlag.

[91] Hagis, P. "Outline of a Proof that Every Odd Perfect Number has At Least Eight Prime Factors," *Mathematics of Computation* 35(151) (July 1980), 1027–1032.

[92] Hardy, G.H. "An Introduction to the Theory of Number," *Bulletin of the American Mathematical Society* 35 (1929), 778–818.

[93] Hardy, G.H. and Littlewood, J.E. "Some Problems of 'Partitio Numerorum,' III: On the Expression of a Number as a Sum of Primes," *Acta Mathematica* 44 (1923), 1–70.

[94] Hardy, G.H. and Wright, E.M. *An Introduction to the Theory of Numbers* 5th ed. (1979). Oxford: Clarendon Press.

[95] Heath-Brown, D.R. "The Divisor Function at Consecutive Integers," *Mathematika* 31 (1984), 141–149.

[96] Heath-Brown, D.R. "Artin's Conjecture for Primitive Roots," *Quarterly Journal of Math* 37(1) (March 1986), 27–38.

[97] Hilbert, D. "Beweis fur die Darstellbarkeit der ganzen Zahlen durch eine feste Anzahl n-ter Potenzen (Waringshes Problem)," *Mathematische Annalen* 67(3), (1909), 281–300.

[98] Hinz, A.M. "Pascal's Triangle and the Tower of Hanoi," *American Mathematical Monthly* 99(6) (June-July 1992), 538–544.

[99] Hoggatt, V.E. *Fibonacci and Lucas Numbers* (1969). Boston: Houghton Mifflin.

[100] Holzer, L. "Minimal Solutions of Diophantine Equations," *Canadian Journal of Mathematics* 2 (1950), 238–244.

[101] Honsberger, R. *Mathematical Gems II* (1976). Mathematical Association of America.

[102] Honsberger, R. *Mathematical Gems III* (1985). Mathematical Association of America.

[103] Honsberger, R. *More Mathematical Morsels* (1991). Mathematical Association of America.

[104] Hua, L.K. *Introduction to Number Theory* (1982). New York: Springer-Verlag.

[105] Huxley, M.N. "Exponential Sums and Lattice Points III," *Proceedings of the London Mathematical Society* 87(3), (2003), 591–609.

[106] Iannucci, D.E. "The Second Largest Prime Divisor of an Odd Perfect Number Exceeds Ten Thousand," *Mathematics of Computation* 68(228), (1999), 1749–1760.

[107] Iannucci, D.E. "The Third Largest Prime Divisor of an Odd Perfect Number Exceeds One Hundred," *Mathematics of Computation* 69(230), (2000), 867–879.

[108] Ingham, A.E. "On the Difference between Consecutive Primes," *Quarterly Journal of Mathematics* os-8(1), (1937), 255–266.

[109] Iwaniec, H. and Mozzochi, C.J. "On the Divisor and Circle Problem," *Journal of Number Theory* (May 1988), 60–93.

[110] Ireland, K. and Rosen, M. *A Classical Introduction to Modern Number Theory* (1982). New York: Springer-Verlag.

[111] Jaeschke, G. "The Carmichael Numbers to 10^{12}," *Mathematics of Computation* 55(191) (July 1990), 383–389.

[112] Jaeschke, G. "On Strong Pseudoprimes to Several Bases," *Mathematics of Computation* 61(204) (October 1993), 915–926.

[113] Jameson, G.J.). *The Prime Number Theorem* (2003). Cambridge: Cambridge U. Press.

[114] Jenkins, P. "Odd Perfect Numbers have a Factor that Exceeds 10^7," *Mathematics of Computation* 72 (2003), 1549–1554.

[115] Jones, G. "6/π2," *Mathematics Magazine* 66(5) (December 1993), 290–298.

[116] Keller, W. "The 17th Prime of the Form 5·2n + 1," *Abstracts AMS* 6 (1985), 121.

[117] Keller, W. "Prime Factors k·2n +1 of Fermat Numbers," on-line Encyclopedia of Integer Sequences ((May 18, 2025)).

[118] Khinchin, A.Y. *Three Pearls of Number Theory* (1952). Rochester: Graylock Press.

[119] Knuth, D. *The Art of Computer Programming.* 3 vols (1997). Reading, MA: Addison-Wesley.

[120] Koblitz, N. *A Course in Number Theory and Cryptography* 2d ed. (1987). New York: Springer Verlag.

[121] Kolata, G. "The Assault on 114, 381,625,757,888,669,235,779,976,146,612,010,218,296,721,242,362,562,561,842,935,706,935,245,733, 897,830,597,123,563,958,705,058,989,075,147,599,290,026,879,543,541." *The New York Times*, March 22, 1994.

[122] Kolata, G. " ... While a Mathematician Calls Classic Riddle Solved." *The New York Times*, October 27, 1994.

[123] Lander, L.J. and Parkin, T.R. "Counterexample to Euler's Conjecture on Sums of Like Powers," *Bulletin of the American Mathematical Society* 72 (1966), 1079.

[124] Larson, L.C. *Problem-Solving Through Problems* (1983). New York: Springer-Verlag.

[125] Lebesgue, V.A. "note in Nouv," Annals of Mathematics 2(11), (1872), 516–519.

[126] Leech, J. "Some solutions of Diophantine Equations," *Proceedings of the Cambridge Philosophical Society* 53 (1967), 778–780.

[127] Legendre, A.M. "Recherches d'analyse indeteminee," Histoire de l'Academie royale des sciences avec les Memoires de Annee (1785), 465–559.

[128] Legendre, A.-M. *Essai sur la Theorie des Nombres* 1st ed. (1798). Paris: Duprat, an VI.

[129] Lehmer, D.H. and Powers, R.E. "On Factoring Large Numbers," *Bulletin of AMS* 37 Iss (10), (1931), 770–776.

[130] Lehmer, D.H. "Mathematical Methods in Large-Scale Computing Units." *Proceedings of 2nd Sumposium on Large-Scale Digital Calculating Machinery* (1951): 141–146.

[131] Lenstra, H.W. "Factoring Integers with Elliptic Curves," *Annals of Math* 126 (1987), 649–673.

[132] Lenstra, A.K. and Lenstra, H.W., (eds.) *The Development of the Number Field Sieve*, Lecture Notes in Math (2006). Berlin: Springer.

[133] Lenstra, A.K., Lenstra, H.W., Manasse, M.S. and Pollard, J.M. "The Factorization of the Ninth Fermat Number," *Mathematics of Computation* 61(203) (July 1993), 319–349.

[134] LeVeque, W.J. ed. *Reviews in Number Theory.* 6 vols (1974). American Mathematical Society.

[135] LeVeque, W.J. *Fundamentals of Number Theory* (1977). Reading, MA: Addison- Wesley.

[136] Levinson, N. "A Motivated Account of an Elementary Proof of the Prime Number Theorem," *American Mathematical Monthly* (March 1969), 225–245.

[137] Long, C.T. *Elementary Introduction to Number Theory* 2d ed. (1972). Lexington, MA: D.C. Heath.

[138] Lucas, E. *Theorie des Nombres* (1891). Paris: Tome Premier, Gauthier-Villars.

[139] Mahler, K. "On the Fractional Parts of the Powers of a Rational Number II," *Mathematics of Computation* 4(2), (1957), 122–124.

[140] Maillet, E. Bull. Soc. Math (1895). 23, 45, France.

[141] Maynard, J. "Small Gaps between Primes," *Annals of Math* 181(1), (2015), 383–413.

[142] Mazur, B. and Stein, W. *Prime Numbers and the Riemann Hypothesis* (2016). Cambridge: Cambridge U. Press.

[143] McCarthy, P.J. *Introduction to Arithmetical Functions* (1985). New York: Springer-Verlag.

[144] Meissner, W. Uber die Teilbarkeit von 2p – 2 durch das Quadrat der Primzahl p = 1093 Sitzungsber, D.K., Akad, P. and Wiss, D. (1913). 663–667, Berlin.

[145] Mertens, F. "Ein Beitrag zur Analytischen Zahlentheorie," *Journal für die reine und angewandte Mathematik* 78 (1874), 46–62.

[146] Miller, G.L. "Riemann's Hypothesis and Tests for Primality." *Proceedings of the Seventh Annual ACM Symposium on the Theory of Computing*, 234–239.

[147] Mills, W.H. "A Prime-Representing Function," *Bulletin of the American Mathematical Society* 53 (1947), 604.

[148] Mobius, A.F. "Uber eine besondere Art von Umkehrung der Reihen," *Journal für die reine und angewandte Mathematik* 9 (1832), 105–123.

[149] Mollin, R.A. *An Introduction to Cryptography* (2001). Boca Raton: Chapman and Hall/CRC.

[150] Mordell, L.J. "On Numbers that can be Expressed as a Sum of Powers", Abhandlungen aus Zahlentheorie und Analysis sur Erinnerung an Edmund Landau (1968). 219–221, Berlin: VEB Deutscher Verlag der Wissenschaften.

[151] Mordell, L.J. *Diophantine Equations* (1969). London and New York: Academic Press.

[152] Moree, P. and Mullen, G.L. "Dickson Polynomial Discriminators." In: Report 95-177 (May 1995). NSW 2409 Australia: Macquarie University.

[153] Mozzochi, C.J. "On the Difference between Consecutive Primes," *Journal of Number Theory* 24 (1986), 181–187.

[154] Nielsen, P.P. "Odd Perfect Numbers have at least Nine Distinct Prime Factors," *Mathematics of Computation* 76(260), (2007), 2109–2126.

[155] Nielsen, P.P. "Odd Perfect Numbers, Diophantine Equations, and Upper Bounds," *Mathematics of Computation* 84(295), (2015), 2549–2567.

[156] Niven, I. "A Simple Proof That π is Irrational," *Bulletin of the American Mathematical Society* 53 (1947), 509.

[157] Niven, I. *"Irrational Numbers," Mathematical Association of America* (1956).

[158] Niven, I., Zuckerman, H.S. and Montgomery, H.L. *An Introduction to the Theory of Numbers* 5th ed. (1991). New York: Wiley.

[159] Norrie, R., in Univ. of St. Andrews five-hundredth anniversary memorial volume (1911), 89.

[160] Ochem, P. and Rao, M. "Odd Perfect Numbers are Greater than 10^{1500}," *Mathematics of Computation* 81 (2012), 1869–1877.

[161] Odlyzko, A.M. and te Riele, H.J.J. "Disproof of the Mertens Conjecture," Journal für die reine und angewandte Mathematik 357 (1985), 138–160.

[162] Olds, C.D. *"Continued Fractions," Mathematical Association of America* (1963).

[163] Ore, O. *Number Theory and Its History* (1988). New York: Dover.

[164] Parady, B.K., Smith, J.F. and Zarantonello, S.E. "The Largest Known Twin Primes," *Mathematics of Computation* 21(99) (July 1990), 381–382.

[165] Parkin, T.R. and Shanks, D. "On the Distribution of Parity in the Partition Function," *Mathematics of Computation* 21(99) (July 1967), 466–480.

[166] Patterson, S.J. *An Introduction to the Theory of the Riemann Zeta-Function* (1987). Cambridge: Cambridge University Press.

[167] Perelli, A. and Pintz, J. "On the Exceptional Set for Goldbach's Problem in Short Intervals," *Journal of the London Mathematical Society* 2(47), (1993), 41–49.

[168] Pillai, S.S. "On m Consecutive Integers I," Proceedings of the Indian Academy of Sciences, Section A 11 (1940), 6–12.

[169] Pinch, R.G.E. "The Carmichael Numbers up to 10^{15}," *Mathematics of Computation* 61(203) (July 1993), 381–391.

[170] Pisano, L. *"Fibonacci's Liber Abaci (Book of Calculation),"* Biblioteca a Nazionale di Firenze (1202).

[171] Platt, D.J. and Trudjian, T. "The Riemann Hypothesis is True up to 3·1012," *Bull. of London Math. Society* 53(3) (January 2021), 792–797.

[172] Poinsot, L. "Reflexions sur les principes fondamentaux de la theorie des nombres," Journal de Mathématiques Pures et Appliquées 10 (1845), 1–101.

[173] Pollard, J.M. "Theorems on Factorization and Primality Testing," *Proceedings of the Cambridge Philosophical Society* 76 (1974), 521–528.

[174] Pollard, J.M. "A Monte-Carlo Method of Factorization," *Nordisk Tidskrift for Informationsbehandling* (BIT) 15 (1975), 331–334.

[175] Pomerance, C. "On the Distribution of Pseudoprimes," *Mathematics of Computation* 37(156) (October 1981), 587–593.

[176] Pomerance, C. "Analysis and Comparison of Some Integer Factoring Algorithms." In: Lenstra, H.W. and Tijdeman, R. *Computational Methods in Number Theory*, Part I (1982), 89–139, Amsterdam: es. Math. Centre Tract 154.

[177] Pomerance, C. *Lecture Notes on Primality Testing and Factoring*. Notes by Gagola, S.M. (1984). Mathematical Association of America.

[178] Pomerance, C., Selfridge, J.L. and Wagstaff, S.S. "The Pseudoprimes to $25 \cdot 10^9$," *Mathematics of Computation* 35(151) (July 1980), 1003–1026.

[179] Pomerance, C. and Spicer, C. "Proof of the Sheldon Conjecture," *The American Mathematical Monthly* 126 (2019), 688–698.

[180] Pratt, K., Robles, N., Zaharescu, A. and Ziendler, D. "More than five-twelfths of the Zeros of the Riemann zeta function are on the Critical Line." In: *Research in the Mathematical Sciences J* 3 (2019). Springer Science and Business.1-60.

[181] Pritchard, P.A. "Long Arithmetic Progressions of Primes: Some Old, Some New," *Mathematics of Computation* 45(171) (July 1985), 263–267.

[182] Pritchard, P.A., Moran, A. and Thyssen, A. "Twenty-two Primes in Arithmetic Progression," *Mathematics of* Computation 64(211) (July 1995), 1337–1339.

[183] Proth, F. "Theoremes sur les nombres premiers," *Comptes Rendus de l'Academie des Sciences de Paris* 87 (1878), 926.

[184] Rabin, M.O. "Digitalized Signatures and Public-Key Functions as Intractable as Factorization." In: *MIT Laboratory for Computer Science Technical Report LCS/TR-212* (1979). Cambridge, MA.

[185] Rabung, J.R. "On Applications of Van der Waerden's Theorem." *Mathematics Magazine* (May-June 1975): 142–148.

[186] Ramanujan, S. "Highly Composite Integers." *Proceedings of London Math. Society*, (1915)

[187] Ribenboim, P. *The Book of Prime Number Records* 2d ed. (1989). New York: Springer-Verlag.

[188] Richert, H.E. "Uber Zerfallungen in Ungleiche Primzahlen," *Mathematische Zeitschrift* 52 (1949), 342–343.

[189] Riesel, H. *Prime Numbers and Computer Methods for Factorization* 2d ed. (1994). Boston: Birkhauser.

[190] Rivest, R.L., Shamir, A. and Adelman, L.M. "A Method for Obtaining Digital Signatures and Public-Key Cryptosystems," *Communications of the ACM* 21 (1978), 120–126.

[191] Rosen, K. *Elementary Number Theory and Its Applications* 3d ed. (1993). Reading, MA: Addison-Wesley.

[192] Roth, K.F. "On Certain Sets of Integers," *Journal of the London Mathematical Society* 28 (1953), 104–109.

[193] Roth, K.F. "Rational Approximations to Algebraic Numbers," Mathematika 2(1), (1955), 1–20.

[194] Rotkiewicz, A. "Note on the Diophantine Equation $1 + x + x2 + ... + xn= ym$," *Elementary Mathematics* 42 (1987), 76.

[195] Sarkozy, A. "On Divisors of Binomial Coefficients, I," *Journal of Number Theory* 20(1) (February 1985), 70–80.

[196] Schafly, A. and Wagon, S. "Carmichael's Conjecture is valid below 10^10,000,000," Mathematics of Computation 63 (1994), 415–419.

[197] Scharlau, W. and Opalka, H. *From Fermat to Minkowski* (1984). New York: Springer-Verlag.

[198] Schnirelmann, L. "On the Additive Properties of Numbers," (orig. in Russian), Proceedings of Don Polytechnic Inst. in Novocherkassk XIV (1930), 3–27.

[199] Schroeder, M.R. *Number Theory in Science and Communication* (1984). Berlin: Springer-Verlag.

[200] Schumer, P.D. *Introduction to Number Theory* (1996). Boston, MA: PWS.

[201] Schumer, P.D. "The Josephus Problem: Once More Around," *Mathematics Magazine* 75(1) (February 2002), 12–17.

[202] Seelhoff, P. "Prime Factors k·2n +1 of Fermat Numbers," on-line Encyclopedia of Integer Sequences (May 18, 2025).

[203] Serret, J.A. *Cours d'Algebre superieure* Tome 1, Gauthiers-Villars.

[204] Shanks, D. *Solved and Unsolved Problems in Number Theory* 3d ed. (1985). New York: Chelsea.

[205] Shapiro, H. *Introduction to the Theory of Numbers* (1983). New York: Wiley.

[206] Shiu, D.K.L. "Strings of Congruent Primes," Journal of the London Mathematical Society 61 (Issue 2), (April 2020), 359–373.

[207] Sierpinski, W. *Elementary Theory of Numbers* (1964). Warszawa: Panstwowe Wydawnictwo Naukowe.

[208] Sinisalo, M.K. "Checking the Goldbach Conjecture up to $4 \cdot 10^{11}$," *Mathematics of Computation* 61(204) (October 1993), 931–934.

[209] Sprague, R. "Uber Zerlegungen in Ungleiche Quadratzahlen," *Mathematische Zeitschrift* 51 (1949), 289–290.

[210] Stemmler, R.M. "The Ideal Waring Theorem for Exponents," *Mathematics of Computation* 18(85), (1964), 144–146.

[211] Stieltjes, T.J. "Sur la theorie des nombres," Annales de la Faculté des Sciences de Toulouse 4 (1890), 1–103.

[212] Stinson, D.R. *Cryptography: Theory and Practice* (2002). Boca Raton: Chapmand and Hall/CRC.

[213] Stopple, J. *A Primer of Analytic Number Theory: From Pythagoras to Riemann* (2003). Cambridge: Cambridge U. Press.

[214] Sylvester, J.J. "On a point in the theory of vulgar fractions," American Journal of Mathematics 3(4), (1880), 332–335.

[215] Sylvester, J.J. "Problem 7382 in Mathematical Questions with their Solutions." In: Miller, W.J.C. (Ed.) Education Times 41 (1884). 21.

[216] Szemeredi, E. "On Sets of Integers Containing No Arithmetic Progressions of Length 4," *Acta Mathematica Hungarica* 18(4), (1967), 479–483.

[217] Taylor, R. and Wiles, A. "Ring-theoretic Properties of Certain Hecke Algebras," *Annals of Mathematics* 141(3) (May 1995), 553–572.

[218] Te Riele, H.J.J. "Computation of all Amicable Pairs Below 10^{10}," *Mathematics of Computation* 47(175) (July 1986), 361–368.

[219] Thue, A. "Uber Annaherungswerte Algebraischer Zahlen," Journal für die reine und angewandte Mathematik 135 (1909), 284–305.

[220] Trappe, W. and Washington, L.C. *Introduction to Cryptography with Coding Theory* 3d ed. (2006). New York: Chelsea, 1985. 2d ed. Upper Saddle River, NJ: Pearson.

[221] Tunnell, J.B. "A Classical Diophantine Problem and Modular Forms of Weight 3/2," *Invention Math* 72 (1983), 323–334.

[222] Van de lune, J., Te Riele, H.J.J. and Winter, D.T. "On the Zeros of the Riemann Zeta Function in the Critical Strip, IV," *Mathematics of Computation* 46(174) (April 1986), 667–681.

[223] Vanden Eynden, C. *Elementary Number Theory* 2d ed. (2001). Long Grove, IL: Waveland Press.

[224] Vandermonde, A.T. Memoire sur des irrationnelles de differents orders avec une application au cercle (1772). French Academy of Sciences.

[225] Van der Waerden, B.L. "Wie der Beweis der Vermutung von Baudet gefunden wurde," *Abhandlungen aus dem Mathematischen Seminar der Universität Hamburg* 28 (1965), 6–15.

[226] Van der Waerden, B.L. *A History of Algebra* (1985). Berlin: Springer-Verlag.

[227] Vardi, I. *Computational Recreations in Mathematica* (1991). Redwood City, CA: Addison-Wesley.

[228] Vaughan, R. and Wooley, T. "Further Improvements in Waring's Problem, IV: Higher Powers," *Acta Arithmetica* 94(3), (2000), 203–285.

[229] Vinogradov, I.M. "A New Method in Analytic Number Theory," Travaux Inst. Math. Stekloff 10 (1937), 5–122.

[230] Von Neumann, J. "Various Techniques Used in Connection with Random Digits." In: Householder, A. S., Forsythe, G.E. and Germond, H.H. (Eds.) *Monte Carlo Method, National Applied Mathematics Series* 12 (1951). 36–38.

[231] Wagon, S. "Editor's Corner: The Euclidean Algorithm Strikes Again," *American Mathematical Monthly* 97(2) (February 1990), 125–129.

[232] Wagon, S. *Mathematica in Action* (1991). New York: W.H. Freeman.

[233] Wallis, J. letter to Collins, J., 25 January 1671, in Cambridge U. Library, The Macclesfield Collection, letters and papers of Barrow, Collins, Flamsteed, Gregory, Halley, Strobe, and Wallis.

[234] Wantzel, P.L. "Recherches sur le moyens de reconnaitre si un Probleme de Geometrie peut se resoudre avec la regle et le compas," *Journal de Mathématiques Pures et Appliquées* 2 (1837), 366–372.

[235] Waring, E. *Meditationes Arithmeticae* (1770)(2023). English trans Creative Media Partners LLC

[236] Widynski, W. "Squares: A Fast Counter-based RNG," In arXiv: 2004.06278v7 [cs.DS] (13 March, 2022).

[237] Wieferich, A. "Zum letzten Fermat'schen Theorem," *Journal für die reine und angewandte Mathematik* 136(3/4), (1909), 293–302.

[238] Wiles, A. "Modular Elliptic Curves and Fermat's Last Theorem," *Annals of Mathematics* 141(3) (May 1995), 443–551.

[239] Williams, H.C. and Dubner, H. "The Primality of R1031," *Mathematics of Computation* 47(176) (October 1986), 703–711.

[240] Weil, A. *Number Theory: An Approach through History* (1984). Boston: Birkhauser.

[241] Wolstenholme, J. "On Certain Properties of Prime Numbers," The Quarterly Journal of Pure and Applied Mathematics 5 (1862), 35–39.

[242] Wunderlich, M.C. and Kubina, J.M. "Extending Waring's Conjecture to 471,600.000," *Mathematics of Computation* 55(192) (October 1990), 815–820.

[243] Young, J. and Buell, D. "The Twentieth Fermat Number is Composite," *Mathematics of Computation* 50(181) (January 1988), 261–263.

[244] Zhang, Y. "Bounded Gaps between Primes," *Annals of Mathematics* 179 (2014), 1121–1174.

Index

https://doi.org/10.1515/9783111579283-013

www.ingramcontent.com/pod-product-compliance
Lightning Source LLC
Chambersburg PA
CBHW080902220326
41598CB00034B/5449